Programming Video Games for the Evil Genius

Evil Genius Series

Bionics for the Evil Genius: 25 Build-it-Yourself Projects

Electronic Circuits for the Evil Genius: 57 Lessons with Projects

Electronic Gadgets for the Evil Genius: 28 Build-it-Yourself Projects

Electronic Games for the Evil Genius

Electronic Sensors for the Evil Genius: 54 Electrifying Projects

50 Awesome Auto Projects for the Evil Genius

50 Model Rocket Projects for the Evil Genius

Fuel Cell Projects for the Evil Genius

Mechatronics for the Evil Genius: 25 Build-it-Yourself Projects

MORE Electronic Gadgets for the Evil Genius: 40 NEW Build-it-Yourself Projects

101 Spy Gadgets for the Evil Genius

123 PIC® Microcontroller Experiments for the Evil Genius

123 Robotics Experiments for the Evil Genius

PC Mods for the Evil Genius: 25 Custom Builds to Turbocharge Your Computer

Programming Video Games for the Evil Genius

Solar Energy Projects for the Evil Genius

25 Home Automation Projects for the Evil Genius

Programming Video Games for the Evil Genius

IAN CINNAMON

New York Chicago San Francisco Lisbon London Madrid
Mexico City Milan New Delhi San Juan Seoul
Singapore Sydney Toronto

Cataloging-in-Publication Data is on file with the Library of Congress.

Programming Video Games for the Evil Genius

1 2 3 4 5 6 7 8 9 0 QPD/QPD 0 1 4 3 2 1 0 9 8

ISBN 978- 0-07-149752-7

MHID 0-07-149752-8

This book is printed on acid-free paper.

McGraw-Hill books are available at special quality discounts to use as premiums and sales promotions, or for use in corporate training programs. To contact a representative please visit the Contact Us pages at www.mhprofessional.com.

Sponsoring Editor
Judy Bass

Acquisitions Coordinator
Rebecca Behrens

Editorial Supervisor
David Fogarty

Project Manager
Imran Mirza

Copy Editor
John Bremer

Proofreader
Ian Ross

Indexer
Pauline Davies

Production supervisor
Richard Ruzycka

Composition
Keyword Group Ltd.

Art Director, Cover
Jeff Weeks

Contents

Foreword

Years ago, Ian Cinnamon attended iD Tech Camps at our UCLA location. Ian received programming instruction in C++ and Java. Year after year, Ian continued to attend camp, and his programming prowess grew steadily—but it became apparent that he was outpacing his peers and needed new challenges.

His instructors commented that Ian was a "sponge," and grasped the programming curriculum quickly—as if he would go home and study and read and solve problems on his free time. It turns out, he was doing just that.

I got the opportunity to meet Ian two summers ago at our Stanford University location. Ian is a fine young man with great manners, excellent social skills, and, obviously, some serious programming talent. He is one of the best programmers I have seen at iD Tech Camps, which is an impressive statistic considering the thousands of programmers who have graduated from iD Tech Camps over the years.

Ian, now just 15 years old, has become a luminary for his generation. His book, *Programming Video Games for the Evil Genius,* offers a step-by-step approach to programming video games—a dream for many young kids.

I hear it all the time... "I wish I knew how to program my own game," and "I don't know where to start." My suggestion is to attend iD Tech Camps and to grab a copy of Ian's book. The crucial steps are knowing where to go and diving in to get started. Ian is empowering his generation and demystifying the code behind games. I can't wait to see where Ian will be in the next five years... ten years. Just watch out.

Pete Ingram-Cauchi

President and CEO, iD Tech Camps,
internalDrive, Inc.

ABOUT THE AUTHOR

Ian Cinnamon is a 15-year old phenom who has been programming for over 7 years, and is certified in both Java and C++. He is currently a sophomore at Harvard-Westlake School in Los Angeles, California.

Introduction

Master your gaming universe

What's better than playing video games? Creating your own video games! Games you design, games you control ... games played by *your* rules.

When you buy an off-the-shelf game at your local big box store, it is often the result of months, sometimes years, of effort by a small army of professionals using highly complex programming aided by the newest, most powerful computers. But the one thing they don't have to make the perfect game is *you*, the Game Creator.

You are the master of your gaming universe. You create the world you want. You set background colors, levels of difficulty, the shape of the game board/playing field, and cheat codes. You invent the characters, you design the challenge, you choose how points are scored and how players win ... or lose.

If you can think it, you can program it.

Your personal video game arcade

Do you ever get bored playing the same game over and over? Not anymore. You can build an assortment of games with endless variations. In these pages, you'll find the secrets to building racing games, board games, shoot-'em-up games, strategy games, retro games, and brain buster games.

Racing games—Get your adrenaline pumping! Construct games in which you race against time, an opponent, or even yourself. Speed and precision rule the road as you zoom around in cars, rockets, broomsticks, skis—whatever mode of transportation your evil genius mind can conjure.

Board games—Mental minefields to drive you crazy! Games and mazes that make you outthink, outwit, outmaneuver your adversaries! Frustrate your opponents and bring them to their knees. Think Tic-Tac-Toe on steroids.

Shoot-'em-up games—Games of lightning reflex and nerve-wracking action! Transform into a soldier, a snake handler, an alien warrior, or a stone-throwing Neanderthal as you take aim within the world you create.

Strategy games—Trap your opponents in an escape-proof box or diffuse a bomb before it can destroy the earth! Either way, sweat the challenge. Cool heads and fast thinking required.

Retro games—Have the classics your way! Make variations to Mario and Pac-Man by programming new twists and turns that make these old games new again.

Brain Buster games—Do you have a good memory? Do you perform well under pressure? Hope so, because in these games, you live or die by what you recall. Make it as simple or as complex as your courage allows.

Programming: the language of games

In music, there are notes; in mathematics, there are equations; in language, there are words; and in the video game world, there are commands which bring life to your games. This is called programming. This is how you tell the computer what you want it to do.

All the games you create will be written in Java, the most universal of programming languages.

What if you know absolutely nothing about programming? What if you have no idea how a computer turns your ideas into actions and images?

No worry! I will lead you step by step through the basics of this very versatile language. Once you have experienced *Programming Video Games for the Evil Genius,* you'll not only have a library of awesome, personalized games, but you'll be a skilled game creator, as well.

The building blocks to game creation

Programming is made up of several building blocks. You'll learn them all easily as we go through each one step-by-step. Screen shots are your guide as you master these critical tools. It's fool-proof. And, like riding a bicycle, once you know how, you never forget.

If you are new to programming, Section 1 offers a speed-of-light review. If you have previous programming experience, you may want to proceed ahead to Section 2.

Some of the programming building blocks used for game creation include:

Statements**:** Command central, the backbone of all games.

Comments**:** Allows you to mark your code, so you know what each line is doing.

Flow control**:** Allows you to repeat code. This is great when you want to retry the game.

Variables**:** This is how you keep track of a player's score, name, level, etc.

"If" statements**:** Lets you test your variables with conditionals. For example, *if* you kill an enemy, your score goes up.

JOptionPane**:** Want to display a player's score? Want to get a player's name for a high score list? This is how you get input and output.

Random numbers**:** The gateway to artificial intelligence. If you want to make an enemy move randomly, you're at the right building block.

Pausing**:** Allows your game to refresh so that your graphics remain crisp and clear.

Arrays and ArrayLists**:** Saves time by grouping similar objects together (enemies, power-ups, etc.).

File IO (Input/Output)**:** Allows you to save the game ... handy when the boss comes in unexpectedly and you've been playing video games instead of working.

Acknowledgments

You would not be reading this book if it weren't for the dedication and support from the following people:

My parents – their unwavering love, encouragement, help, and even jokes have been vital to writing this book. They've always been there for me and I know they always will. I love you, Mom and Dad.

My little sister, Molly – she is my unofficial publicist, telling everyone she meets to buy this book.

Judy Bass, my editor at McGraw-Hill – her enthusiasm for this project and her faith in me from the very beginning will always be valued.

Pete Ingram-Cauchi, the CEO of iD Tech Camps – his courses at UCLA and Stanford ignited my enthusiasm for all things programming.

The Compiler

Getting your computer to listen to you

You only need three things to start making your own games—a computer (PC, Mac, or Linux), this book ... and a compiler. This software program translates your code into a language your computer can understand. If you don't already have a compiler, it can be downloaded FREE through the Internet.

To install Java and its compiler, go to java.sun.com and click on "Java SE" to the right under "Popular Downloads."

Click "Get the JDK 6 with NetBeans 5.5"; JDK stands for *Java Development Kit*. This will allow the computer to understand the code. NetBeans is an IDE (Integrated Development Environment) which makes code writing easier.

Install NetBeans by clicking on the setup icon.

Now, the interface: Click "File">"New Project"

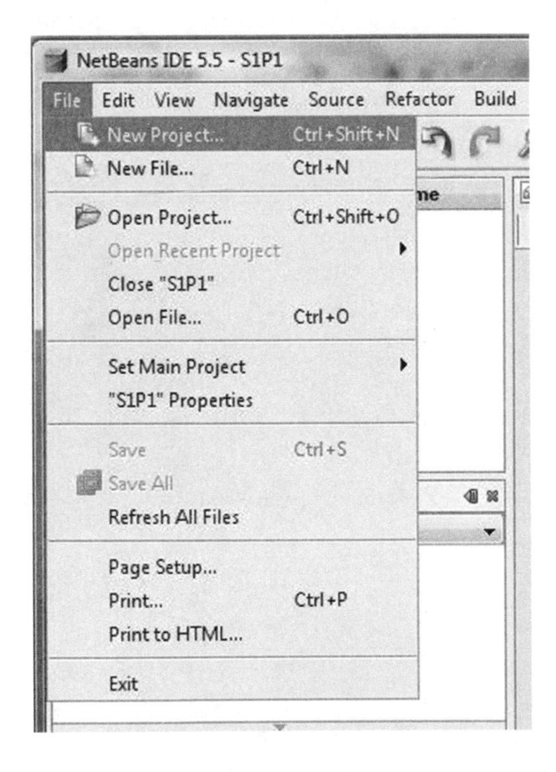

Click "Java Application"

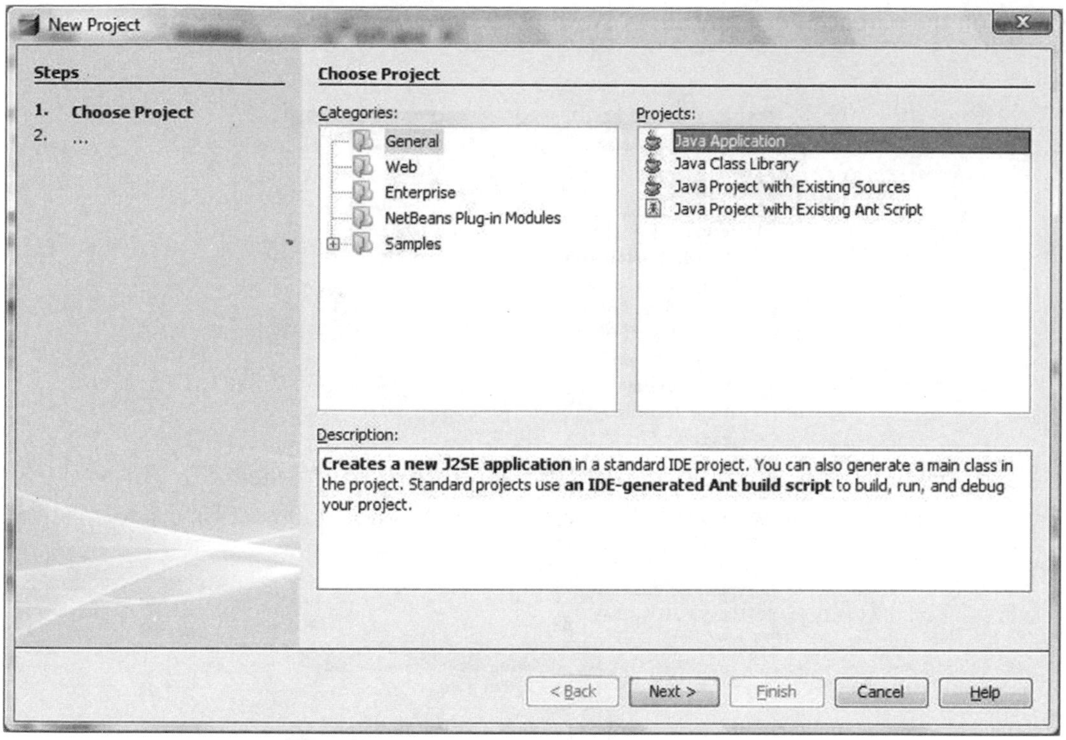

New Project

Steps

1. Choose Project
2. ...

Choose Project

Categories:

General
Web
Enterprise
NetBeans Plug-in Modules
Samples

Projects:

Java Application
Java Class Library
Java Project with Existing Sources
Java Project with Existing Ant Script

Description:

Creates a new J2SE application in a standard IDE project. You can also generate a main class in the project. Standard projects use **an IDE-generated Ant build script** to build, run, and debug your project.

< Back | Next > | Finish | Cancel | Help

Click "Next" and type in the Project name. It should begin with an uppercase letter and have no spaces.

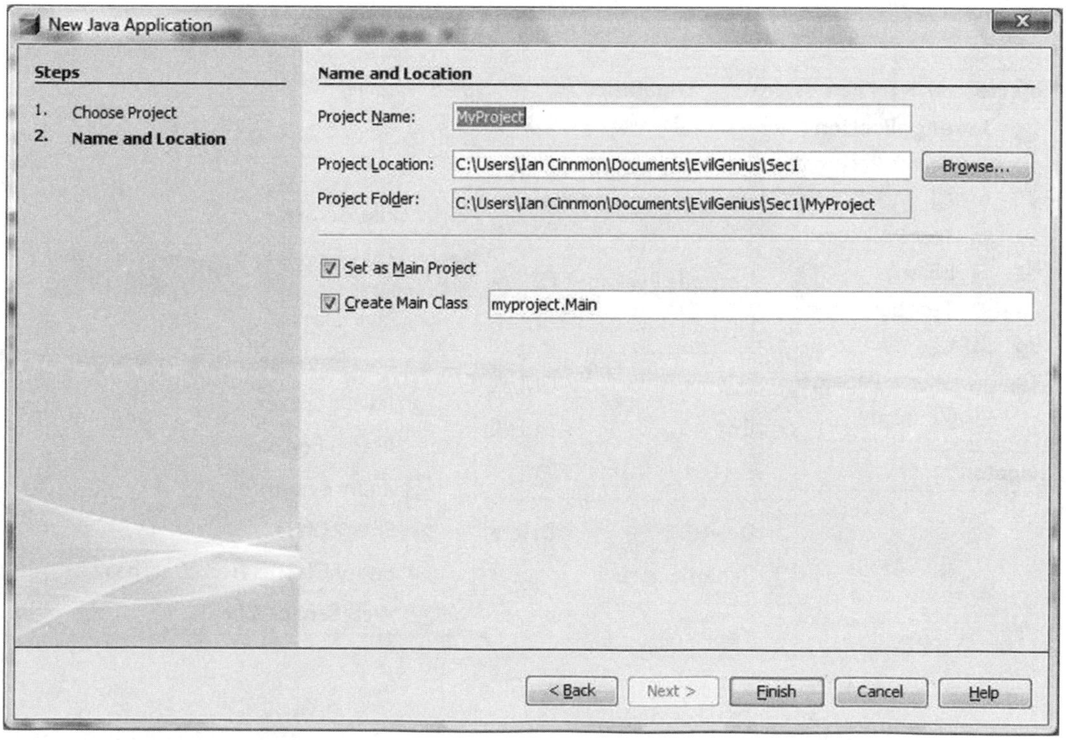

New Java Application

Steps

1. Choose Project
2. **Name and Location**

Name and Location

Project Name: MyProject

Project Location: C:\Users\Ian Cinnmon\Documents\EvilGenius\Sec1 | Browse...

Project Folder: C:\Users\Ian Cinnmon\Documents\EvilGenius\Sec1\MyProject

Set as Main Project

Create Main Class myproject.Main

< Back | Next > | Finish | Cancel | Help

Click "Finish."

In order to create an environment in which you can write your own code, start by deleting "Main.java," then right click on the parent folder, and click "New" > "Empty Java File ..."

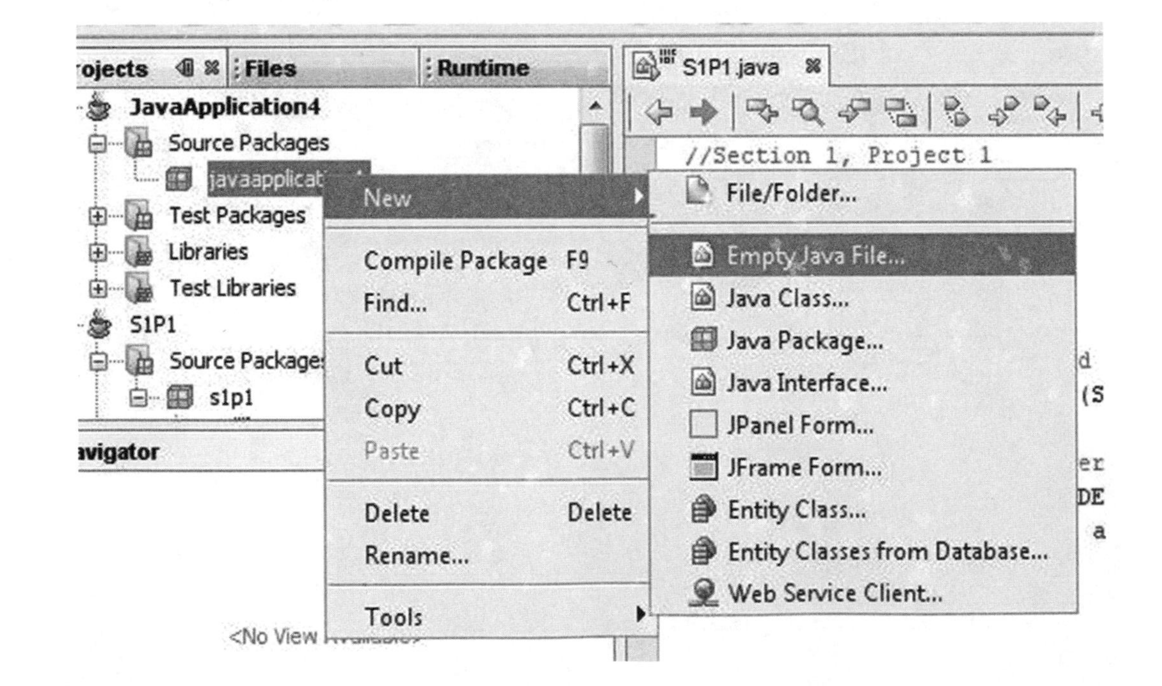

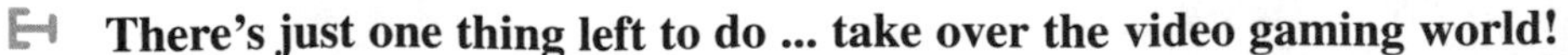

There's just one thing left to do ... take over the video gaming world!

Section One

Java Jumpstart

Project 1: The Repeater

Project

Type a message, press two buttons, and the computer will repeat it.

New building blocks

Classes, Statements, Variables

When first learning Java, there are four essential types of code to know: Comments, Statements, Declarations, and Flow Control.

Comments

Comments always begin with //

Compilers ignore comments.

They are only used so the programmer can document the code. The documentation lets you edit the code later with ease. Once the code increases to hundreds of lines long, it is extremely difficult to remember the function of each line. This will help.

Statements

These let you do things! Like printing to the screen ...

They ***always*** end in semicolons.

Declarations

Declarations use statements to create or modify variables.

Variables are the same as they are in algebra ($2x = 4$), except they store more than just numbers. We'll learn more about them in the next project.

Flow control

This lets you manipulate what statements to use. More about this later (Section 1, Project 4).

Every Java program exists within a container called a Class. A class is created with the following code:

```
public class <class name>
```

The class name should always match the file name, which should always begin with a capital letter and have no spaces.

To show that certain code belongs to a class, use the {character after the class name and} after the code.

Inside a class, Java will search for the main method. A method is a group of code that can be run multiple times. The main method is a special method—Java always calls it first (runs the code in it). This is what it looks like:

```
public static void main (String[] args)
```

For now, just know that this line of code must be present for the main method. We'll learn what each word means later.

Within a class (outside of a method) there can *only* be declarations and comments. These declarations are called class variables. Within a

```
System.out.println("Whatever you type here... will be repeated!");
```

Figure 1-1 *This code outputs text to the screen*

class and a method, there can be declarations, comments, flow control, and statements. Class variables can be accessed by all methods; method variables can only be accessed by the method in which they are created.

Here is a sample statement, which is also illustrated in Figure 1-1, that lets you print to the screen:

```
System.out.println("This stuff in quotes
is displayed.");
```

This code is a method that Java makes for us. All we need to do is tell it what to display (the words in quotes).

Within quotes, there are escape sequences. These sequences let you manipulate the text in cool ways. Just add them inside the quotes:

Escape Sequence	*Result*
`\n`	Creates a new line
`\t`	Indents the line (creates a tab)
`\"`	Lets you quote within a quote.

Making the game

So ... if you want to make a program to repeat the text you have entered, you must create a class, write a main method, and add the `System.out.println()` code. Whatever you put in the quotes, the computer will repeat.

Try this on your own. If it works, you win! Proceed to the next project.

If you need help, the completed code is below:

```
//create a class named S1P1
public class S1P1
{
  //this is the main method
  public static void main (String[ ] args)
  {
    //this prints whatever is below.
    System.out.println("Whatever you type here ... will be repeated!");
  }
}
```

Click "Build" > "Build Main Project" > "OK"

This compiles your game, as shown in Figure 1-2.

Click "Run" > "Run Main Project"

This runs the game, as shown in Figure 1-3.

Figure 1-4 illustrates the output of your first game!

In the next project, you will learn how to store important information (such as the "current score") and take input from the user!

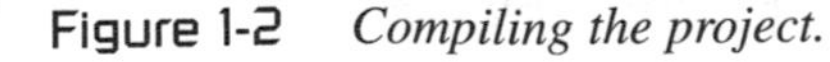

Figure 1-2 *Compiling the project.*

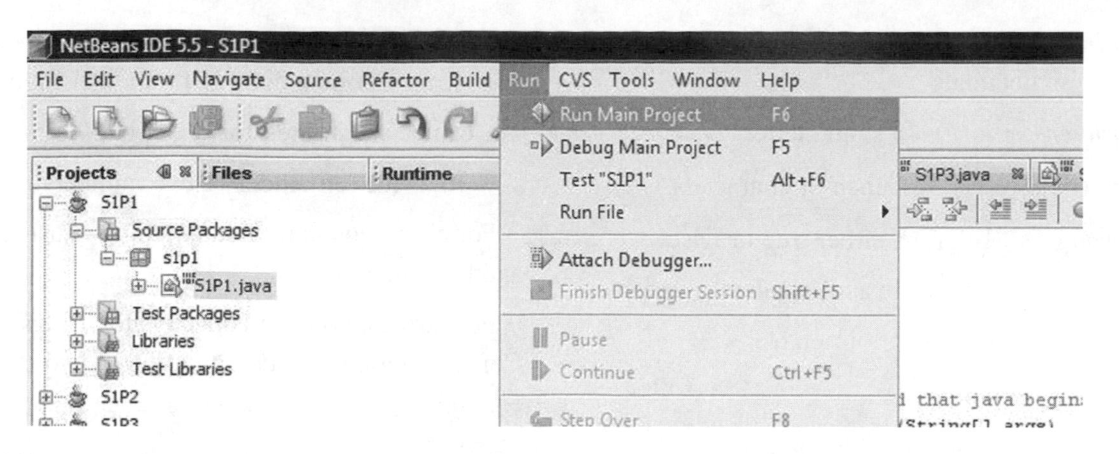

Figure 1-3 *Clicking this button starts the game.*

```
Output
Debugger Console   S1P1 (run)
init:
deps-jar:
Compiling 1 source file to C:\Users\Ian Cinnmon\Documents\evilGenius\Sec1\S1P1\build\classes
compile:
run:
Whatever you type here... will be repeated!
BUILD SUCCESSFUL (total time: 0 seconds)
Finished building S1P1 (run).
```

Figure 1-4 *Game output.*

Project 2: Guess the Password

Project

Program the computer to verify the correct password. In this two player exercise, one player gives hints while the other types in password guesses. The computer will let the player know when he/she is correct.

New building blocks

Variables (Creation and Use), If-statements, Casting, and Input

So far, you know how to create a simple program that outputs messages to the screen.

Now, you'll learn how to create variables, manipulate them, and test them.

Variable creation

This *always* ends in semicolons!!!!

Here are the basic variable types that you will use most often:

Integer (`"int"`)—a positive or negative whole number

Double ("double")—a positive or negative number with decimals

Character ("char")—a single letter

String ("string")—a bunch of characters

Boolean ("boolean")—either true or false

How to create (or declare and initialize) a variable:

For "int," "double," or "char:"

```
int <name> = <value>;
```

"int" can be replaced with "double" or "char"

<name> is replaced with any name starting with a lower case letter (no numbers or symbols)

For ints, <value> can be any number (e.g. - 10, - 9999, 298)

For chars, <value> should be the letter that is contained within single quotes (e.g. `char letter = 'l' ;`)

For Strings, <value> can be anything as long as it is contained in quotes (e.g. `String text = "I am a string.";`)

You can also create a variable and not set its value (this is called a Declaration). To do this, just end the line early (with a semicolon):

```
int number;
```

Or, if you have already declared a variable and want to change (or set) its value, use the following code (this is called initialization).

```
number = 1;
```

Or, to change the value of an int/double use the following:

```
number++;
```

This adds one to "number"

```
number- -;
```

This subtracts one from "number"

```
number+=5;
```

Add 5 (or any value/variable) to "number"

```
number = <variable> + <variable>;
```

This sets number equal to the sum of the variables. Variable can be a number or another int/double.

+ can be substituted for -,*, ?, or %.

% finds the remainder of two numbers

For chars, you can't add anything—just set it to another value

For Strings, you can concatenate Strings. Concatenation is the combination of two strings, e.g. `String txt = "This is added to" + "this."`

How to test your variables

Use "if-statements" to test variables with the following code:

```
if(<variable>==<variable>){}
```

<variable> can be replaced with any variable.

If the variable is a String, do not use ==. Instead, do the following (pretend <String> and <String2> are the names of two String variables)

```
if(<String>.equals(<String2>))
```

To test if it is not equal ...

```
if(!<String>.equals(<String2>))
```

Instead of = =, which tests for equality, you can use:

! = means not equals (for ints/doubles/chars)

> means greater than (for ints/doubles)

< means less than (for ints/doubles)

>= means greater than or equal to (for ints/doubles)

<= means less than or equal to (for ints/doubles)

So ... "if-statements" run the code in curly braces when the conditional is true. What if you want some code to run when the conditional is false? Or what if you want to execute code when the "if-statement" doesn't run???

Java can do that! Here's how:

```
if(<question>)
{
  //code for true
```

```
}
else if (<question2>)
{
  //if this is true
}
else
{
  //if none of the above are true
}
```

You can use any of the code above (or leave parts out) as long as the "if" is first and "else" (if you have one) is last.

There are only two ways to use booleans with if-statements (pretend boolean b exists):

1: `if(b)`

This means "if b has the value of true"

2: `if(!b)`

This means "if b has the value of false"

Booleans can be set like other variables:

```
boolean b = true;
```

Or ...

```
boolean b = false;
```

Input

Now that you know how to compare values, you need to learn how to access what the player is typing ... through input. First, add code to the very beginning of the program (before the class declaration). The code is below. Just memorize it for now. It will be explained later.

```
import javax.swing.*;
```

Now, insert the following code where you want to ask the player for input:

```
String input = JOptionPane.showInputDialog
("What's your name?");
```

This will create a dialog with an input area that asks "What's your name?" (see Figure 2-1).

Input is the value that the user entered. The value is always a string, but can be turned into other values.

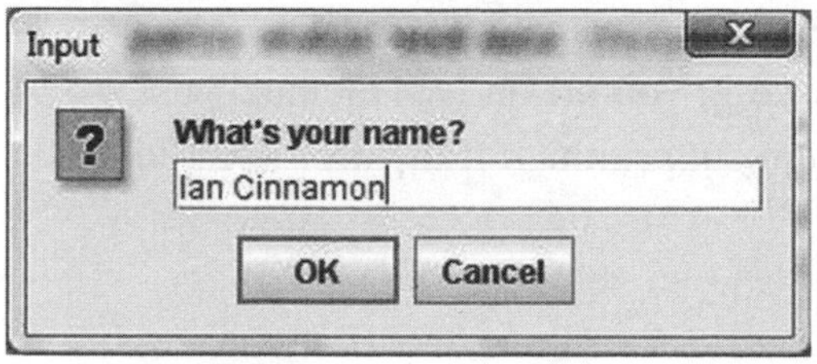

Figure 2-1 *Input dialog box.*

Note that input lessens the need for hard-coding. Hard-coding is when you actually program the value of a variable. Therefore, by getting input, you can set the variables after the code is compiled.

Casting

Casting allows you to turn one type of variable (such as a String) into another type of variable (such as an int). To do this, use the following commands (pretend you already have a String named input)

```
int inputInIntForm = Integer.parseInt(input);
double inputInDoubleForm =
Double.parseDouble(input);
```

You can now manipulate input!!

Making the game

This is a text-based game that models the final level of an RPG (role-playing game).

Set a password by using a String. Ask the user for the password. If he/she gets it right, display a celebratory message (e.g. "Congratulations! You guessed the password"). Now, let's get started on the details of the game.

First, create the class. Next, create the main method. Then, create a String variable. Call it "input." Set "input" equal to the user's input (use the JOptionPane code). Test the "input" variable against your secret password with the following code:

```
if(input.equals("secret password"))
```

Display a positive message (e.g. "You guessed it: you are an Evil Genius!" using System.out.println()) if the user guessed the correct password. If incorrect, display a fun insult (e.g. "LoSeR").

```
//first, allow for input getting
import javax.swing.*;

//create a class named S1P2
public class S1P2
{
  //main method
  public static void main (String[ ] args)
  {
    //this String will hold the user's input
    String input;

    //get input now
    input = JOptionPane.showInputDialog("Enter the secret message.");

    //test for correctness, "Evil Genius" is my secret message!
    if(input.equals("Evil Genius"))
    {
      //user got it right, so tell him/her!
      System.out.println("YOU GOT THE SECRET MESSAGE!!!");
    }
    //if user got it wrong ...
    else
    {
      //tell him/her ...
      System.out.println("WRONG!!! Hahaha!");
    }
  }
}
```

Figures 2-2 through 2-5 illustrate the game play of Guess the Password.

Want to pause the game or create random numbers to make artificial intelligence simpler? Just go on to the next project.

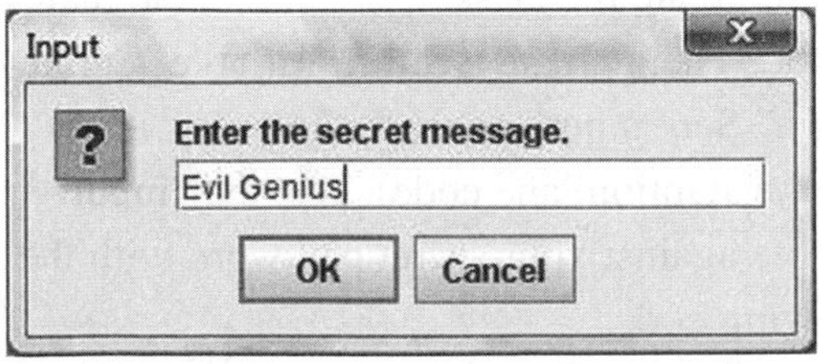

Figure 2-2 *Guessing the secret password.*

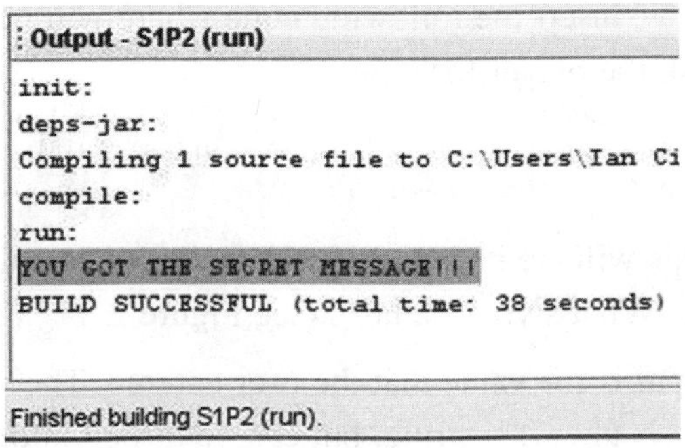

Figure 2-3 *Correct guess!*

Input

Enter the secret message.

I don't know!!!

OK Cancel

Figure 2-4 *Guessing a different password.*

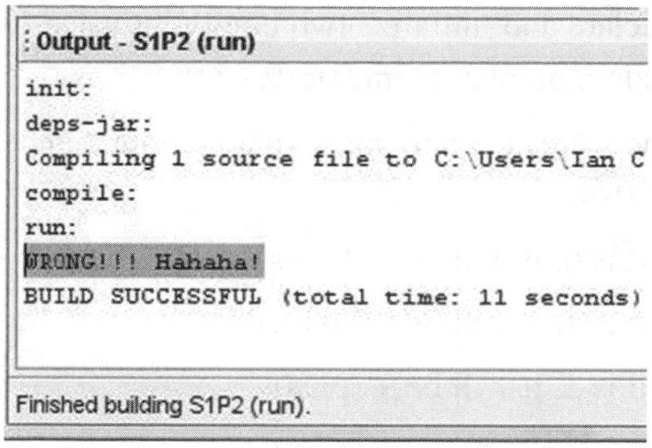

```
Output - S1P2 (run)
init:
deps-jar:
Compiling 1 source file to C:\Users\Ian C
compile:
run:
WRONG!!! Hahaha!
BUILD SUCCESSFUL (total time: 11 seconds)
Finished building S1P2 (run).
```

Figure 2-5 *Incorrect guess.*

Project 3: Number Cruncher

Project

The program displays a math equation.
The player must solve the problem before the computer counts down from three and displays the correct answer. Can you beat the computer?

New building blocks

Random numbers, Pausing

Random numbers

All random numbers are of type double, because they are between 0 (inclusive) and 1 (exclusive). There is a simple command to create random numbers.

To create a random double between the value of 0 and 1, use the following code:

```
Math.random()
```

This returns a value, similar to the way JOptionPane returns values. However, this returns a double instead of a String.

So ... if you want to create a random 1 digit number, try the following:

```
int rand =
(int)(Math.round(Math.random()*10));
```

(int) makes sure that it is int variable format

`Math.round()` rounds it to the nearest whole number

Pausing

You can make the computer pause for a given period by using a very simple command:

```
Thread.sleep(100);
```

100 is the number of milliseconds you want to sleep (this can also be an int variable).

Also, you must add something to the last part of the main method:

After (String[] args), add "throws Exception"

The entire line should look like:

```
public static void main (String[ ] args)
throws Exception
```

Making the Game

OK ... you can now make a game that can be used in math competitions around the world!

Create a math equation with two numbers using any of these operations: adding, subtracting, multiplying, dividing, or modding.

Declare and initialize two empty int values with a random number from 0 to 9.

Then, create a new int with a random number form 0–4.

If it is 0, it will be * (times)

If it is 1, it will be / (divided by)

If it is 2, it will be + (plus)

If it is 3, it will be – (minus)

If it is 4, it will be % (mod)

Create a new variable to hold the solution of the math problem.

Now, pause the program (to let the user answer the question).

Display the variable that holds the solution.

If the player calculates the answer correct before the computer displays it, the player wins!

Here's the code:

```
  //first, allow for input
import javax.swing.*;

//create a class named S1P3
public class S1P3
{
  //main method (throws Exception) added for Thread.sleep()
  public static void main (String[ ] args) throws Exception
  {
    //random numbers for the equaton
    int num1 = (int)(Math.round(Math.random()*10));
    int num2 = (int)(Math.round(Math.random()*10));
    //random number for the sign
    int sign = (int)(Math.round(Math.random()*3));
    //will store the answer
    int answer;

    //make stuff noticable:
    System.out.println("\n\n*****");

    if(sign==0)
    {
      //tell user and calculate answer
      System.out.println(num1+ " * "+num2);
      answer = num1*num2;
    }
    else if(sign==1)
    {
      //tell user and calculate answer
      System.out.println(num1+" / "+num2);
      answer = num1/num2;
    }
    else if(sign==1)
    {
      //tell user and calculate answer
      System.out.println(num1+" + "+num2);
      answer = num1+num2;
    }
    else if(sign==1)
    {
      //tell user and calculate answer
      System.out.println(num1+" - "+num2);
      answer = num1-num2;
    }
    else
    {
      //tell user and calculate answer
```

```
            System.out.println(num1+" % "+num2);
            answer = num1%num2;
        }
        //make it easier to read ...
        System.out.println("*****\n");
        //count down from 3
        System.out.println("3 ...");
        Thread.sleep(1000);
        System.out.println("2 ...");
        Thread.sleep(1000);
        System.out.println("1...");
        Thread.sleep(1000);
        //print the answer
        System.out.println("ANSWER: "+answer);
    }
}
```

An equation is displayed in Figure 3-1.

The computer counts down in Figure 3-2.

And the answer is displayed in Figure 3-3!

In the next project, you will start using loops. This technique allows you to repeat the code that runs the game. The screen refreshes itself so that the images are clear and move smoothly.

```
Output - S1P3 (run)

*****
1 % 7
*****
```

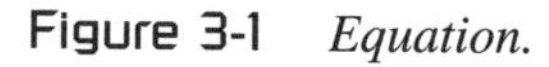
Figure 3-1 *Equation.*

```
Output S1P3 (run)
*****
1 % 7
*****

3...
2...
1...
```

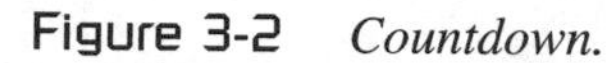
Figure 3-2 *Countdown.*

```
Output
Debugger Console  S1P3 (run)
*****
9 / 3
*****

3...
2...
1...
ANSWER: 3
BUILD SUCCESSFUL (total time: 3 seconds)
```

Figure 3-3 *Answer.*

Project 4: Number Cruncher Extreme

Project

This is an add-on to the previous exercise, Number Cruncher. The player can either opt to play again or change the difficulty level by increasing or decreasing the time allowed to calculate the answer.

New building blocks

Loops

Loops

It's time to learn some flow control techniques. Flow control lets you repeat and execute different parts of the code at different times, depending on a condition (e.g. x==4). There are three types of flow control, called loops:

"for" loops

`For loops` allow you to execute the code within the curly brace until a given condition is satisfied. Here's the format:

```
for(<variable init>; <condition>;
<iterator>) { //repeated code goes here}
```

`<Variable init>` is a standard variable declaration. (e.g. int i = 0;)

`<Condition>` can be i<0 or i >0 or i>=0 or i<=0 or anything else that can be inside the if statement's parentheses.

`<Iterator>` can be (and usually is) `i++`.

The process of a "for loop": When the JRE (Java Runtime Environment) reaches the variable declaration, it creates the variable (which exists only in the for statement). If the condition is true, the code runs once. Then, the iterator code runs. The condition is checked again and the process repeats.

Sample for loop:

```
for(int i = 0; i <10; i++) {}
```

Whatever is in the parentheses will execute ten times.

"while" loops

While loops are like for loops, except they do not directly include the variable declaration and iteration. They only include the condition.

A sample `while loop`:

```
int i = 0;
while(i<10) { i++;}
```

The process of a "while loop": When the compiler encounters a while loop, the condition is checked. If true, it will enter the loop and continue to recheck the condition after every iteration.
If false, it stops looping.

"do ... while" loops

Do ... while loops are almost identical to while loops.

Here's the format:

```
int i = 0;
do
{
i++;
}
while(i <10);
```

The Process of a "do ... while loop": `Do ... while loops` are the same as while loops, except the code in braces always executes once (before the condition is checked).

Special keywords

The most important keyword when working with loops is "break," which immediately exits the loop. Here's an example:

```
for(int i = 0; i <10; i++)
{
  if(i==1)
  break;
}
```

Making the game

Change the delay time and degree of difficulty of the previous exercise, Number Cruncher (1 = evil genius, 10 = mathematically challenged).

Let the user input the difficulty level and count down from that level to zero. The higher the level, the more time the player is given to answer. The lower the level, the less time is given to answer.

You can ask the user if he/she wants the program run again (using a `do ... while` loop).

Occasionally, the program may try to divide by zero, which will cause it to "throw an exception" (quit). To fix this bug, put the second number random code inside a `do ... while` loop. If the number is zero, run it again!

You may have realized that the answers are not always exact ... they are ints, not doubles. With ints, the equation $^3/_4$ produces 1, not 0.75.

Hint:

To make the solution value more exact, change all the variables of type "int" (except the random sign variable) to type "double."

Here's the code:

```
//first, allow for input getting
import javax.swing.*;

//create a class named S1P3
public class S1P4
{

//main method (throws Exception) added for Thread.sleep()
  public static void main (String[ ] args) throws Exception
  {
    //this will be how many 1/2 seconds the user gets
    int difficulty;
    difficulty = Integer.parseInt(JOptionPane.showInputDialog("How good are you?\n"+
               "1 = evil genius...\n"+"10 = evil, but not a  genius"));
    //this will tell the loop whether to continue or not:
    boolean cont = false;

    //the contents of the main method are about to be enclosed in a do
    //while loop...
    do
    {
      //reset cont to false
      cont = false;

      //random numbers for the equaton
      double num1 = (int)(Math.round(Math.random()*10));

      //this do..while loop prevents exceptions
      //num 2 must be declared outside of the do while so
      //the "while" part can see it. It will still be initialized
      //inside of the do part, though.
      double num2;
      do
      {
        //init num2
        num2 = (int)(Math.round(Math.random()*10));

      }
```

```
while(num2==0.0); //if it is 0, do it again!
//random number for the sign
int sign = (int)(Math.round(Math.random()*3));
//will store the answer
double answer;
//make stuff noticable:
System.out.println("\n\n*****");
if(sign==0)
{
  //tell user and calculate answer
  System.out.println(num1+" times "+num2);
  answer = num1*num2;
}
else if(sign==1)
{
  //tell user and calculate answer
  System.out.println(num1+" divided by "+num2);
  answer = num1/num2;
}
else if(sign==1)
{
//tell user and calculate answer
  System.out.println(num1+" plus "+num2);
  answer = num1+num2;
}
else if(sign==1)
{
  //tell user and calculate answer
  System.out.println(num1+" minush "+num2);
  answer = num1-num2;
}
else
{
  //tell user and calculate answer
  System.out.println(num1+" % "+num2);
  answer = num1%num2;
}
//make it easier to read...
System.out.println("*****\n");
//count down from difficulty... use a for loop!!!
for(int i = difficulty; i >= 0; i- -)
{
  //count down at double speed!
  System.out.println(i+"...");
  //instead of waiting a second,
  //this time only wait 1/2 second
  //per difficulty level.
  Thread.sleep(500);
}
//print the answer
System.out.println("ANSWER: "+answer);
//ask the user if he/she wants to play again
String again;
again = JOptionPane.showInputDialog("Play again?");
//if the user says yes, set cont to true.
if(again.equals("yes"))
  cont = true;
}
```

```
      while(cont); //keep going until continue is false
    }
}
```

Figures 4-1 through 4-3 illustrate the game in play.

Turn the page to learn how to save information to files. This process allows you to call up a player's progress in any game you create.

Figure 4-1 *Inputting the difficulty level*

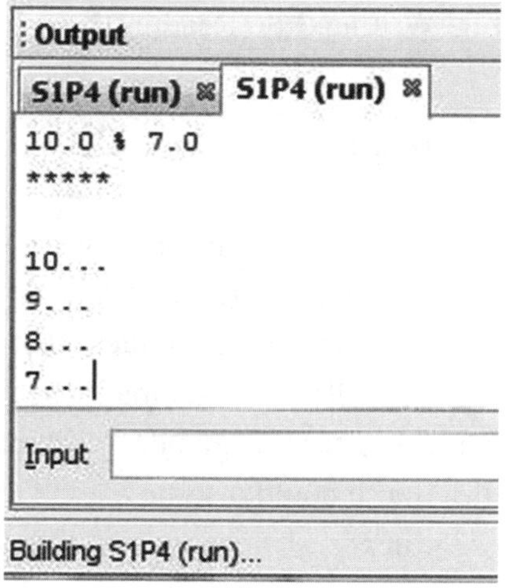

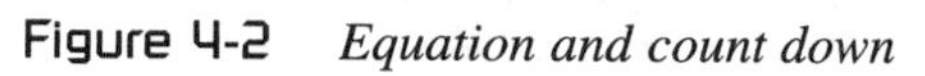

Figure 4-2 *Equation and count down*

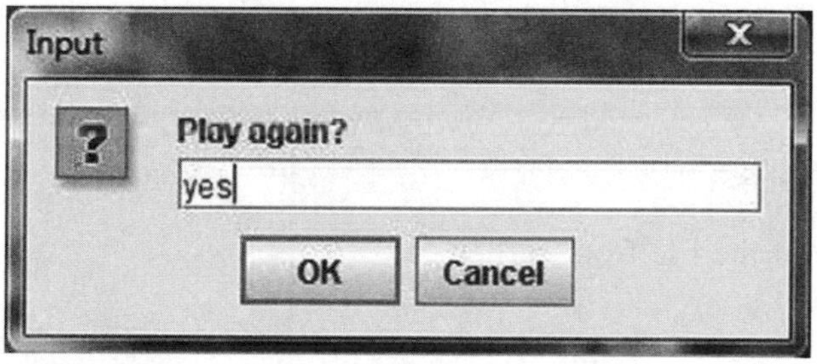

Figure 4-3 *Play again?*

Project 5: Crack the Code

Project

This is an add-on to Project 2, Guess the Password. Instead of hard-coding the password, the user can set the password through the pop-up windows. Also, the password will be permanent: it will be saved to a file.

New building blocks

File IO (File Writing, File Reading)

Creating files

Every type of file is saved the same way. The only difference is the extension (.doc, .txt, .avi, .jpg, etc). Extensions exist to tell the computer what program to use to open the file. For example, when you store the password for this game, you'll make a .psswrd file. But, it could also be a .evilGenius or .<anything> file.

First, create a File by using the following code:

```
File file = new File("password.psswrd");
```

So far, this does nothing. It simply holds a space for a file named "password.psswrd." The file is shown in Figure 5-1.

Now, you'll learn how to actually save the above file to your computer. First, you must write the following code:

```
FileOutputStream outStream = new
FileOutputStream(<file name>);
```

"`<File name>`" is the name of the file from earlier (in this case, "file")

"`outStream`" is the variable name; it can be renamed anything

```
PrintWriter out = new
PrintWriter(outStream);
```

"`outStream`" is the name of the `FileOutputStream` from above

"`out`" is also a variable name

So far, you have designated space for a file and prepared the computer to save it. Use the following code to do the fun part—actually save the file:

```
out.println(<this text is written to the
file>);
```

"`<This text is written to the file>`" is usually of type String

After you have completed making the file, you must tell the computer that you are done. Use the following code:

```
out.flush();
out.close();
outStream.close();
```

Good job! Now you know how to write and save files.

Accessing files

Once again, create a File object:

```
File file = new File("password.psswrd");
```

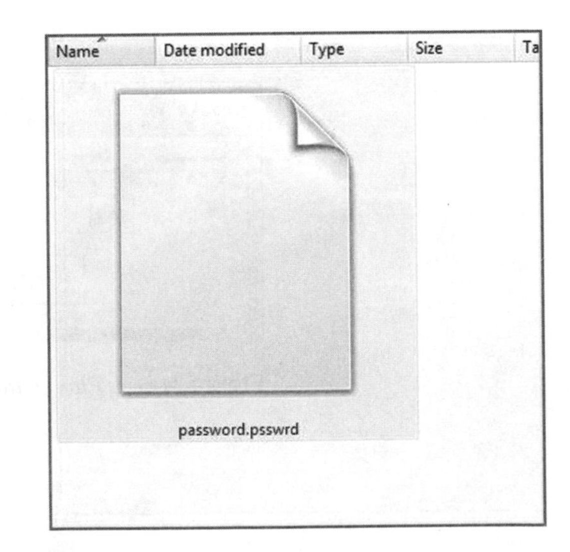

Figure 5-1 *Your file.*

This time, the String (right now "password.psswrd") is the name of the file you will be opening.

Next, create a FileReader. A FileReader tells the computer to open the file and prepare to read the text.

```
FileReader fr = new FileReader(file);
```

`"file"` is the File to access (from above)

`"fr"` is a variable name

Now, create a BufferedReader. A BufferedReader tells the computer to read the text from the opened file.

```
BufferedReader buffer = new
BufferedReader(fr);
```

`"fr"` is the name of the FileReader

`"buffer"` is a variable name

To access the first line of your file, use the following:

```
String line = buffer.readLine();
```

Once you have finished accessing the file, you must tell the computer you are done:

```
buffer.close();
fr.close();
```

Remember to `"throws Exception"` ***and*** `"import java.io.*;"`

Making the game

When the player first opens the game, two options will be displayed: play the game or reset the game.

If the player chooses to reset the game, a new password must be set (which will be saved to a file).

If the player chooses to play the game, he/she will be allowed to try to guess the password.

To write the game, use a JOptionPane to offer the player the above options.

If the player wants to enter a password (reset the game), use your file writing code and save the new password to the file.

If the player opts to try to crack the password, access the file and check the player's guess with the text in the File.


```
//first, allow for input getting
import javax.swing.*;
import java.io.*;

//create a class named S1P2
public class S1P2
{
  //main method
  public static void main (String[ ] args) throws Exception
  {
    //this String will hold the user's input
    String input;

    //get input now
    input = JOptionPane.showInputDialog("1 to set password,\n"+"2 to unlock the message");
    //this is the file that will be set and opened
    File file = new File("password.psswrd");

    //test for entering or setting the password
    if(input.equals("1"))
    {
      //setting the password...

      //get the password
      String p = JOptionPane.showInputDialog("Enter the password to set");

      //these are the two lines we learned about...
      FileOutputStream outStream = new FileOutputStream(file);
```

```
    PrintWriter out = new PrintWriter(outStream);

    //set the password
    out.println(p);

    //close it all
    out.flush();
    out.close();
    outStream.close();
  }
  //if user wants to test the password
  else
  {
    //first, we must get the password:
    FileReader fr = new FileReader(file);
    BufferedReader buffer = new BufferedReader(fr);

    //this is the password in the file
    String password = buffer.readLine();

    //get the user's attempted password
    String userPass;
    userPass = JOptionPane.showInputDialog("Enter your guess...");

    //test the password:
    if(password.equals(userPass))
    {
      //if correct
      JOptionPane.showMessageDialog(null,"CORRECT!!!!");
    }
    else
    {
      //if incorrect:
      JOptionPane.showMessageDialog(null,"WRONG =(");
    }
  }
 }
}
```

Figures 5-2 through 5-8 illustrate the game play of Crack the Code.

Figure 5-2 *Set or guess the password.*

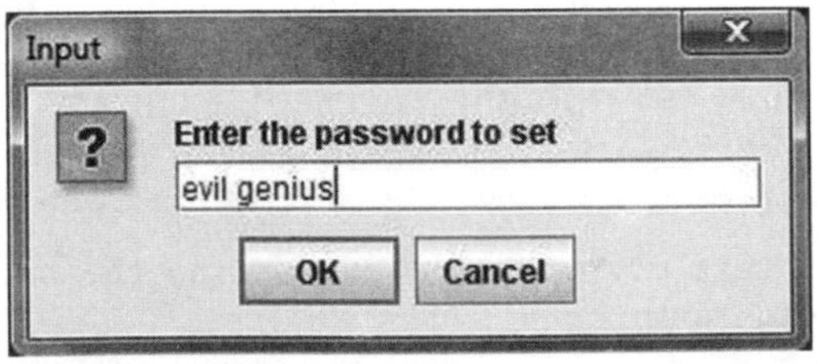

Figure 5-3 *Enter the password.*

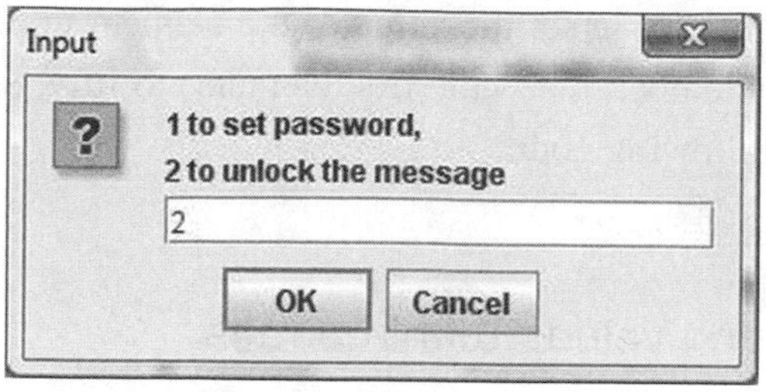

Figure 5-4 *Guess the number.*

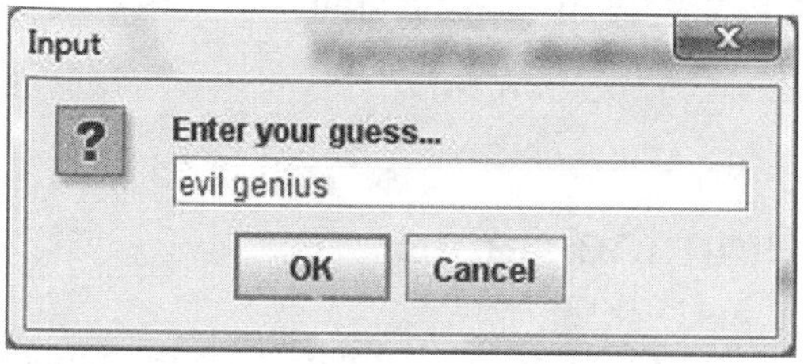

Figure 5-5 *Correct guess;*

Figure 5-6 *Congrats!*

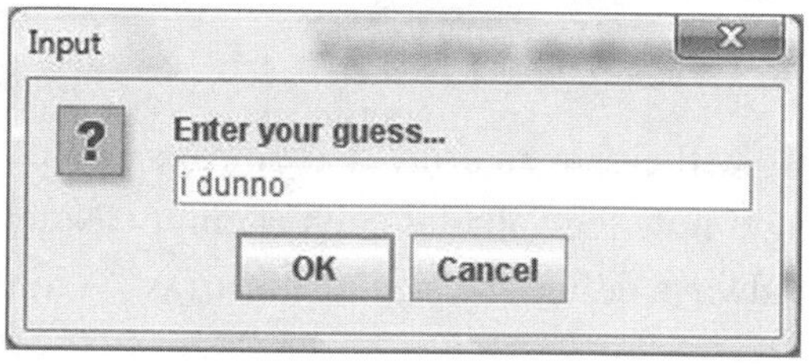

Figure 5-7 *Replay the game.*

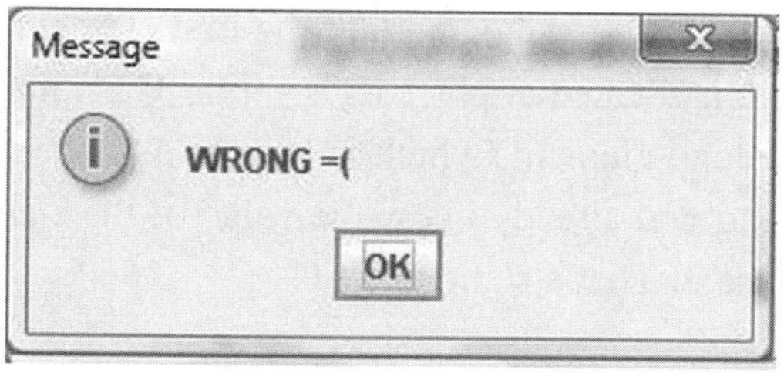

Figure 5-8 *Incorrect password.*

In the next project, you will learn how to hold large amounts of similar objects. A very useful tool in game play when tracking enemies or power ups.

Project 6: Virtual Game Library

Project

Keep an archive of the games you have created.

New building blocks

Arrays

Now that you know how to make some cool basic games, you need to know how to store them for easy access. This is accomplished with data structures.

The data structure that will be covered in this project is called an "array." An "array" lets you store many variables, as long as they are the same type (e.g. int, int, int or double, double, double). This way, you can look up the value of a variable in the array based on its position.

The downside of arrays? It is difficult to add new variables, which are called elements when in an array. But have no fear! A different data structure that allows you to easily add new elements will be introduced in the next project.

Arrays

To simultaneously create and initialize an array, use the following code:

```
int bunchOfInts[ ] = {-981,45,-6,7};
```

This creates an array called "bunchOfInts" with the arbitrary values –981, 45, –6, and 7.

To just create an array without initializing the elements, use the following code:

```
int sixInts[ ] = new int[ 6];
```

This will create an array that holds six ints, but right now the value of each is empty. Note: you must always define the size of the array — in this case, six.

To access or edit a value/element in an array, remember one important fact: The element count starts at ***zero***. For example, in the array `"bunchOfInts,"` the value of the first element (which is located at place 0) is –981. The value of the second element (which is located at place 1) is 45. And you already know the value of the element in place 2. That's right! It's –6.

Setting values within arrays

```
<arrayName>[ <element position to edit>] =
<value>;
```

If you want to set the value of the element in place 0 (remember, that's the first element) to fifty, use the following code:

```
bunchOfInts[ 0] = 50;
```

Getting values within arrays

```
<variable> = [ <element position to edit>]
=<value>;
```

If you want to find the value of the element in place 0, use the following code:

```
int num = bunchOfInts[ 0];
```

Making the game

Create a virtual library that can store the names of the games you have created. The Virtual Game Library allows you to access the games by entering the element number.

First, put everything in a `do ... while` loop so the program can be run again if the user desires.

Then, create and initialize an array (at the same time) of type String with the values being the titles of the previous programs/games.

Next, use a JOptionPane to get input (the element number).

By using the new JOptionPane output code display the name of the game.

```
//this import stuff lets you use JOptionPane
import javax.swing.*;
public class S1P5
{
  public static void main (String[ ] args)
  {
    //this will be used in the while part of the do...while
    boolean cont = false;
    do
    {
      cont = false; //reset cont
      //this is the array of Strings with the game names
      String names[ ] = {"Define: \"Games\"","The Dungeon Defender",
      "Regional Math-a-thon","National Math-a-thon"};
      //now, we'll ask the user which name to return
```

```
      int element = Integer.parseInt(JOption Pane. showInputDialog("Which element?"));
      //this will be outputted in the output JOptionPane
      String output = "The Name of the Game is:\n";

      //concat! And get the element
      output+=names[ element] ;

      //this is the output JOptionPane
      JOptionPane.showMessageDialog(null,output);

      //get input for repeating
      String repeat =JOptionPane.showInput Dialog("Again?");
      if(repeat.equals("yes"))
        cont = true;
    }
    while(cont); //while cont is true, repeat
  }
}
```

Figures 6-1 through 6-3 illustrate the use of arrays in Virtual Game Library.

The next step is learning about ArrayLists. ArrayLists are similar to arrays, except they do more and are easier to use. Continue on!

Figure 6-1 *Element number of the game is entered.*

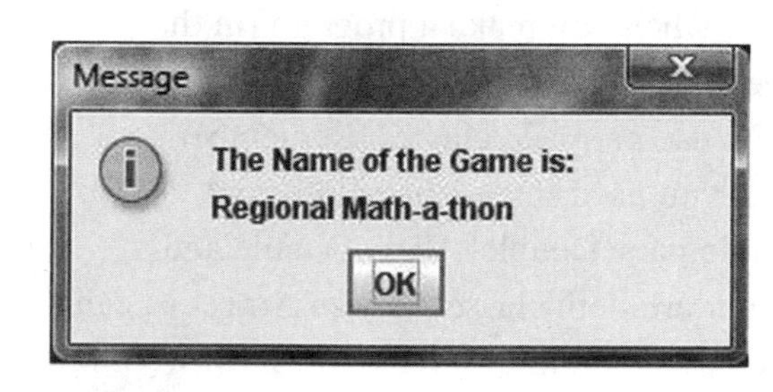

Figure 6-2 *Title is displayed.*

Figure 6-3 *Program repeats.*

Project 7: Virtual Game Library Pro Edition

Project

This is an add-on to Virtual Game Library. In the previous program, the games in the library had to be hard-coded in. Now you can add new games without having to alter the code.

New building blocks

ArrayLists

ArrayLists (java.util.*; must be imported)

ArrayLists hold classes (also known as objects). Yes ... this is the same as the class you create when you make a program (in the code "public class"). For now, however, you'll use premade classes. Remember when you used Integer.parseInt() and Double.parseDouble? Well, Double and Integer are both classes that an ArrayList can hold. To keep things simple, for now, we'll only use these two classes.

All classes (Integer and Double) belong to a hierarchy. The highest member of every class is called an "Object." Therefore, an ArrayList always returns an "Object," because it can be cast into the specific class (Integer or Double) you originally passed in.

```
//import ...
import javax.swing.*;
import java.util.*;

public class S1P6
{
  public static void main (String[ ] args)
  {
    //the ArrayList
```

Here is the code to make an ArrayList (notice how you do not need to specify a type):

```
ArrayList structure = new ArrayList();
```

You cannot initialize values on the same line. To add a value, use the following code:

```
structure.add(new Integer(5));
```

You can add as many elements as you like.

To get a value (if you know the element (which ***always*** begins at place 0), use the following code:

```
Object tempObj = structure.get(0);
Integer tempInt = (Integer)tempObj;
int finalNum = tempInt.intValue();
```

The first line retrieves the Object from the ArrayList. The second line turns the object into an Integer (to turn it into a Double, replace "Integer" with "Double"). The last line turns the Integer into an int (if you are working with Doubles, change `"intValue"` to `"doubleValue"` and `"int"` to `"double."`).

Making the game

Use the previous projects's code. Only a few changes are needed:

Ask the user to enter either 1 to add a new game or 2 to access a game. Next, turn the array into an ArrayList. Then, if the user enters 2, use the previous project's code to return the value.
If the user enters 1, add the new String to the ArrayList.

```
    ArrayList games = new ArrayList();

      //this will be used in the while part of the do...while
    boolean cont = false;
    do
    {
      cont = false; //reset cont

      //what do you want to do?
      int choice = Integer.parseInt(JOptionPane.
              showInputDialog(“Enter\n”+“1 to add a new game\n”+“2 to access games”));

      if(choice==1)
      {
        //get the name
        String name;
        name = JOptionPane.showInputDialog(“Game name?”);

        //add it!
        games.add(name);
      }
      if(choice==2)
      {
        //now, we' ll ask the user which name to return
        int element = Integer.parseInt(JOptionPane. showInputDialog(“Which element?”));

        //this will be outputted in the output JOptionPane
        String output = “The Name of the Game is:\n”;

        //concat! And get the element
        output+=((String)games.get(element));

        //this is the output JOptionPane
        JOptionPane.showMessageDialog(null,output);
      }
      //get input for repeating
      String repeat =
      JOptionPane.showInputDialog(“Again?”);
      if(repeat.equals(“yes”))
        cont = true;

    }
    while(cont); //while cont is true, repeat
  }
}
```

Figures 7-1 through 7-6 illustrate the use of ArrayLists in Virtual Game Library Pro Edition.

Proceed to the next project and create an exciting number guessing game utilizing the many skills and concepts you have learned.

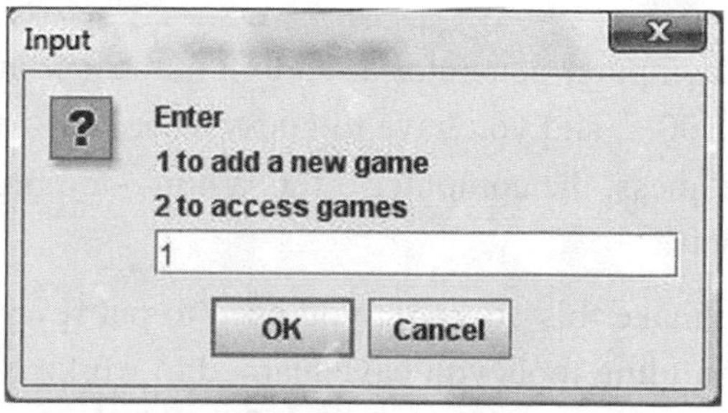

Figure 7-1 *Either access existing games or add new games.*

Input
Game name?
National Math-a-thon
OK Cancel

Figure 7-2 *Add a new game.*

Input
Enter
1 to add a new game
2 to access games
2
OK Cancel

Figure 7-3 *Access existing game.*

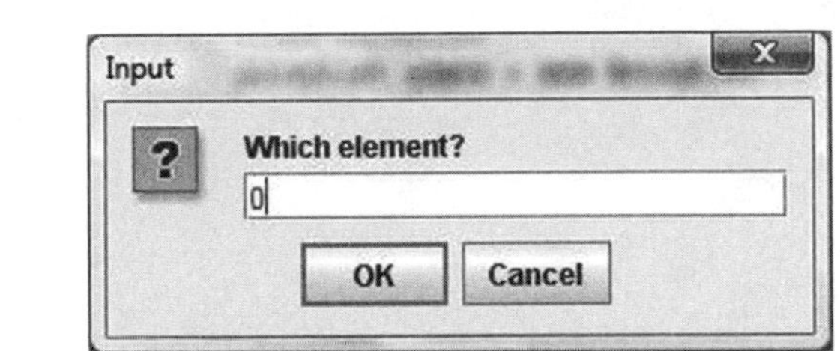

Figure 7-4 *Enter game number.*

Message
The Name of the Game is:
National Math-a-thon
OK

Figure 7-5 *Name of the game.*

Input
Again?
no
OK Cancel

Figure 7-6 *Rerun the program.*

Project 8: Number Guesser

Project

The computer generates a random number from 0 to 100 ... and you have to guess what it is! After each guess, the computer offers a hint—too high or too low.

To make this program, you need to put several of the building tools you have learned to work: Classes, Statements, Variables, Input, Loops, Comments, Output, Casting, and If-Statements

Making the game

Start by generating the random number. Then, use a `do ... while` loop. Get input (the user's guess) and compare it with the correct number using if-statements. This will provide hints for the player. The `do ... while` will exit if the guess is correct.

```
import javax.swing.*;

public class S1WrapUp
{
  public static void main (String[ ] args)
  {
    //this will hold the user's guess'
    int guess = -1;

    //number of user guesses
    int count = 0;

    //create the number:
    int num = (int) (Math.random()*100);

    //this is the loop to ask the user:
    do
    {
      guess = Integer.parseInt(JOptionPane. showInputDialog("Guess a number between 0 and
            100!"));
      if(guess>num)
        JOptionPane.showMessageDialog(null,"Too high");

      if(guess>num)
        JOptionPane.showMessageDialog(null,"Too low");

      count++;
    }
    //keep going until the user gets the number
    while(num!=guess);

    JOptionPane.showMessageDialog(null,"You guessed the number - "+num+" - in "
    +count+" guess(es)!!!");

  }
}
```

Figures 8-1 through 8-12 depict the game play of Number Guesser.

If you have trouble writing this code, don't despair! Just review the previous projects to refresh your knowledge of the concepts.

If you are able to do this on your own, consider yourself well on your way to becoming a true Evil Genius!

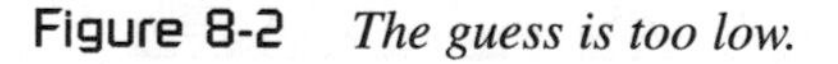

Figure 8-2 *The guess is too low.*

Figure 8-1 *Guess of 50.*

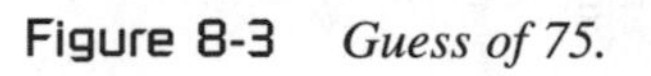

Figure 8-3 *Guess of 75.*

Figure 8-4 *The guess is too low.*

Figure 8-5 *Guess of 88.*

Figure 8-6 *88 is too low.*

Figure 8-7 *Guess of 94.*

Figure 8-8 *Even 94 is too low.*

Figure 8-9 *Guess of 97.*

Figure 8-10 *Still too low!*

Figure 8-11 *Guess of 99.*

Figure 8-12 *99 is CORRECT!!!*

Java Jumpstart quick review

Comments Comments let you document code so you know what each part of your game does.

Statements Statements are commands. They are the backbone of all games.

Flow control Loops let you repeat code. This is useful if the user wants to retry the game if he/she has lost.

Variables Variables allow you to keep track of the player's score, points, etc.

If-statements If-statements let you test variables. For example, you can determine whether a player has any lives left ... or if he's stone cold dead.

JoptionPane JOptionPanes let you get input and output. Output is used to display a player's score. Input is used to access the user's name for a high-scores list.

Random numbers Random numbers allows for artificial intelligence (make an enemy move randomly, etc.).

Pausing Pausing allows the screen to refresh so the game play graphics will be at optimum clarity.

Arrays Arrays keep track of the game environment (e.g. a list of cars in a racing game, a list of bullets in a shoot 'em up).

ArrayLists Arraylists are used to store players' names and scores.

File IO FileIO allows for a permanent high-score list on any game.

Section Two

Racing Games

Project 9: Radical Racing—The Track

Radical Racing

Race cars you build around a track you design with exciting graphics and sound effects!

Project

Create the environment—a virtual racetrack using a Graphical User Interface (GUI).

New building blocks

JFrames, Methods

JFrames

A JFrame is one of the simplest ways to create a Graphical User Interface. It creates a pop-up box (similar to a JOptionPane) of a specified size. You can add graphics, buttons, text fields, labels, etc.

A JFrame, before any graphics or images are added, is shown in Figure 9-1.

To create a JFrame, first add "extends JFrame" at the end of the line "public class ..." For example, if you have a class called Game, the line would look like this:

```
public class Game extends JFrame
```

Now, change the main method so it contains only one line of code:

```
<class name> <var name> = new <class name>();
```

For example, in "public class Game" the only code inside the main method would look like this:

```
Game g = new Game();
```

This line of code executes the class's constructor. A constructor is similar to the main method, except it is automatically called when you create a new instance of the class (by using the code above). Therefore, all the code necessary to run your game should be in the constructor, not the main method.

A constructor is programmed by using the following code:

```
public <class name>()
{
  //code goes here
}
```

Using the "class Game" example, the code would look like this:

```
public Game(){}
```

Inside the constructor, you need to add the following four lines of code so that you can create the JFrame, set the title, and set the size:

```
super("title goes here");
setSize(400,400);
setVisible(true);
setDefaultCloseOperation(JFrame.EXIT_ON_CLOSE);
```

Good going! You have now created your first GUI! And remember, a GUI is like a blank canvas: you can easily add your own shapes, images, buttons, textfields, or anything else you can imagine! Read on to learn how to include threatening enemies, daring heroes, or even disgusting aliens.

Figure 9-1 *Simple JFrame.*

Type the following code outside of the constructor, but within the class:

```
public void paint(Graphics g)
{
  super.paint(g);
}
```

In this method, you can draw any shape you want — rectangles, squares, circles, arcs, or even giant letters. For now, however, we will focus on two commands: setting the color and drawing a rectangle.

To set the color, use the following code (you must import java.awt.*;):

```
g.setColor(Color.<any color>);
```

"<any color>" can be replaced with any color you like. In NetBeans, a list of colors to choose from will pop up when you begin typing "Color." The list is illustrated in Figure 9-2.

To draw a Rectangle, use the following code:

```
g.fillRect(<x>,<y>,<width>,<height>);
```

<x>, <y>, <width>, and <height> are all variables of type "int."

Let's draw a green rectangle like the one shown in Figure 9-3.

Now you know how to draw a simple racetrack!

Hint: Eventually, you are going to need to check if the car goes out of bounds. Plan ahead and save the points in an object called a Rectangle. This makes collision detection easy to program. You can then draw the Rectangle by getting its values.

To make and draw a Rectangle, use the following code:

```
Rectangle r1 = new Rectangle(100, 100, 300, 400);
g.fillRect(r1.x, r1.y, r1.width, r1.height);
```

And, if you want to re-draw the JFrame, simply type the following code anywhere in your program:

```
repaint();
```

Remember—the "public void paint ..." code is always executed when the program is first run.

Methods

Methods hold code and allow it to be run multiple times. The primary use of methods is to allow the programmer to repeat code more easily.

You can call (aka run) methods from the constructor. They can be called from the

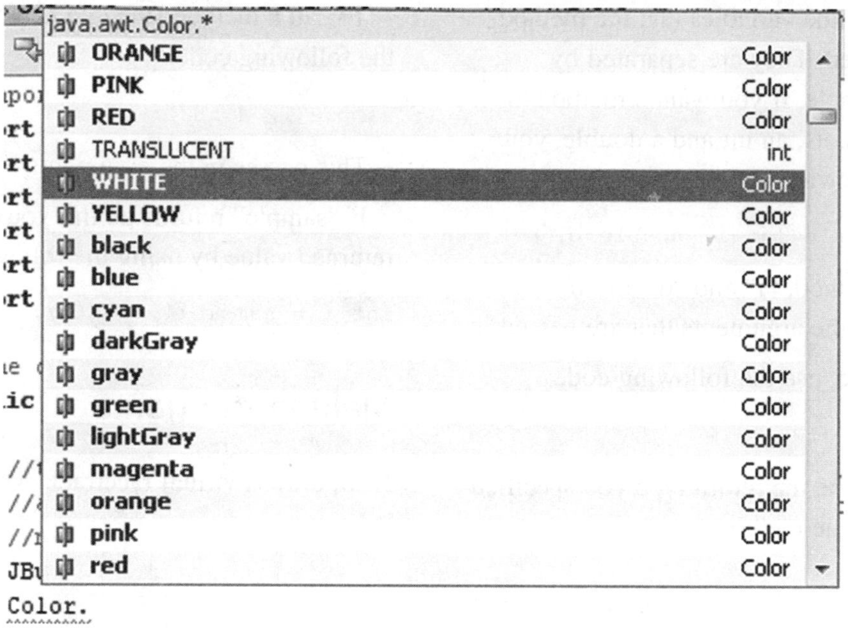

Figure 9-2 *Some of the many colors available.*

constructor or from another method (but not the main method). A method can take in a variable (also known as an argument), return (send back) a variable, do both, or do neither.

To create a method, use the following code:

```
public <return type> <name> (arguments)
{
    //code
}
```

"`<return type>`" should be "void" if the method returns nothing. If it returns an int, it should be "int." If it returns an Integer, it should say "Integer." It can be any primitive (int, double, char etc.) or object.

"`<name>`" is the name of a method. Name it anything, as long as it starts with a letter.

Figure 9-3 *Simple green rectangle inside of a JFrame.*

"arguments" are the variables that the method takes in when called. They are separated by commas. For example, if you want a method to take in two arguments, an int and a double, you would use the following code:

```
public void sample(int i, double d){}
```

Inside of the method, "i" and "d" are the variable names of the arguments that are passed in.

To return a value, use the following code:

```
return <variable>;
```

`<variable>` must be the return type you specified when you created the method.

To call a method (the one named "sample"), use the following code:

```
sample(1, 4.563);
```

This passes in the arguments "1" and "4.563."

If "sample" returns an int, you can get the returned value by using the following code:

```
int i = sample(1, 4.563);
```

Making the game

Now, draw a virtual racetrack. The completed code is below:

```
//import everything:
import javax.swing.*;
import javax.swing.event.*;
import java.awt.*;
import java.awt.event.*;
//this creates the class where you use JFrame
public class G1P1 extends JFrame
{
  //this is the constand that will hold the screen size
  final int WIDTH = 900, HEIGHT = 650;
  /*THE FOLLOWING ARE ALL OF THE RECTANGLES THAT WILL BE DRAWN*/
  /*
  *The following code (creating the Rectangles) may seem complicated at
  *first, but it only seems that way because it creates the pieces
  *based on the WIDTH and HEIGHT. In your version, you could just hard
  *code values.
  */
  //create rectangles that will represent the left, right, top, bottom,
  //and center
  Rectangle left = new Rectangle(0,0,WIDTH/9,HEIGHT);
  Rectangle right = new Rectangle((WIDTH/9)*9,0,WIDTH/9,HEIGHT);
  Rectangle top = new Rectangle(0,0,WIDTH, HEIGHT/9);
  Rectangle bottom = new Rectangle(0,(HEIGHT/9)*9,WIDTH,HEIGHT/9);
  Rectangle center = new
     Rectangle((int)((WIDTH/9)*2.5),(int)
            ((HEIGHT/9)*2.5), (int)((WIDTH/9)*5),(HEIGHT/9)*4);
  //these obstacles will obstruct the path and make navigating harder
  Rectangle obstacle = new
     Rectangle(WIDTH/2,(int)((HEIGHT/9)*7),WIDTH/10,HEIGHT/9);
  Rectangle obstacle2 = new
     Rectangle(WIDTH/3,(int)((HEIGHT/9)*5),WIDTH/10,HEIGHT/4);
  Rectangle obstacle3 = new
     Rectangle(2*(WIDTH/3),(int)((HEIGHT/9)*5),WIDTH/10,HEIGHT/4);
  Rectangle obstacle4 = new Rectangle(WIDTH/3,HEIGHT/9,WIDTH/30,HEIGHT/9);
  Rectangle obstacle5 = new Rectangle(WIDTH/2,(int)((HEIGHT/9)*1.5),WIDTH/30,HEIGHT/4);
```

```
  //the following rectangle is the finish line for both players
  Rectangle finish = new Rectangle(WIDTH/9,(HEIGHT/2)-HEIGHT/9, (int)((WIDTH/9)*1.5),
           HEIGHT/70);

  //the following rectangle is the start line for the outer player
  Rectangle lineO = new
     Rectangle(WIDTH/9,HEIGHT/2,(int)((WIDTH/9)*1.5)/2,HEIGHT/140);
  //the following rectangle is the start line for the inner player
  Rectangle lineI = new Rectangle(((WIDTH/9)+((int)((WIDTH/9)*1.5)/2)),
           (HEIGHT/2)+(HEIGHT/10),
           (int)((WIDTH/9)*1.5)/2, HEIGHT/140);

  //the constructor:
  public G1P1()
  {
    //the following code creates the JFrame
    super("Radical Racing");
    setSize(WIDTH,HEIGHT);
    setDefaultCloseOperation(JFrame.EXIT_ON_CLOSE);
    setVisible(true);
  }

  //this will draw the cars and the race track
  public void paint(Graphics g)
  {
    super.paint(g);

    //draw the background for the racetrack
    g.setColor(Color.DARK_GRAY);
    g.fillRect(0,0,WIDTH,HEIGHT);

    //when we draw, the border will be green
    g.setColor(Color.GREEN);

    //now, using the rectangles, draw it
    g.fillRect(left.x,left.y,left.width,left.height);
    g.fillRect(right.x,right.y,right.width,right.height);
    g.fillRect(top.x,top.y,top.width,top.height);
    g.fillRect(bottom.x,bottom.y,bottom.width,bottom.height);
    g.fillRect(center.x,center.y,center.width,center.height);
    g.fillRect(obstacle.x,obstacle.y,obstacle.width,obstacle.height);
    g.fillRect(obstacle2.x,obstacle2.y,obstacle2.width,obstacle2.height);
    g.fillRect(obstacle3.x,obstacle3.y,obstacle3.width,obstacle3.height);
    g.fillRect(obstacle4.x,obstacle4.y,obstacle3.width,obstacle4.height);
    g.fillRect(obstacle5.x,obstacle5.y,obstacle5.width,obstacle5.height);
    //set the starting line color to white
    g.setColor(Color.WHITE);
    //now draw the starting line
    g.fillRect(lineO.x,lineO.y,lineO.width,lineO.height);
    g.fillRect(lineI.x,lineI.y,lineI.width,lineI.height);
    //set the color of the finish line to yellow
    g.setColor(Color.YELLOW);
    //now draw the finish line
    g.fillRect(finish.x,finish.y,finish.width,finish.height);
  }

  //this starts the program by calling the constructor:
  public static void main (String[ ] args)
  {
    new G1P1();
  }
}
```

A screenshot of the completed track is displayed in Figure 9-4.

In the next project, you will make Radical Racing come to life by creating two cars that accelerate forward. Vroom! Vrooooom!

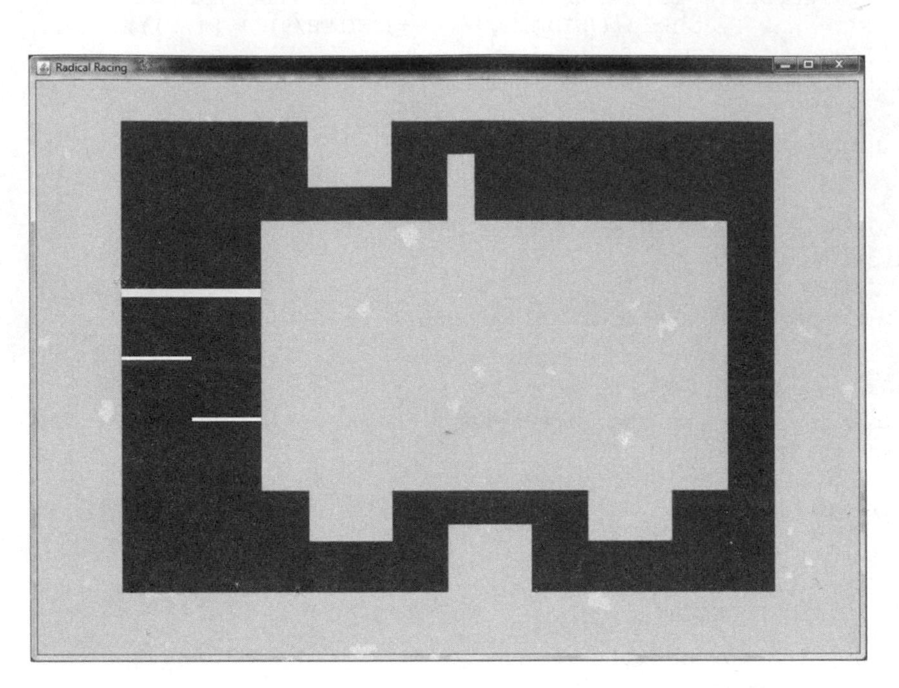

Figure 9-4 *Radical Racing track.*

Project 10: Radical Racing—The Cars

Project:

Now that you can create a GUI, you are going to make it come to life. Using Threads (which let multiple things happen at once), you will make two cars accelerate in a forward motion.

New building blocks:

Threads/Inner Classes

Threads/inner classes

To create a Thread, you must first program an "inner class." This is how you do it: type the following code inside the main class but outside of any methods or constructors:

```
public class <name> extends Thread
{
}
```

"<name>" can be anything, as long as it is different from the main class's name.

Inside of the thread, you must make a mandatory method (it must exist in order for the program to compile). Use the following code:

```
public void run() {}
```

Inside this method, we will want one of the cars to move forward. Therefore, there should be an infinite while loop:

```
while(true)
{
}
```

Inside the while loop, you should insert the following code.

```
try
{
  //code
}
catch(Exception e) {break;}
```

This code checks for errors inside the "try" block. If one is found, the computer executes the code in the "catch" block. In this case, if an error is found, "break" is called, which exits the infinite while loop.

The controls that make the car move should be in place of "//code"

Once you are done making your Thread, call it from the constructor using the following code (in this example, Move is the name of your inner class):

```
Move m = new Move();
m.start();
```

Making the game

First, draw a Rectangle to represent the car (you will learn to customize your own speed-demon racers images in Project 12). Then, program a Thread (using an inner class). Insert the run method, add the while loop, and add the "try ..." code. Next, create a global variable to keep track of the first car's speed.

Inside of the "try ..." code, slowly increase the speed until the car reaches its maximum velocity (use 4). Then, make the car move forward by altering the "y" value of its rectangle by adding it to the car's speed. Finally, refresh the screen and add in a delay (using `Thread.sleep(75);`).

If you can get that working, try adding the second car.

The completed code is below.

```
//import everything:
import javax.swing.*;
import javax.swing.event.*;
import java.awt.*;
import java.awt.event.*;

//this creates the class where you use JFrame
public class G1P2 extends JFrame
{
  //this is the constand that will hold the screen size
  final int WIDTH = 900, HEIGHT = 650;

  //these will keep track of each player's speed:
  double p1Speed =.5, p2Speed =.5;

  //create rectangles that will represent the left, right, top, bottom, and
  //center
  Rectangle left = new Rectangle(0,0,WIDTH/9,HEIGHT);
  Rectangle right = new Rectangle((WIDTH/9)*8,0,WIDTH/9,HEIGHT);
  Rectangle top = new Rectangle(0,0,WIDTH,HEIGHT/9);
  Rectangle bottom = new Rectangle(0,(HEIGHT/9)*8,WIDTH,HEIGHT/9);
  Rectangle  center  =  newRectangle((int)((WIDTH/9)  *2.5),(int)((HEIGHT/9)*2.5),(int)
          ((WIDTH/9)*5), (HEIGHT/9)*4);
  //these obstacles will obstruct the path and make navigating harder
  Rectangle obstacle = new Rectangle(WIDTH/2,(int)((HEIGHT/9)*7),
          WIDTH/10,HEIGHT/9);
  Rectangle obstacle2 = new
       Rectangle(WIDTH/3,(int)((HEIGHT/9)*5),WIDTH/10,HEIGHT/4);
  Rectangle obstacle3 = new
```

```
   Rectangle(2*(WIDTH/3),(int)((HEIGHT/9)*5),WIDTH/10,HEIGHT/4);
Rectangle obstacle4 = new Rectangle(WIDTH/3,HEIGHT/9,WIDTH/30,HEIGHT/9);
Rectangle obstacle5 = new Rectangle(WIDTH/2,(int) ((HEIGHT/9)*1.5),WIDTH/30,HEIGHT/4);

//the following rectangle is the finish line for both players
Rectangle finish = new Rectangle(WIDTH/9,(HEIGHT/2)-HEIGHT/9,
          (int) ((WIDTH/9)*1.5),HEIGHT/70);

//this is the rectangle for player 1's (outer) car:
Rectangle p1 = new Rectangle(WIDTH/9,HEIGHT/2, WIDTH/30,WIDTH/30);

//this is the rectang;e for player 2's (inner) car:
Rectangle p2 = new Rectangle(((WIDTH/9)+((int)((WIDTH/9)*1.5)/2)),(HEIGHT/2)+ (HEIGHT/10),
          WIDTH/30,WIDTH/30);

//the constructor:
public G1P2()
{
  //the following code creates the JFrame
  super("Radical Racing");
  setSize(WIDTH,HEIGHT);
  setDefaultCloseOperation(JFrame.EXIT_ON_CLOSE);
  setVisible(true);

  //start the inner class (which works on its own, because it is a
          //Thread)
  Move1 m1 = new Move1();
  Move2 m2 = new Move2();
  m1.start();
  m2.start();
}

//this will draw the cars and the race track
public void paint(Graphics g)
{
  super.paint(g);

  //draw the background for the racetrack
  g.setColor(Color.DARK_GRAY);
  g.fillRect(0,0,WIDTH,HEIGHT);

  //when we draw, the border will be green
  g.setColor(Color.GREEN);

  //the following rectangle is the start line for the outer player
  Rectangle lineO = new
    Rectangle(WIDTH/9,HEIGHT/2,(int)((WIDTH/9)*1.5)/2,HEIGHT/140);

  //the following rectangle is the start line for the inner player
  Rectangle lineI = new Rectangle(((WIDTH/9)+((int)((WIDTH/9)*1.5)/2)),
          (HEIGHT/2)+(HEIGHT/10),(int)((WIDTH/9)*1.5)/2,HEIGHT/140);

  //now, using the rectangles, draw it
  g.fillRect(left.x,left.y,left.width,left.height);
  g.fillRect(right.x,right.y,right.width,right.height);
  g.fillRect(top.x,top.y,top.width,top.height);
  g.fillRect(bottom.x,bottom.y,bottom.width,bottom.height);
  g.fillRect(center.x,center.y,center.width,center.height);
  g.fillRect(obstacle.x,obstacle.y,obstacle.width,obstacle.height);
  g.fillRect(obstacle2.x,obstacle2.y,obstacle2.width,obstacle2.height);
  g.fillRect(obstacle3.x,obstacle3.y,obstacle3.width,obstacle3.height);
  g.fillRect(obstacle4.x,obstacle4.y,obstacle3.width,obstacle4.height);
  g.fillRect(obstacle5.x,obstacle5.y,obstacle5.width,obstacle5.height);
  //set the starting line color to white
```

```
    g.setColor(Color.WHITE);
    //now draw the starting line
    g.fillRect(lineO.x,lineO.y,lineO.width,lineO.height);
    g.fillRect(lineI.x,lineI.y,lineI.width,lineI.height);
    //set the color of the finish line to yellow
    g.setColor(Color.YELLOW);
    //now draw the finish line
    g.fillRect(finish.x,finish.y,finish.width,finish.height);

    //set the color to blue for p1
    g.setColor(Color.BLUE);
    //now draw the actual player
    g.fill3DRect(p1.x,p1.y,p1.width,p1.height,true);

    //set the color to red for p2
    g.setColor(Color.RED);
    //now draw the actual player
    g.fill3DRect(p2.x,p2.y,p2.width,p2.height,true);
  }

  private class Move1 extends Thread
  {
    public void run()
    {
      //now, this should all be in an infinite loop, so the process
        //repeats
      while(true)
      {
        //now, put the code in a "try" block. This will let the
          //program exit
        //if there is an error.
        try
        {

          //first, refresh the screen:
          repaint();
          //increase speed a bit
          if(p1Speed<=5)
              p1Speed+=.2;

          p1.y-=p1Speed;

          //this delays the refresh rate:
          Thread.sleep(75);
        }
        catch(Exception e)
        {
          //if there is an exception (an error), exit the loop.
          break;
        }
      }
    }
  }

  private class Move2 extends Thread
  {
    public void run()
    {
      //now, this should all be in an infinite loop, so the process
        //repeats
      while(true)
      {
        //now, put the code in a "try" block. This will let the
```

```
        //program exit
      //if there is an error.
      try
      {
        //first, refresh the screen:
        repaint();
        //increase speed a bit
        if(p2Speed<=5)
          p2Speed+=.2;

        p2.y-=p2Speed;

        //this delays the refresh rate:
        Thread.sleep(75);
      }
      catch(Exception e)
      {
        //if there is an exception (an error), exit the loop.
        break;
      }
    }
  }
}
//this starts the program by calling the constructor:
public static void main (String[ ] args)
{
  new G1P2();
}
}
```

Figures 10-1 and 10-2 illustrate the game play of Radical Racing.

Go on to the next project to learn how to add collision detection. And that's not all! Your keyboard will be turned into your own steering wheel.

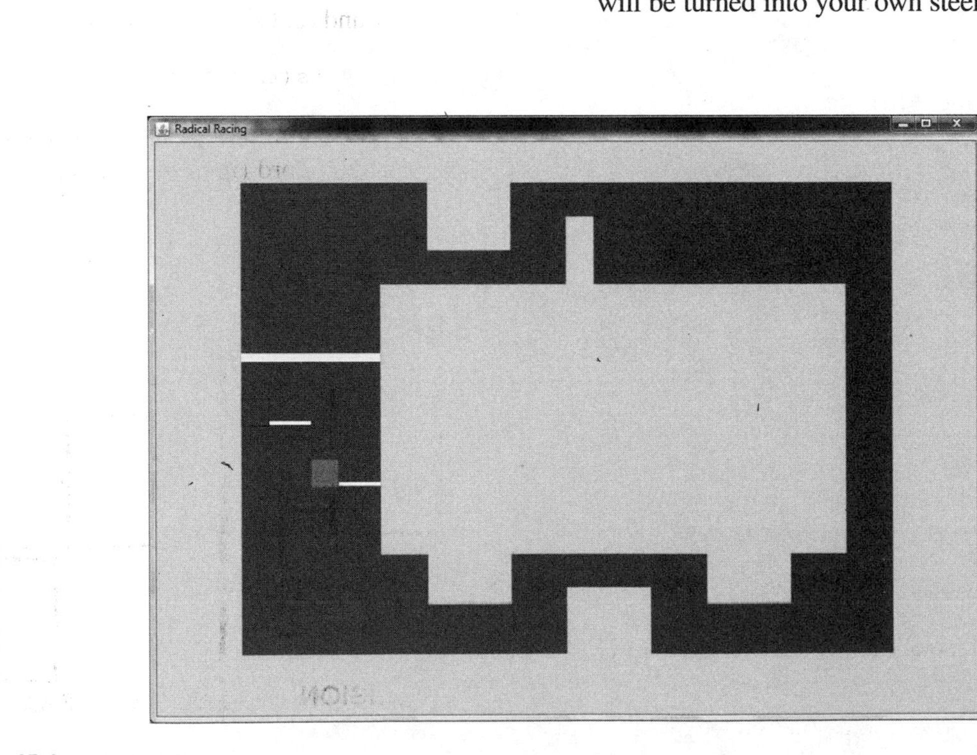

Figure 10-1 *Radial Racing.*

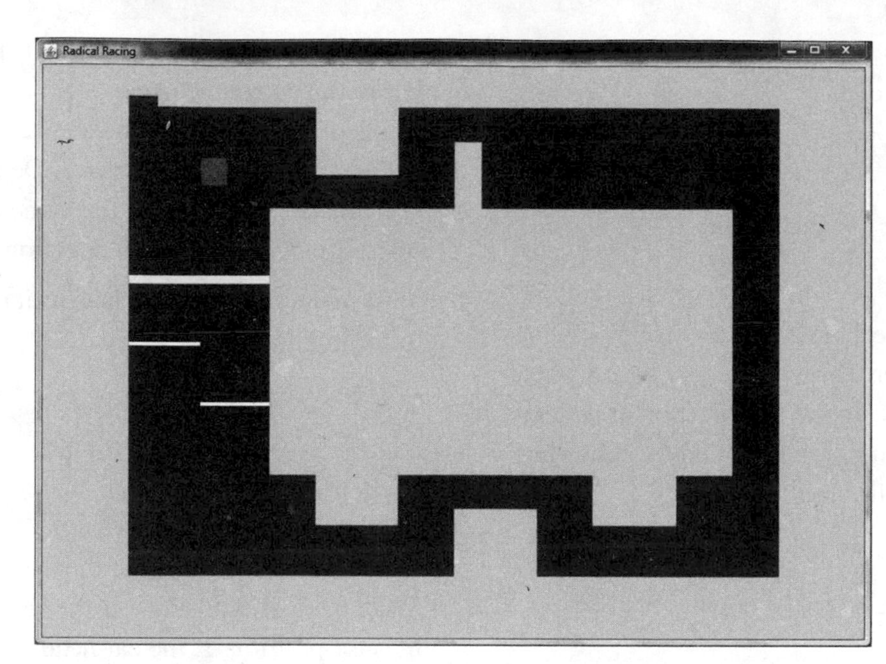

Figure 10-2 *Cars accelerating forward.*

Project 11: Radical Racing—Collision!

Project

Program the cars to bounce back and slow down if they run into each other or the boundaries. Plus, transform your keyboard into a steering wheel by assigning specific keys to direct the racers.

New building blocks

Collision detection and KeyListener

Collision detection

Collision Detection (checking to see if objects hit each other, as illustrated in Figure 11-1) is easy in Java. Because you created all of your graphics with Rectangles, we can easily check for collisions: Rectangle has a built-in "intersects" method.

Let's say you want to check for a collision between Rectangles r1 and r2. Use the following code:

```
if(r1.intersects(r2))
```

To check for multiple collisions, you can use the "and" or "or" keyword (in if-statements). "And" is "&&" and "or" is "||"

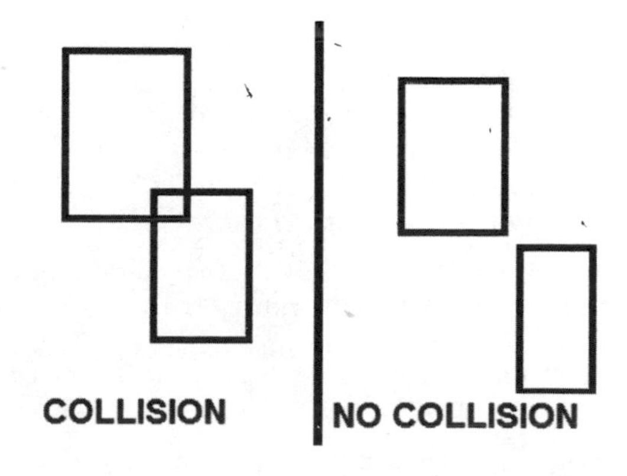

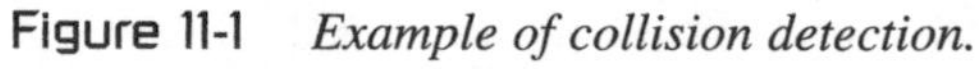

Figure 11-1 *Example of collision detection.*

For example, to see if Rectangle r1 hit r2 *or* r3, use the following code:

```
if(r1.intersects(r2) || r1.intersects(r3))
```

KeyListener

The KeyListener lets the computer receive keyboard input from the user. This will be very useful when you let the players control the direction of their cars. To use a KeyListener, you must add the following code to your "public class ..." line:

```
implements KeyListener
```

This can also be added to the inner classes (Threads).

Then, inside the constructor (or "run" method, if you are dealing with Threads), type the following code:

```
addKeyListener(this);
```

This line tells the keyboard to "wake up" and start listening for keyboard commands.

Next, you must add three methods to your code:

```
public void keyPressed(KeyEvent e){}
public void keyReleased(KeyEvent e){}
public void keyTyped(KeyEvent e){}
```

Each of these methods are automatically called when a key is typed. "keyPressed" is called when the key is pushed down. "keyReleased" is called when you let go of a key after pressing it. "keyTyped" is called after a key is pressed and released. You will be using "keyTyped" most often. Inside of "keyTyped" you can use an if-statement to check what key was pressed. Depending on the key, you can then change the game by altering such variables as the car's direction.

To figure out what key has been pressed, use the following line:

```
e.getKeyChar()
```

Making the game

Create global variables for the car's direction. Make them of type "int." If the value is 0, the car travels up; if it is 1, the car heads left; etc. To keep things simple, create final variables. For example, initialize an int called "UP" and set its value to zero. Then, if you need to see if the car is going up, compare it to "UP" instead of 0.

Next, use the KeyListener to change the direction of the player. Also, make the direction change by using if statements to test the direction of the car and alter "x" if it is going left/right and alter "y" if it is going up/down.

Let's check to see if the player has collided with a wall or the other car. If so, set the player's speed to a negative value so the car moves backwards before picking up speed.

The completed code is below:

```
//RADICAL RACING
//Project 11
//By Ian Cinnamon

//import everything:
import javax.swing.*;
import javax.swing.event.*;
import java.awt.*;
import java.awt.event.*;

//this creates the class where you use JFrame
public class G1P3 extends JFrame
{
  //this is the constand that will hold the screen size
  final int WIDTH = 900, HEIGHT = 650;

  //these will keep track of each player's speed:
  double p1Speed =.5, p2Speed =.5;
```

```
//these are ints that represent directions:
final int UP = 0, RIGHT = 1, DOWN = 2, LEFT = 3;

//these will keep track of the player's directions (default = up)
int p1Direction = UP;
int p2Direction = UP;

//create rectangles that wil represent the left, right, top, bottom, and center
Rectangle left = new Rectangle(0,0,WIDTH/9, HEIGHT);
Rectangle right = new Rectangle((WIDTH/9)*8,0,WIDTH/9,HEIGHT);
Rectangle top = new Rectangle(0,0,WIDTH,HEIGHT/9);
Rectangle bottom = new Rectangle(0,(HEIGHT/9)*8,WIDTH,HEIGHT/9);
Rectangle center = new Rectangle((int)((WIDTH/9) *2.5),(int)((HEIGHT/9)*2.5),
        (int)((WIDTH/9)*5),(HEIGHT/9)*4);
//these obstacles will obstruct the path and make navigating harder
Rectangle obstacle = new Rectangle(WIDTH/2,(int)((HEIGHT/9)*7),WIDTH/10,HEIGHT/9);
Rectangle obstacle2 = new Rectangle(WIDTH/3,(int)((HEIGHT/9)*5),WIDTH/10,HEIGHT/4);
Rectangle obstacle3 = new Rectangle(2*(WIDTH/3),(int)((HEIGHT/9)*5),WIDTH/10,HEIGHT/4);
Rectangle obstacle4 = new Rectangle(WIDTH/3,HEIGHT/9,WIDTH/30,HEIGHT/9);
Rectangle obstacle5 = new Rectangle(WIDTH/2,(int)((HEIGHT/9)*1.5),
                        WIDTH/30,HEIGHT/4);

//the following rectangle is the finish line for both players
Rectangle finish = new Rectangle(WIDTH/9,(HEIGHT/2)-HEIGHT/9,
                   (int)((WIDTH/9)*1.5),HEIGHT/70);

//this is the rectangle for player 1's (outer) car:
Rectangle p1 = newRectangle(WIDTH/9,HEIGHT/2, WIDTH/30,WIDTH/30);

//this is the rectang;e for player 2's (inner) car:
Rectangle p2 = new Rectangle(((WIDTH/9)+((int)((WIDTH/9)*1.5)/2)),(HEIGHT/2)+
        (HEIGHT/10),WIDTH/30,WIDTH/30);
//the constructor:
public G1P3()
{
  //the following code creates the JFrame
  super("Radical Racing");
  setSize(WIDTH,HEIGHT);
  setDefaultCloseOperation(JFrame.EXIT_ON_CLOSE);
  setVisible(true);

  //start the inner class (which works on its own, because it is a Thread)
  Move1 m1 = new Move1();
  Move2 m2 = new Move2();
  m1.start();
  m2.start();
}

//this will draw the cars and the race track
public void paint(Graphics g)
{
  super.paint(g);

  //draw the background for the racetrack
  g.setColor(Color.DARK_GRAY);
  g.fillRect(0,0,WIDTH,HEIGHT);

  //when we draw, the border will be green
  g.setColor(Color.GREEN);

  //the following rectangle is the start line for the outer player
  Rectangle lineO = new Rectangle(WIDTH/9,HEIGHT/2,(int)((WIDTH/9)*1.5)/2,HEIGHT/140);
  //the following rectangle is the start line for the inner player
  Rectangle lineI = new Rectangle(((WIDTH/9)+((int)((WIDTH/9)*1.5)/2)),
          (HEIGHT/2)+(HEIGHT/10), (int)
          ((WIDTH/9)*1.5)/2,HEIGHT/140);
```

```
    //now, using the rectangles, draw it
    g.fillRect(left.x,left.y,left.width,left.height);
    g.fillRect(right.x,right.y,right.width,right.height);
    g.fillRect(top.x,top.y,top.width,top.height);
    g.fillRect(bottom.x,bottom.y,bottom.width,bottom.height);
    g.fillRect(center.x,center.y,center.width,center.height);
    g.fillRect(obstacle.x,obstacle.y,obstacle.width,obstacle.height);
    g.fillRect(obstacle2.x,obstacle2.y,obstacle2.width,obstacle2.height);
    g.fillRect(obstacle3.x,obstacle3.y,obstacle3.width,obstacle3.height);
    g.fillRect(obstacle4.x,obstacle4.y,obstacle3.width,obstacle4.height);
    g.fillRect(obstacle5.x,obstacle5.y,obstacle5.width,obstacle5.height);
    //set the starting line color to white
    g.setColor(Color.WHITE);
    //now draw the starting line
    g.fillRect(lineO.x,lineO.y,lineO.width,lineO.height);
    g.fillRect(lineI.x,lineI.y,lineI.width,lineI.height);
    //set the color of the finish line to yellow
    g.setColor(Color.YELLOW);
    //now draw the finish line
    g.fillRect(finish.x,finish.y,finish.width,finish.height);

    //set the color to blue for p1
    g.setColor(Color.BLUE);
    //now draw the actual player
    g.fill3DRect(p1.x,p1.y,p1.width,p1.height,true);

    //set the color to red for p2
    g.setColor(Color.RED);
    //now draw the actual player
    g.fill3DRect(p2.x,p2.y,p2.width,p2.height,true);

  }

  private class Move1 extends Thread implements KeyListener
  {
    public void run()
    {
    //add the code to make the KeyListener "wake up"
    addKeyListener(this);

    //now, this should all be in an infinite loop, so the process repeats
    while(true)
    {
      //now, put the code in a "try" block. This will let the program exit
      //if there is an error.
      try
      {
        //first, refresh the screen:
        repaint();

        //check to see if car hits the outside walls.
        //If so, make it slow its speed by setting its speed
        //to -4.
        if(p1.intersects(left) || p1.intersects(right) ||
           p1.intersects(top) || p1.intersects(bottom) ||
           p1.intersects(obstacle) || p1.intersects(obstacle2)||
           p1.intersects(p2) || p1.intersects(obstacle3) ||
           p1.intersects(obstacle4) || p1.intersects(obstacle5))
        {
          p1Speed = -4;
        }

        //if the car hits the center, do the same as above
        //but make the speed -2.5.
```

```
        if(p1.intersects(center))
        {
          p1Speed = -2.5;
        }

        //increase speed a bit
        if(p1Speed<=5)
          p1Speed+=.2;

        //these will move the player based on direction
        if(p1Direction==UP)
        {
          p1.y-=(int)p1Speed;
        }
        if(p1Direction==DOWN)
        {
          p1.y+=(int)p1Speed;
        }
        if(p1Direction==LEFT)
        {
          p1.x-=(int)p1Speed;
        }
        if(p1Direction==RIGHT)
        {
          p1.x+=(int)p1Speed;
        }

        //this delays the refresh rate:
        Thread.sleep(75);
      }
      catch(Exception e)
      {
        //if there is an exception (an error),
              exit the loop.
        break;
      }
    }
  }

  //you must also implement this method from KeyListener
  public void keyPressed(KeyEvent event)
  {

  }
    //you must also implement this method from KeyListener
  public void keyReleased(KeyEvent event)
  {

  }
    //you must also implement this method from KeyListener
    public void keyTyped(KeyEvent event)
    {
      if(event.getKeyChar()=='a')
      {
        p1Direction = LEFT;
      }
      if(event.getKeyChar()=='s')
      {
        p1Direction = DOWN;
      }
      if(event.getKeyChar()=='d')
      {
        p1Direction = RIGHT;
```

```
      }
      if(event.getKeyChar()==' w' )
      {
        p1Direction = UP;
      }
    }
  }

  private class Move2 extends Thread implements KeyListener
  {
    public void run()
    {
      //add the code to make the KeyListener "wake up"
      addKeyListener(this);

      //now, this should all be in an infinite loop, so the process repeats
      while(true)
      {
        //now, put the code in a "try" block. This will let the program exit
        //if there is an error.
        try
        {
          //first, refresh the screen:
          repaint();

          //check to see if car hits the outside walls.
          //If so, make it slow its speed by setting its speed
          //to -4.
          if(p2.intersects(left) || p2.intersects(right) ||
              p2.intersects(top) || p2.intersects(bottom) ||
            p2.intersects(obstacle) || p2.intersects(obstacle2) ||
              p1.intersects(p2))
          {
            p2Speed = -4;
          }

          //if the car hits the center, do the same as above
          //but make the speed -2.5.
          if(p2.intersects(center))
          {
            p2Speed = -2.5;
          }

          //increase speed a bit
          if(p2Speed<=5)
            p2Speed+=.2;

          //these will move the player based on direction
          if(p2Direction==UP)
          {
            p2.y-=(int)p2Speed;
          }
          if(p2Direction==DOWN)
          {
            p2.y+=(int)p2Speed;
          }
          if(p2Direction==LEFT)
          {
            p2.x-=(int)p2Speed;
          }
          if(p2Direction==RIGHT)
          {
            p2.x+=(int)p2Speed;
```

```
        }
        //this delays the refresh rate:
        Thread.sleep(75);
      }
      catch(Exception e)
      {
        //if there is an exception (an error), exit the loop.
        break;
      }
    }
  }
  //you must also implement this method from KeyListener
  public void keyPressed(KeyEvent event)
  {

  }

    //you must also implement this method from KeyListener
  public void keyReleased(KeyEvent event)
  {

  }
  //you must also implement this method from KeyListener
  public void keyTyped(KeyEvent event)
  {
    if(event.getKeyChar()== 'j' )
    {
      p2Direction = LEFT;
    }
    if(event.getKeyChar()=='k' )
    {
      p2Direction = DOWN;
    }
    if(event.getKeyChar()=='l' )
    {
      p2Direction = RIGHT;
    }
    if(event.getKeyChar()=='i' )
    {
      p2Direction = UP;
    }
  }
}
//this starts the program by calling the constructor:
public static void main (String[ ] args)
{
  new G1P3();
}
}
```

In Figures 11-2 through 11-4, the mobility of the cars and their collision detection are illustrated.

In Project 12, make the game come to life with bold graphics and sounds!

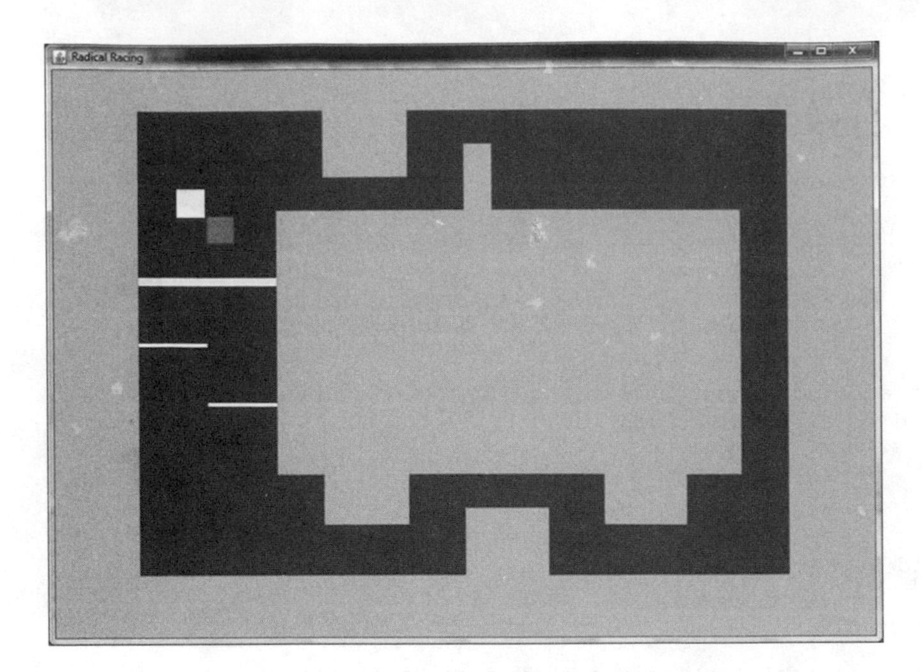

Figure 11-2 *Cars are about to collide.*

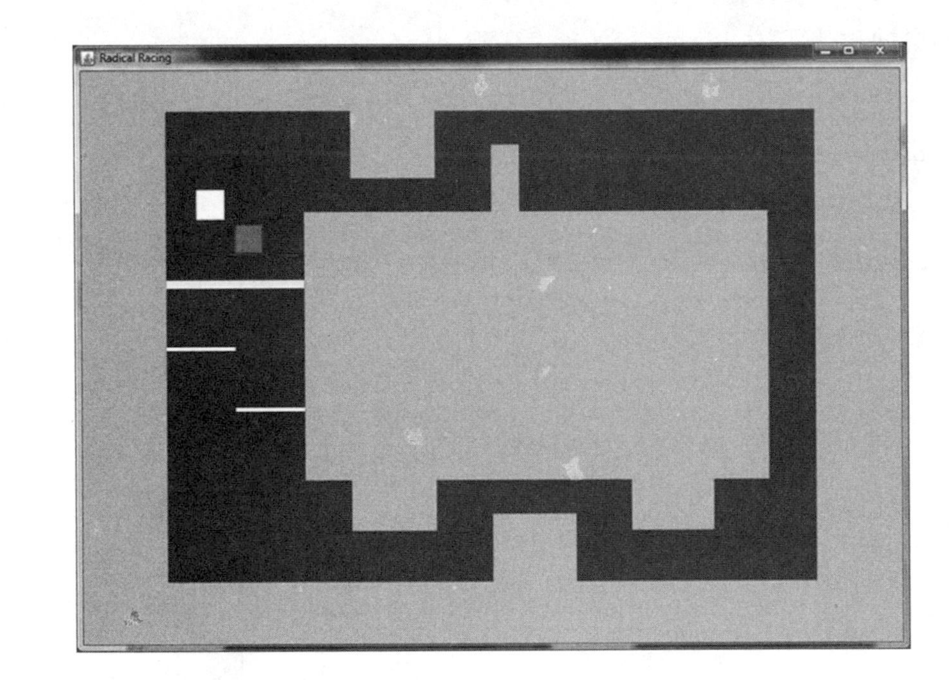

Figure 11-3 *Cars move backwards after collision.*

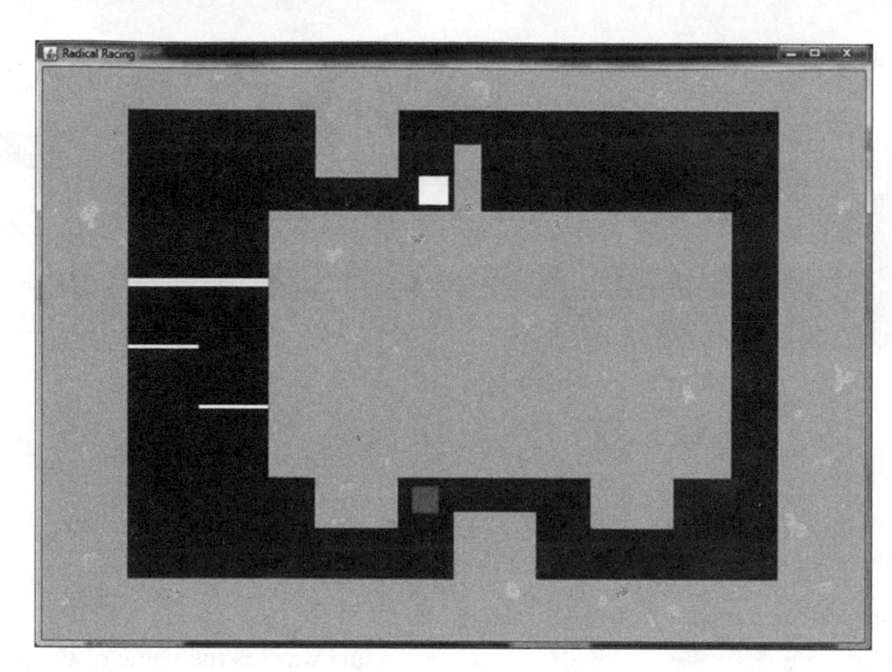

Figure 11-4 *Cars turn and complete the track.*

Project 12: Radical Racing—Customizing

Project

Add modifications to the game with images and sounds. Create a "welcome" screen. Add a lap tracking feature that announces the winner after three laps.

New Building Blocks

Images, Sound

Images

Want to change the graphics in your game? It's not difficult. You can turn your racecar into a spaceship, a boat, a plane, a cheetah, even a John Deere tractor ... anything that moves!

First, you have to draw the image yourself or find it on the internet. Once you have chosen your mode of transportation, save it to `<directory of Java project>`/build/classes. Figures 12-1

Figure 12-1 *Race car.*

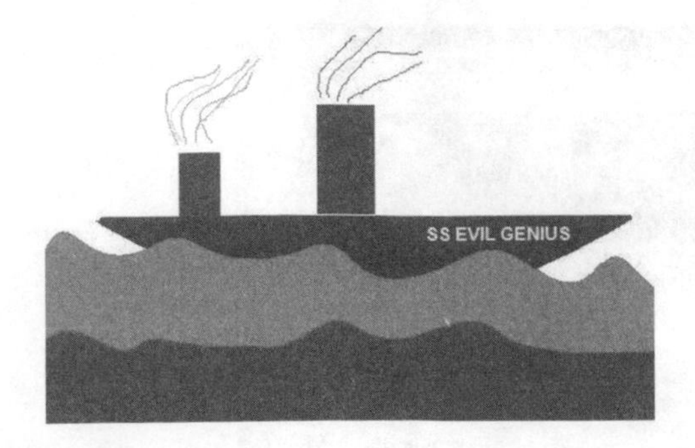

Figure 12-2 *Boat.*

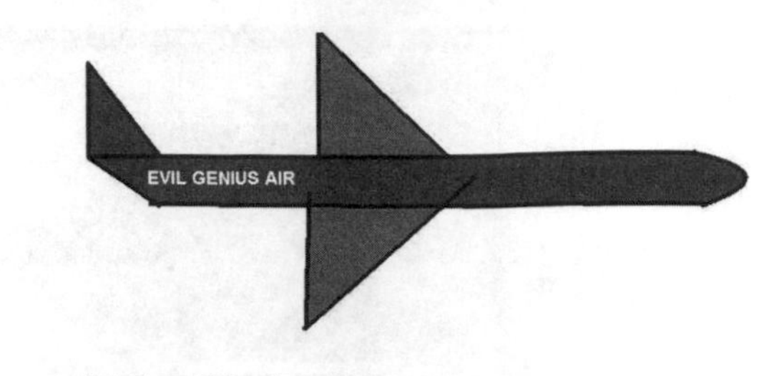

Figure 12-3 *Airplane.*

through 12-3 represent some possible means of transportation that can be used in the game.

Add the following code before you draw the image:

```
Image img = null;
try
{
  URL url =
    this.getClass().getResource ("name");
  img =
    Toolkit.getDefaultToolkit().getImage(url);
}
catch(Exception e){}
```

"name" should be the name of the image you want to use (including the extension).

Then, replace the code that draws the shapes with the following:

```
g.drawImage(img, x, y, this);
```

"x" and "y" should be the x and y from the Rectangle.

One more thing — don't forget to import the following:

```
import java.io.*;
import java.net.*;
```

All of the images used in the games in this book are available for free from www.mcgraw-hill.com/authors/cinnamon

Sound

You can quickly add sound effects to your games with the following code:

```
URL eng = this.getClass().getResource
("engine.wav");
AudioClip snd = JApplet.newAudioClip(eng);
snd.loop();
```

"engine.wav" is the name of the sound file.

"`snd.loop()`" will loop the sound file. If you don't want to loop it, you can replace that code with:

```
snd.play()
```

This will play the sound once.

Don't forget to `import java.applet.AudioClip;`

The sounds used in the games in this book are all available for free from www.mcgraw-hill.com/authors/cinnamon

Making the game

Add a "welcome" screen using a JOptionPane. You can also write code to keep track of laps by checking the number of collisions with the "finish" line.

Draw images to replace the plain, rectangular cars. Make sure you orient the image to the direction the racer is traveling (when it is traveling right, the image should point to the right).

The completed code is below:

```
//import everything:
import javax.swing.*;
import javax.swing.event.*;
import java.awt.*;
import java.awt.event.*;
import java.io.*;
import java.net.*;
import java.applet.AudioClip;

//this creates the class where you use JFrame
public class G1P4 extends JFrame
{
  //the following are all 8 images:
  URL url1 = null, url2 = null, url3 = null, url4 = null,
     url5 = null, url6 = null, url7 = null, url8 = null;
   Image img1,img2,img3,img4,
     img5,img6,img7,img8;
   //the following URL and Image are for the center title
   URL URLt = null;
   Image title = null;

   //this is the constand that will hold the screen size
   final int WIDTH = 900, HEIGHT = 650;
  //this will be true if someone already won
  boolean winnerChosen = false;

  //these will keep track of each player's speed:
  double p1Speed =.5, p2Speed =.5;

  //this will keep track of how may laps a player has run
  int p1Laps = 0, p2Laps = 0;

  //these are ints that represent directions:
  final int UP = 0, RIGHT = 1, DOWN = 2, LEFT = 3;

  //these will keep track of the player's directions (default = up)
  int p1Direction = UP;
  int p2Direction = UP;
  //create rectangles that will represent the left, right, top, bottom, and center
  Rectangle left = new Rectangle(0,0,WIDTH/9,HEIGHT);
  Rectangle right = new Rectangle((WIDTH/9)*8,0,WIDTH/9,HEIGHT);
  Rectangle top = new Rectangle(0,0,WIDTH,HEIGHT/9);
  Rectangle bottom = new Rectangle(0,(HEIGHT/9)*8,WIDTH,HEIGHT/9);
  Rectangle center = new Rectangle((int)((WIDTH/9)*2.5),(int)((HEIGHT/9)*2.5), (int)
            ((WIDTH/9)*5),(HEIGHT/9)*4);
  //these obstacles will obstruct the path and make navigating harder
  Rectangle obstacle = new Rectangle(WIDTH/2,(int)((HEIGHT/9)*7),WIDTH/10,HEIGHT/9);
  Rectangle obstacle2 = new Rectangle(WIDTH/3,(int)((HEIGHT/9)*5),WIDTH/10,HEIGHT/4);
  Rectangle obstacle3 = new Rectangle(2*(WIDTH/3),(int)((HEIGHT/9)*5),WIDTH/10,HEIGHT/4);
  Rectangle obstacle4 = new Rectangle(WIDTH/3,HEIGHT/9,WIDTH/30,HEIGHT/9);
  Rectangle obstacle5 = new Rectangle(WIDTH/2,(int)((HEIGHT/9)*1.5),WIDTH/30,HEIGHT/4);
  //the following rectangle is the finish line for both players
  Rectangle finish = new Rectangle(WIDTH/9,(HEIGHT/2)-HEIGHT/9, (int)((WIDTH/9)*1.5),
            HEIGHT/70);
  //this is the rectangle for player 1's (outer) car:
  Rectangle p1 = new Rectangle(WIDTH/9,HEIGHT/2,WIDTH/30,WIDTH/30);

  //this is the rectangle for player 2's (inner) car:
  Rectangle p2 = new
            Rectangle(((WIDTH/9)+((int)((WIDTH/9)*1.5)/2)),(HEIGHT/2)+(HEIGHT/10),WIDTH/3
            0,WIDTH/30);
  //the constructor:
  public G1P4()
  {
```

```
//the following code creates the JFrame
super("Radical Racing");
setSize(WIDTH,HEIGHT);
setDefaultCloseOperation(JFrame.EXIT_ON_CLOSE);
setVisible(true);

//load the URLs
try
{
  url1 = this.getClass().getResource("G1P4Img1.jpg");
  url2 = this.getClass().getResource("G1P4Img2.jpg");
  url3 = this.getClass().getResource("G1P4Img3.jpg");
  url4 = this.getClass().getResource("G1P4Img4.jpg");
  url5 = this.getClass().getResource("G1P4Img5.jpg");
  url6 = this.getClass().getResource("G1P4Img6.jpg");
  url7 = this.getClass().getResource("G1P4Img7.jpg");
  url8 = this.getClass().getResource("G1P4Img8.jpg");
  URLt = this.getClass().getResource("title.png");
}
catch(Exception e){}

//attach the URLs to the images
img1 = Toolkit.getDefaultToolkit().getImage(url1);
img2 = Toolkit.getDefaultToolkit().getImage(url2);
img3 = Toolkit.getDefaultToolkit().getImage(url3);
img4 = Toolkit.getDefaultToolkit().getImage(url4);
img5 = Toolkit.getDefaultToolkit().getImage(url5);
img6 = Toolkit.getDefaultToolkit().getImage(url6);
img7 = Toolkit.getDefaultToolkit().getImage(url7);
img8 = Toolkit.getDefaultToolkit().getImage(url8);
title = Toolkit.getDefaultToolkit().getImage(URLt);

//display a welcome dialog (JOptionPane) that includes the rules
JOptionPane.showMessageDialog(null,"WELCOME TO RADICAL RACING!\n\n"+"Game: 2 player
        racing\n"+"Goal: Complete 3 full laps before your
        opponent!\n"+"Controls:\n"+"Player1:\n"+"(BLUE CAR) WASD directional, speed
        is automatic\n"+" Player 2:\n"+" (RED CAR) IJKL directional, speed is
        automatic\n"+"Also, be sure to avoid the green grass. It's slick\n"+"and
        might make you spin out!\n\n"+"Click OK to start");
//start the inner class (which works on its own, because it is a Thread)
Move1 m1 = new Move1();
Move2 m2 = new Move2();
m1.start();
m2.start();

//now, play the sound:
try

{
  URL eng = this.getClass().getResource ("engine.wav");
  AudioClip snd = JApplet.newAudioClip(eng);
  snd.loop();
}
catch(Exception e){}
}

//this will draw the cars and the race track
public void paint(Graphics g)
{
  super.paint(g);

  //draw the background for the racetrack
  g.setColor(Color.DARK_GRAY);
  g.fillRect(0,0,WIDTH,HEIGHT);
```

```
    //when we draw, the border will be green
    g.setColor(Color.GREEN);

    //the following rectangle is the start line for the outer player
    Rectangle lineO = new Rectangle(WIDTH/9, HEIGHT/2,(int)((WIDTH/9)*1.5)/2,HEIGHT/140);
    //the following rectangle is the start line for the inner player
    Rectangle lineI = new Rectangle(((WIDTH/9)+
            ((int)((WIDTH/9)*1.5)/2)),(HEIGHT/2)+(HEIGHT/10),(int)((WIDTH/9)*1.5)/2,
            HEIGHT/140);
    //now, using the rectangles, draw it
    g.fillRect(left.x,left.y,left.width,left.height);
    g.fillRect(right.x,right.y,right.width,right.height);
    g.fillRect(top.x,top.y,top.width,top.height);
    g.fillRect(bottom.x,bottom.y,bottom.width, bottom.height);
    g.fillRect(center.x,center.y,center.width,center.height);
    g.fillRect(obstacle.x,obstacle.y,obstacle.width,obstacle.height);
    g.fillRect(obstacle2.x,obstacle2.y,obstacle2.width,obstacle2.height);
    g.fillRect(obstacle3.x,obstacle3.y,obstacle3.width,obstacle3.height);
    g.fillRect(obstacle4.x,obstacle4.y,obstacle3.width,obstacle4.height);
    g.fillRect(obstacle5.x,obstacle5.y,obstacle5.width,obstacle5.height);
    //set the starting line color to white
    g.setColor(Color.WHITE);
    //now draw the starting line
    g.fillRect(lineO.x,lineO.y,lineO.width,lineO.height);
    g.fillRect(lineI.x,lineI.y,lineI.width,lineI.height);
    //set the color of the finish line to yellow
    g.setColor(Color.YELLOW);
    //now draw the finish line
    g.fillRect(finish.x,finish.y,finish.width,finish.height);

    //this code will draw the title image to the center of the screen:
    g.drawImage(title,center.x+10,center.y+80,this);

    //draw the images for p1
    if(p1Direction==UP)
      g.drawImage(img5,p1.x,p1.y,this);
    if(p1Direction==LEFT)
      g.drawImage(img8,p1.x,p1.y,this);
    if(p1Direction==DOWN)
      g.drawImage(img7,p1.x,p1.y,this);
    if(p1Direction==RIGHT)
      g.drawImage(img6,p1.x,p1.y,this);
    //draw the images for p2
    if(p2Direction==UP)
      g.drawImage(img1,p2.x,p2.y,this);
    if(p2Direction==LEFT)
      g.drawImage(img4,p2.x,p2.y,this);
    if(p2Direction==DOWN)
      g.drawImage(img3,p2.x,p2.y,this);
    if(p2Direction==RIGHT)
      g.drawImage(img2,p2.x,p2.y,this);
}

private class Move1 extends Thread implements KeyListener
{
  public void run()
  {
    //add the code to make the KeyListener "wake up"
    addKeyListener(this);

    //now, this should all be in an infinite loop, so the process repeats
    while(true)
    {
```

```
//now, put the code in a "try" block. This will let the program exit
//if there is an error.
try
{
  //first, refresh the screen:
  repaint();

  //check to see if car hits the outside walls.
  //If so, make it slow its speed by setting its speed
  //to -4.
  if(p1.intersects(left) || p1.intersects(right) || p1.intersects(top) ||
     p1.intersects(bottom) || p1.intersects(obstacle) ||
       p1.intersects(obstacle2)|| p1.intersects(p2) || p1.intersects (obstacle3) ||
     p1.intersects(obstacle4) || p1.intersects(obstacle5))
  {
    p1Speed = -4;
  }
  //if the car hits the center, do the same as above
  //but make the speed -2.5.
  if(p1.intersects(center))
  {
    p1Speed = -2.5;
  }

  //check how many laps:
  if(p1.intersects(finish)&&p1Direction==UP)
  {
    p1Laps++;
  }
  //3 full laps will occur when laps is about 24.
  //so, use an if statement to check it and tell the user.
  //also, if winnerChosen is false, tell them they won and set it
  //to true. Otherwise, tell the user they lost.
  if(p1Laps>=24)
  {
    if(!winnerChosen)
    {
      winnerChosen = true;JOptionPane.showMessageDialog(null, "Player 1 (blue)
      Wins!!!");
      break;
    }
    else
    {
      JOptionPane.showMessageDialog(null, "Player 1 (blue): LOSER!:(\n" + "Player 2
      (red): WINNER!!!:D");
      break;
    }
  }

  //increase speed a bit
  if(p1Speed<=5)
    p1Speed+=.2;
  //these will move the player based on direction
  if(p1Direction==UP)
  {
    p1.y-=(int)p1Speed;
  }
  if(p1Direction==DOWN)
  {
    p1.y+=(int)p1Speed;
  }
```

```
        if(p1Direction==LEFT)
        {
          p1.x-=(int)p1Speed;
        }
        if(p1Direction==RIGHT)
        {
          p1.x+=(int)p1Speed;
        }

        //this delays the refresh rate:
        Thread.sleep(75);
      }
      catch(Exception e)
      {
        //if there is an exception (an error), exit the loop.
        break;
      }
    }
  }

  //you must also implement this method from KeyListener
  public void keyPressed(KeyEvent event)
  {
  }

  //you must also implement this method from KeyListener
  public void keyReleased(KeyEvent event)
  {
  }

    //you must also implement this method from KeyListener
  public void keyTyped(KeyEvent event)
  {
    if(event.getKeyChar()=='a')
    {
      p1Direction = LEFT;
    }
    if(event.getKeyChar()=='s')
    {
      p1Direction = DOWN;
    }
    if(event.getKeyChar()=='d')
    {
      p1Direction = RIGHT;
    }
    if(event.getKeyChar()=='w')
    {
      p1Direction = UP;
    }
  }
}

private class Move2 extends Thread implements KeyListener
{
public void run()
  {
    //add the code to make the KeyListener "wake up"
    addKeyListener(this);

    //now, this should all be in an infinite loop, so the process repeats
    while(true)
    {
      //now, put the code in a "try" block. This will let the program exit
      //if there is an error.
```

```
try
{
  //first, refresh the screen:
  repaint();

  //check to see if car hits the outside walls.
  //If so, make it slow its speed by setting its speed
  //to -4.
  if(p2.intersects(left) || p2.intersects(right) ||
    p2.intersects(top) || p2.intersects(bottom) ||
    p2.intersects(obstacle) ||p2.intersects(obstacle2) ||
    p1.intersects(p2) || p1.intersects(obstacle3) ||
    p1.intersects(obstacle4) || p1.intersects(obstacle5))
  {
    p2Speed = -4;
  }

  //if the car hits the center, do the same as above
  //but make the speed -2.5.
  if(p2.intersects(center))
  {
    p2Speed = -2.5;
  }

  //check how many laps:
  if(p2.intersects(finish)&&p2Direction==UP)
  {
    p2Laps++;
  }
  //3 full laps will occur when laps is about 24.
  //so, use an if statement to check it and tell the user.
  //also, if winnerChosen is false, tell them they won and set it
  //to true. Otherwise, tell the user they lost.
  if(p2Laps>=24)
  {
    if(!winnerChosen)
    {
      winnerChosen = true;
      JOptionPane.showMessageDialog(null, "Player 2 (red) Wins!!!");
      break;
    }
    else
    {
      JOptionPane.showMessageDialog(null,
        "Player 2 (red): LOSER!:(\n" +
        "Player 1 (blue): WINNER!!!:D");
      break;
    }
  }

  //increase speed a bit
  if(p2Speed<=5)
    p2Speed+=.2;
  //these will move the player based on direction
  if(p2Direction==UP)
  {
    p2.y-=(int)p2Speed;
  }
  if(p2Direction==DOWN)
  {
    p2.y+=(int)p2Speed;
  }
  if(p2Direction==LEFT)
```

```
        {
          p2.x-=(int)p2Speed;
        }
        if(p2Direction==RIGHT)
        {
          p2.x+=(int)p2Speed;
        }

        //this delays the refresh rate:
        Thread.sleep(75);
      }
      catch(Exception e)
      {
        //if there is an exception (an error), exit the loop.
        break;
      }
    }
  }

  //you must also implement this method from KeyListener
  public void keyPressed(KeyEvent event)
  {
  }
    //you must also implement this method from KeyListener
  public void keyReleased(KeyEvent event)
  {

  }
//you must also implement this method from KeyListener
  public void keyTyped(KeyEvent event)
  {
    if(event.getKeyChar()=='j')
    {
      p2Direction = LEFT;
    }
    if(event.getKeyChar()=='k')
    {
      p2Direction = DOWN;
    }
    if(event.getKeyChar()=='l')
    {
      p2Direction = RIGHT;
    }
    if(event.getKeyChar()=='i')
    {
      p2Direction = UP;
    }
  }
}

//this starts the program by calling the constructor:
public static void main (String[ ] args)
{
  new G1P4();
}
}
```

Figures 12-4 through 12-6 depict the game with the new graphics.

Customizing the game

Make your own track—make it round, make it square, whatever you want!

Add more detail/colors to both racers and the track.

Program hidden obstacles (potholes) in the middle of the track.

Experiment with different sound effects/music.

Change the rules—give one player a head start while increasing the other racer's speed.

Add a slow-motion section to the track.

Add more cars so more people can play.

Shape-shift: have the racers change vehicles after each lap.

Reverse world: race backwards (change which keys do what).

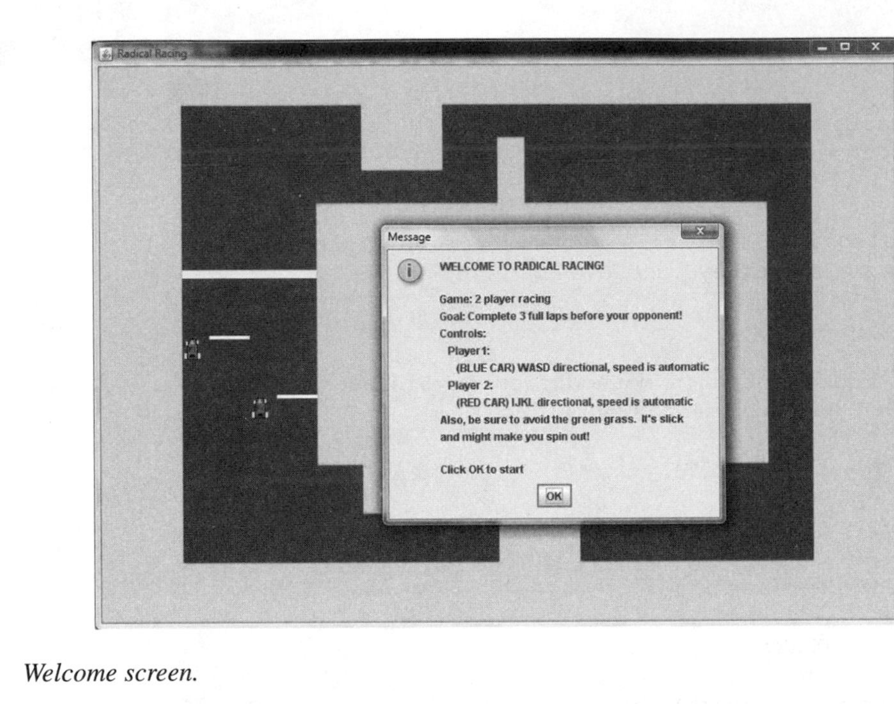

Figure 12-4 *Welcome screen.*

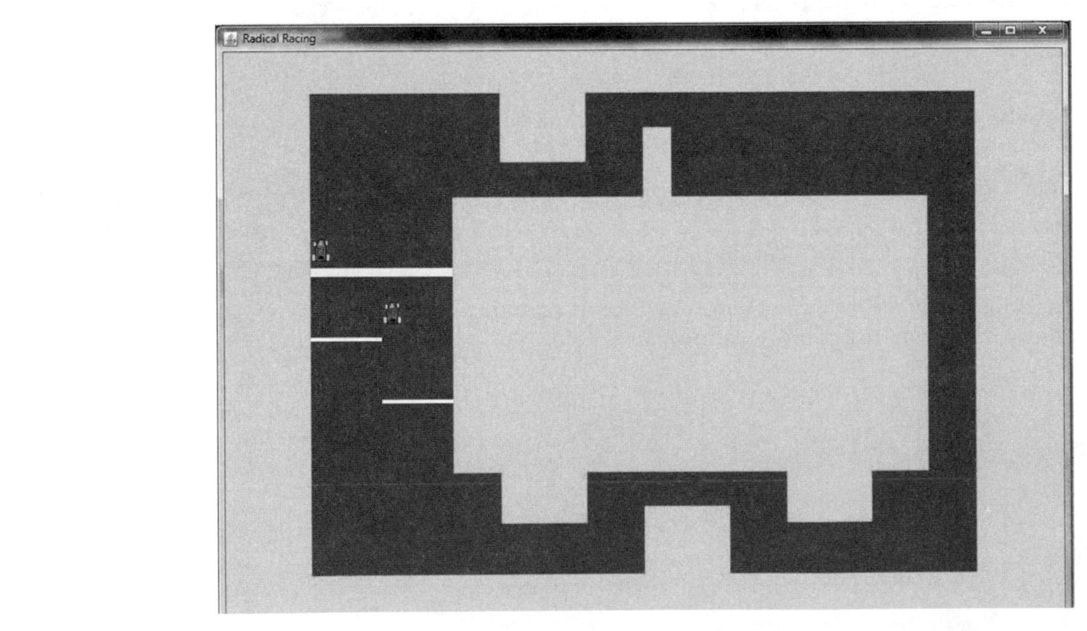

Figure 12-5 *Cars are ready to race!*

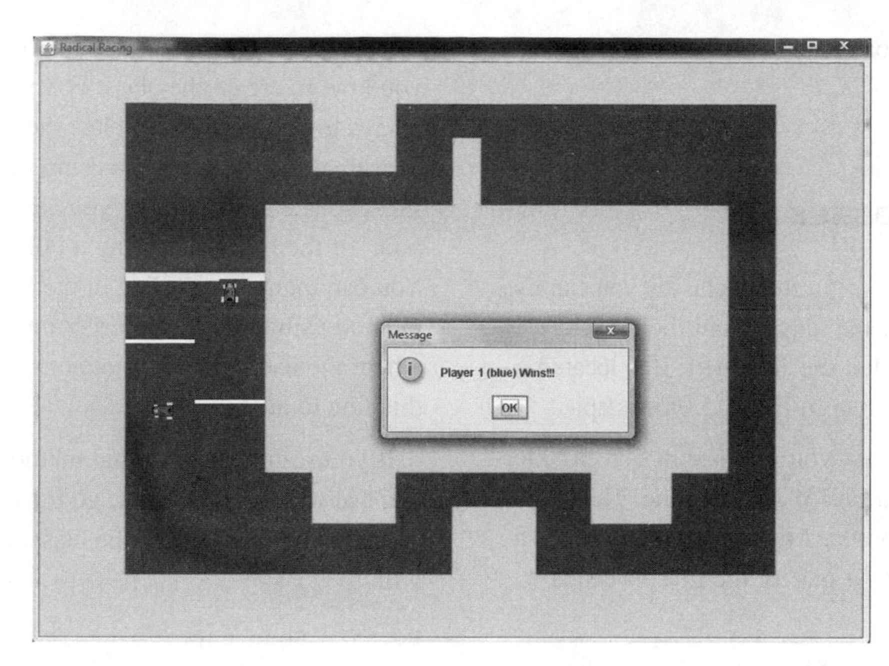

Figure 12-6 *Blue car wins!*

Project 13: Screen Skier—The Slope

Screen skier

It's a slippery slope as you have exactly 45 seconds to ski down the mountain on a course you create. Fail and die or win and grab the glory!

Project

In this project, you will create the slope-drawing environment: is it a smooth and gentle slope, a sheer drop, or a bumpy terrain of boulders and trees? You decide!

New building blocks

MouseListener, Premade Classes

MouseListener

You already know how to program the computer to see what the player is typing. Now, you can get the computer to see where the user is clicking. In KeyListener, you had to add three methods, whether or not you used all of them. You must do the same with MouseListener, except there are five methods. They are:

```
public void mouseClicked(MouseEvent e){}
public void mouseEntered(MouseEvent e){}
public void mouseExited(MouseEvent e){}
public void mousePressed(MouseEvent e){}
public void mouseReleased(MouseEvent e){}
```

In Screen Skier, you will use only the last two. For "`mousePressed`," you will see where the player first pressed the mouse (the first point on the slope). You will then use "`mouseReleased`" to see where the player ended the track.

Of course, the methods are useless if you don't know how to see the coordinates of the player's click.

Here's how:

To get the x coordinate, use the following code:

```
e.getX();
```

To get the y coordinate, use this code:

```
e.getY();
```

Premade classes

Java has hundreds of premade classes you can use. A complete list of the classes and their descriptions is called the Java API. It is located here: http://java.sun.com/j2se/1.5.0/docs/api/

One premade class you will use in Screen Skier holds the information for a single line. The class is called Line2D.Double. Another class that holds a single point (the first part of the line) is called Point2D.Double.

If you look in the API, it will tell you what you must import. In addition, you will see a table labeled "Constructor Summary." This will tell you how to create the class. For a Point2D.Double, it says to pass in two doubles: the x and y coordinates. Therefore, by using the `"mousePressed"` method, you can get the first point of the line by creating a Point2D.Double. You can then create a line in the `"mouseReleased"` method using your previous point and the point where the user released the mouse. You can add this line to an ArrayList.

If you want to know what methods you can use with a given class, go to NetBeans and create a new instance of the class. For example, with the class `Point2D.Double,` do the following:

```
Point2D.Double pnt = new Point2D.Double
(23.5,563.0);
```

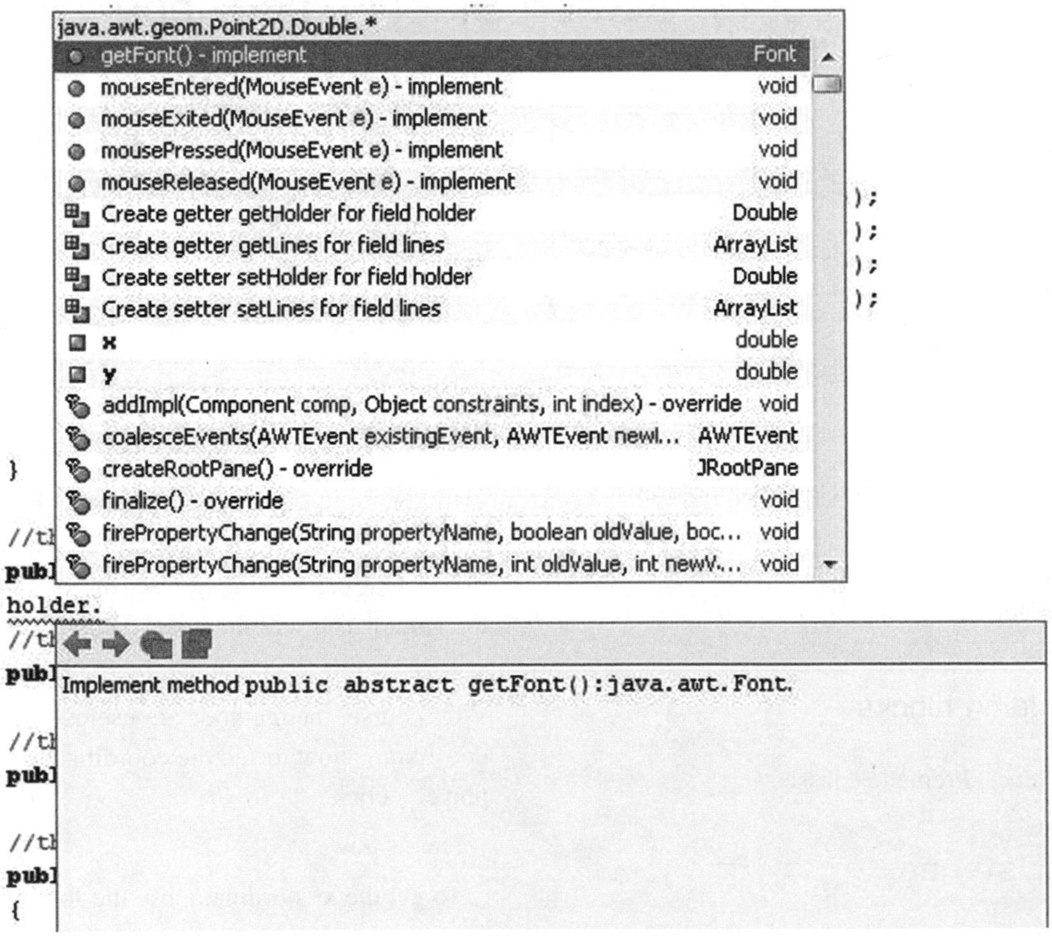

Figure 13-1 *Pop-up box in NetBeans.*

Then, type the following:

```
pnt.
```

When you type the period, NetBeans will create a list of all the methods you can use. Most are self-explanatory, but for the more complex methods, NetBeans will make a description pop-up, as shown in Figure 13-1.

Making the game

It's time to create the first part of Screen Skier: the slope drawing environment. Use a mouseListener and Point2D.Double and Line2D.Double to draw the ski slope. Store the lines in an ArrayList and use a "for loop" to print all of the lines in the ArrayList from the paint method.

```
//import everything:
import javax.swing.*;
import javax.swing.event.*;
import java.awt.*;
import java.awt.event.*;
import java.awt.geom.*;
import java.util.*;
//the actual class:
public class G2P1 extends JFrame implements MouseListener
{
  //this ArrayList holds the lines:
  ArrayList lines = new ArrayList();

  //this will hold the first point of the line
  Point2D.Double holder;

  //the constructor:
  public G2P1()
  {
    //set the title:
    super("Screen Skier - Programming Video Games for the Evil Genius");
    setSize(700,700);
    setVisible(true);
    setDefaultCloseOperation(JFrame. EXIT_ON_CLOSE);
    addMouseListener(this);
  }

  public void paint(Graphics g)
  {
    super.paint(g);

    //set the color to black (for the lines):
    g.setColor(Color.black);

    //this for loop will go thru every line in the ArrayList and draw them:
    for(int i = 0; i <lines.size(); i++)
    {
      //get the line in the ArrayList:
      Line2D.Double temp = (Line2D.Double) lines.get(i);

      //now get all the x's and y's
      int x1 = Integer.parseInt(""+Math.round (temp.getX1()));
      int y1 = Integer.parseInt(""+Math.round (temp.getY1()));
      int x2 = Integer.parseInt(""+Math.round (temp.getX2()));
      int y2 = Integer.parseInt(""+Math.round (temp.getY2()));

      g.drawLine(x1,y1,x2,y2);
    }

  }

  //these are the mouse listener methods:
```

```
    public void mouseClicked(MouseEvent e){}

    //these are the mouse listener methods:
    public void mouseEntered(MouseEvent e){}

    //these are the mouse listener methods:
    public void mouseExited(MouseEvent e){}

    //these are the mouse listener methods:
    public void mousePressed(MouseEvent e)

    {
      //this code will run when the mouse is first pressed, getting the
      //starting point of the line.
      holder = new Point2D.Double(e.getX(),e.getY());
    }

    //these are the mouse listener methods:
    public void mouseReleased(MouseEvent e)
    {
      //this code will run when the mouse is released, completing the
      //line and adding it to the ArrayList.
      Point2D.Double end = new Point2D.Double(e.getX(),e.getY());
      lines.add(new Line2D.Double(holder,end));

      //now, repinat the screen so the line is drawn:
      repaint();

    }

    public static void main (String[ ] args)
    {
      //begin the program:
      new G2P1();
    }
}
```

Figure 13-2 *Blank canvas.*

Figures 13-2 through 13-4 depict the line drawing capabilities of Screen Skier.

Let your mind go wild! In the next project, create any kind of skier you want to glide down slopes of your design.

Figure 13-3 *Single line.*

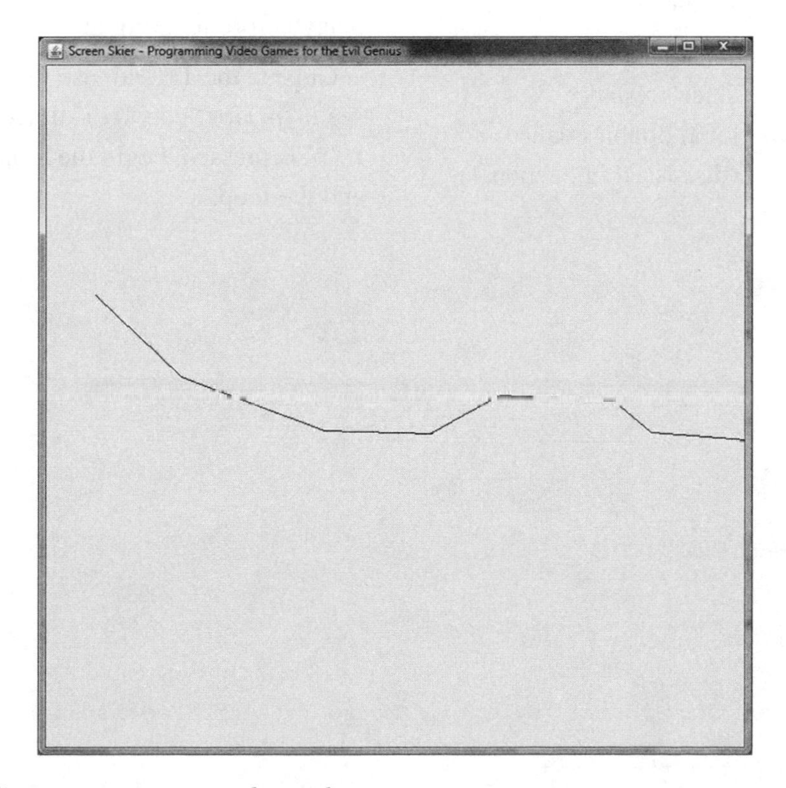

Figure 13-4 *Many lines connect to create the track.*

Project 14: Screen Skier—Practice Run

Project

Create a figure to ski down the track the player has drawn.

Making the game

Start by generating the image and the Rectangle that will represent the skier.

To have the skier race down the slope, initialize a Thread that controls the player's movements. Inside the Thread, make a method that sets up the skier's starting position (a bit to the right and above the first point of the first line), as shown in Figure 14-1.

Call this method from the run method before the `while loop`. In addition, create another method that takes in a boolean as its argument that can end the `while loop`. Inside the `while loop`, check to see if the skier's Rectangle intersects any of the lines (use the "intersects" method).

In order to control the skier's x and y movements, create two global doubles called "velocity" and "gravity" (discussed in Section 1).

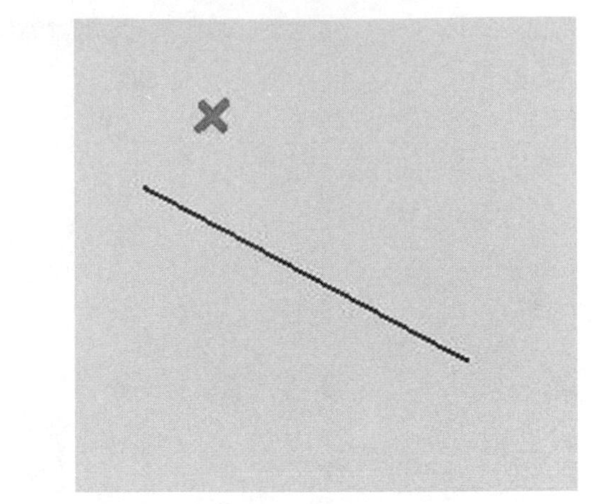

Figure 14-1 *Starting position.*

If the skier is not on the line, slowly increase "gravity." If the skier is on the line, find the difference between the two points of the line and divide by 50 for the y axis and 100 for the x axis. Add this to the respective variables, "gravity" and "velocity." Then, apply these variables to the skier's x and y position. Don't forget to repaint and delay before iterating through the loop again.

Outside the Thread, use the KeyListener to see if 'p' (for "play") or 'q' (for "quit") is pressed. If 'p' is pressed, begin the loop. If 'q' is pressed, end the loop.

```
//import everything:
import javax.swing.*;
import javax.swing.event.*;
import java.awt.*;
import java.awt.event.*;
import java.awt.geom.*;
import java.util.*;
import java.io.*;
import java.net.*;

//the actual class:
public class G2P2 extends JFrame implements MouseListener, KeyListener
{
  //this ArrayList holds the lines:
  ArrayList lines = new ArrayList();
```

```
//this will hold the first point of the line
Point2D.Double holder;

//this is the Thread:
Move move;

//this will hold the character's information:
Rectangle guy = null;

//this will tell repaint whether or not to draw the guy:
boolean drawGuy = false;

//the constructor:
public G2P2()
{
  //set the title:
  super("Screen Skier - Programming Video Games for the Evil Genius");
  setSize(700,700);
  setVisible(true);
  setDefaultCloseOperation(JFrame.EXIT_ON_CLOSE);

  addMouseListener(this);
  addKeyListener(this);
}

public void paint(Graphics g)
{
  super.paint(g);

  //set the color to black (for the lines):
  g.setColor(Color.black);

  //this for loop will go thru every line in the ArrayList and draw them:
  for(int i = 0; i <lines.size(); i++)
  {
    //get the line in the ArrayList:
    Line2D.Double temp = (Line2D.Double) lines.get(i);

    //now get all the x's and y's
    int x1 = Integer.parseInt(""+Math.round (temp.getX1()));
    int y1 = Integer.parseInt(""+Math.round (temp.getY1()));
    int x2 = Integer.parseInt(""+Math.round (temp.getX2()));
    int y2 = Integer.parseInt (""+Math.round(temp.getY2()));

    g.drawLine(x1,y1,x2,y2);
  }

  if(drawGuy)
  {
    try
    {

      //draw the guy:
      URL url = this.getClass().getResource ("guy.png");
      Image img = Toolkit.getDefaultToolkit().getImage(url);
      g.drawImage(img, guy.x, guy.y, this);
    }
    catch(Exception e){}
  }

}

//these are the mouse listener methods:
public void mouseClicked(MouseEvent e){}

//these are the mouse listener methods:
public void mouseEntered(MouseEvent e){}
```

```
//these are the mouse listener methods:
public void mouseExited(MouseEvent e){}

//these are the mouse listener methods:
public void mousePressed(MouseEvent e)
{
  //this code will run when the mouse is first pressed, getting the
  //starting point of the line.
  holder = new Point2D.Double(e.getX(),e.getY());
}

//these are the mouse listener methods:
public void mouseReleased(MouseEvent e)
{
  //this code will run when the mouse is released, completing the
  //line and adding it to the ArrayList.
  Point2D.Double end = new Point2D.Double(e.getX(),e.getY());
  lines.add(new Line2D.Double(holder,end));

  //now, repinat the screen so the line is drawn:
  repaint();
}

//these are the key listener methods:
public void keyPressed(KeyEvent e){}

//these are the key listener methods:
public void keyReleased(KeyEvent e){}

//these are the key listener methods:
public void keyTyped(KeyEvent e)
{
  //if the user presses "p" OR "P," start the Thread
  if(e.getKeyChar()=='p' || e.getKeyChar()=='P')
  {
    //init the Thread:
    move = new Move();
    move.start();
    move.action(true);
  }

  //if the user presses "q" OR "Q," stop the Thread
  if(e.getKeyChar()=='q' || e.getKeyChar()=='Q')
  {
    move.action(false);
    drawGuy = false;
    move = null;
  }
}

//this is the thread that will make the character move
private class Move extends Thread
{

  //these variables will hold the player's speed and gravity
  double velocity;
  double gravity;

  //stops/starts the thread:
  boolean go = false;

  public void run()
  {
    if(go)
    {
      initGuy();
```

```
    velocity = 0;
    gravity = 1;
  }
  while(go)
  {
    try
    {
      //this will hold the line the guy is on (null if none)
      Line2D.Double lineTaken = null;
      //this will say whether the character is even on a line
      boolean onLine = false;
      //gravity needs to be reset when the guy first lands on
      //the line. This will hold that info. It holds the line #
      int firstOnLine = -1;

      //check if he is on a line:
      for(int i = lines.size()-1; i>=0; i--)
      {
        //get the line:
        Line2D.Double temp = (Line2D.Double) lines.get(i);

        if(temp.intersects(guy.x,guy.y,30,30))
        {
          lineTaken = temp;
          onLine = true;

          if(firstOnLine!=i)
          {
            firstOnLine = i;
            gravity = 0;
          }

          break;
        }
      }
      //if there is a line it is on ...
      if(onLine)
      {
        //now, get the new gravity by subtracting the y's and
        //dividing by 20
        double mGrav = (lineTaken.y2-line Taken.y1)/50;

        //now, get the new velocity by subtracting the x's and
        //dividing by 20
        double mVel = (lineTaken.x2-line Taken.x1)/100;

        //set the maximum values
        if(velocity<5)
          velocity+=mVel;
        if(gravity<2.5)
          gravity+=mGrav;
      }
      else
      {
        gravity+=.2;
      }

      //alter the guy's movements:
      guy.x += velocity;
      guy.y += gravity;

      //delay before repainting:
      Thread.sleep(75);
      //repaint:
      repaint();
```

```
        }
        catch(Exception e){  break; }
      }
    }
    public void action(boolean a)
    {
      //stops the thread:
      go = a;
    }
    public void initGuy()
    {

      /*
       * This code will set up the character's position
       */
      //get the first line
      Line2D.Double firstLine = (Line2D.Double) lines.get(0);
      //get the first "x" and "y" of that line:
      int x = Integer.parseInt(""+Math.round (firstLine.x1));
      int y = Integer.parseInt(""+Math.round (firstLine.y1));
      guy = new Rectangle(x+30,y-20,30,30);
      drawGuy = true;
    }
  }
  public static void main (String[ ] args)
  {
    //begin the program:
    new G2P2();
  }
}
```

Figure 14-2 *Slope comes to life.*

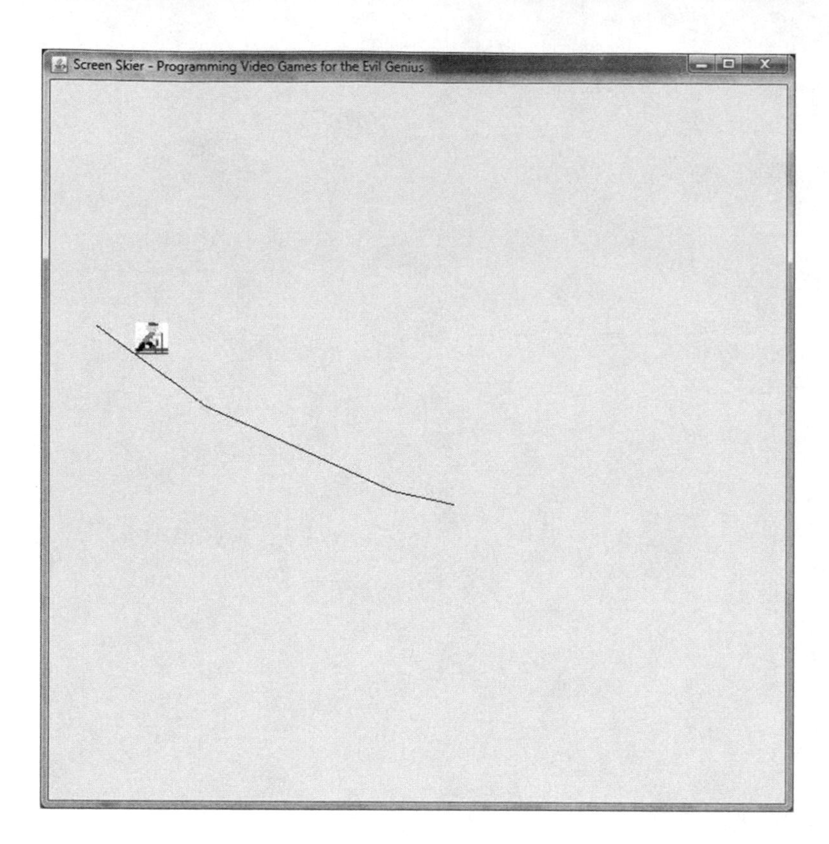

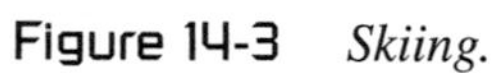
Figure 14-3 *Skiing.*

Figure 14-4 *More skiing.*

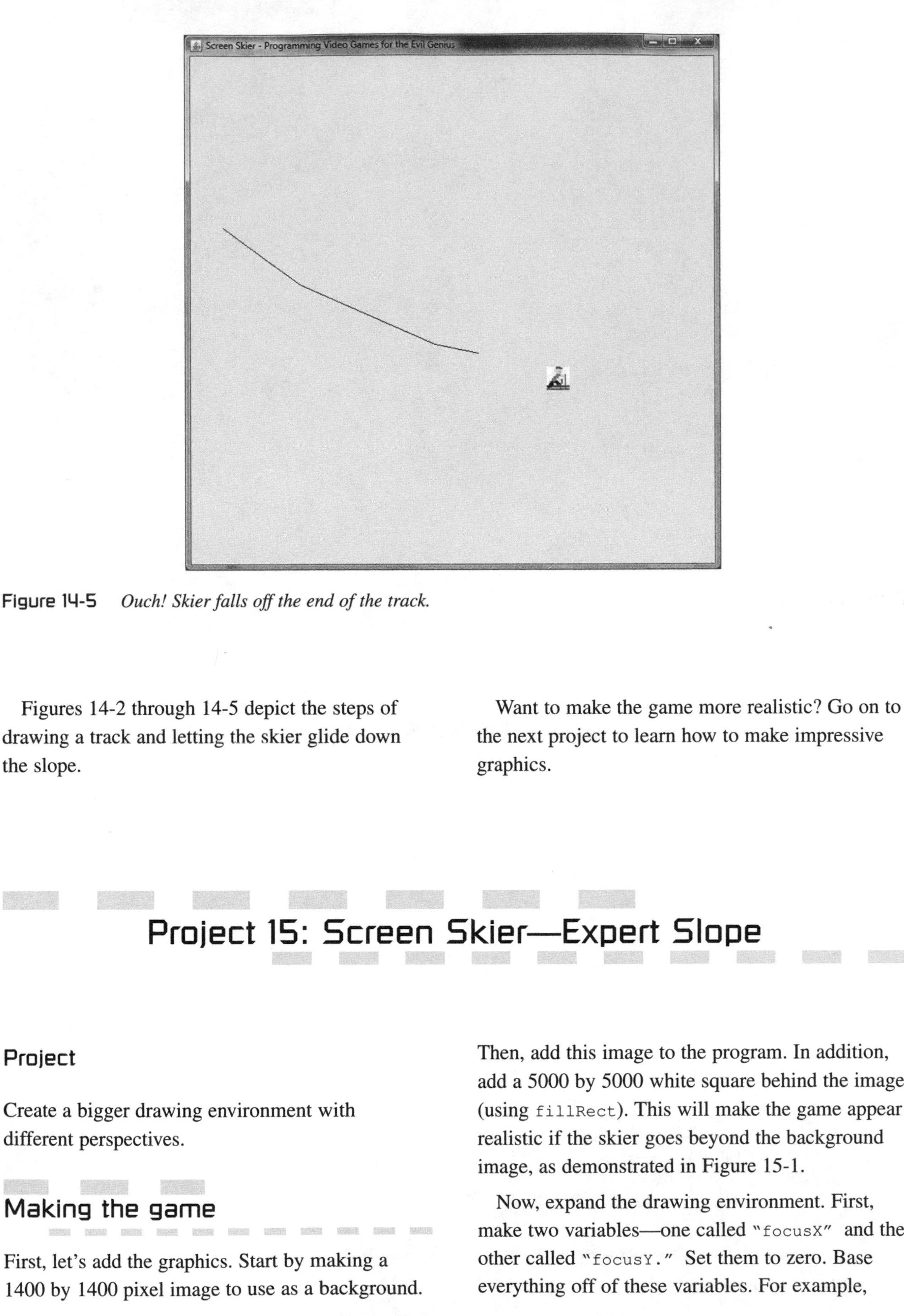

Figure 14-5 *Ouch! Skier falls off the end of the track.*

Figures 14-2 through 14-5 depict the steps of drawing a track and letting the skier glide down the slope.

Want to make the game more realistic? Go on to the next project to learn how to make impressive graphics.

Project 15: Screen Skier—Expert Slope

Project

Create a bigger drawing environment with different perspectives.

Making the game

First, let's add the graphics. Start by making a 1400 by 1400 pixel image to use as a background. Then, add this image to the program. In addition, add a 5000 by 5000 white square behind the image (using `fillRect`). This will make the game appear realistic if the skier goes beyond the background image, as demonstrated in Figure 15-1.

Now, expand the drawing environment. First, make two variables—one called "`focusX`" and the other called "`focusY`." Set them to zero. Base everything off of these variables. For example,

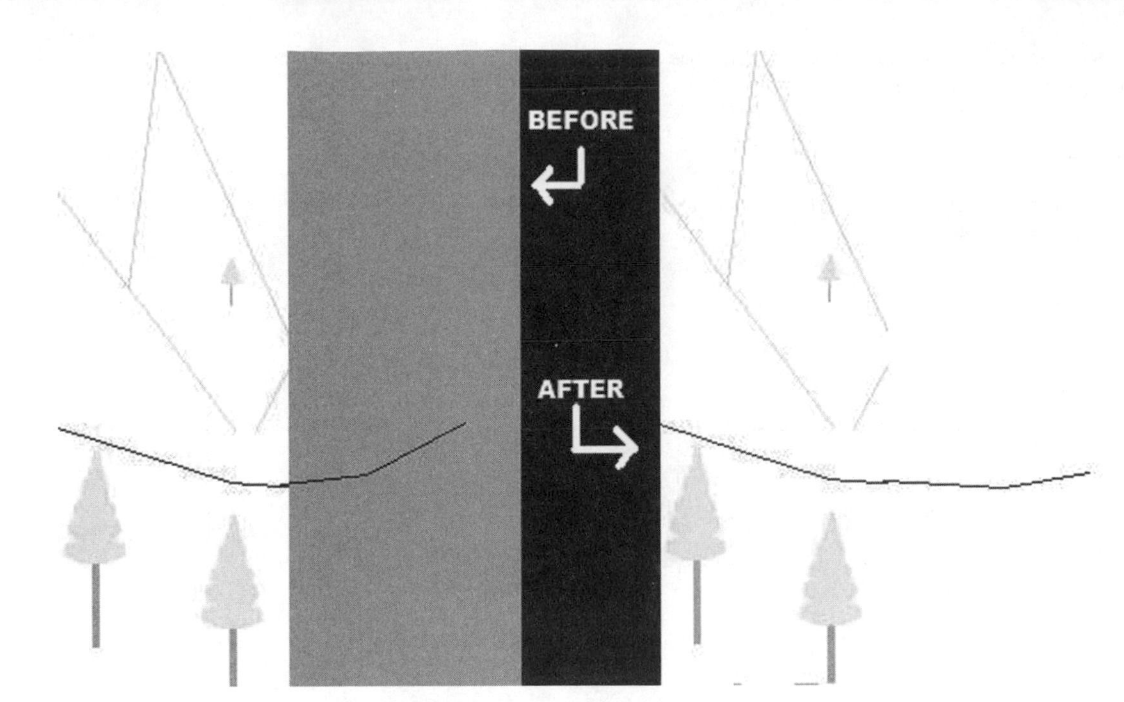

Figure 15-1 *Background—before and after.*

when you get the x coordinate of a click, use the following:

```
e.getX()+focusX
```

When you draw the skier (in the paint method), change `focusX` and `focusY` to `"guy.x-100"` and `"guy.y-100,"` respectively. This will make the screen follow the skier. You can also add keyListeners to alter `focusX` and `focusY` by 100 pixels so that the player can move the focus around when creating the ski slope.

```
//import everything:
import javax.swing.*;
import javax.swing.event.*;
import java.awt.*;
import java.awt.event.*;
import java.awt.geom.*;
import java.util.*;
import java.io.*;
import java.net.*;

//the actual class:
public class G2P3 extends JFrame implements MouseListener, KeyListener
{
  //this ArrayList holds the lines:
  ArrayList lines = new ArrayList();

  //this will hold the first point of the line
  Point2D.Double holder;

  //this is the Thread:
  Move move;
```

```
//this will hold the character's information:
Rectangle guy = null;

//this will tell repaint whether or not to draw the guy:
boolean drawGuy = false;

//this will make the screen focus on the guy:
int focusX = 0;
int focusY = 0;

//the constructor:
public G2P3()
{
  //set the title:
  super("Screen Skier - Programming Video Games for the Evil Genius");
  setSize(700,700);
  setVisible(true);
  setDefaultCloseOperation(JFrame.EXIT_ON_CLOSE);

  addMouseListener(this);
  addKeyListener(this);
}

public void paint(Graphics g)
{
  super.paint(g);

  //draw the background:
  try
  {
    //this is the general, plain white background:
    g.setColor(Color.white);
    g.fillRect(-2000,-2000,5000,5000);
    //this is the image background
    URL url = this.getClass().getResource ("bg.png");
    Image img = Toolkit.getDefaultToolkit(). getImage(url);
    g.drawImage(img, 0-focusX, 0-focusY, this);

  }
  catch(Exception e){}

  //set the color to black (for the lines):
  g.setColor(Color.black);

  //this for loop will go tru every line in the ArrayList and draw them:
  for(int i = 0; i <lines.size(); i++)
  {
    //get the line in the ArrayList:
    Line2D.Double temp = (Line2D.Double) lines.get(i);

    //now get all the x's and y's
    int x1 = Integer.parseInt(""+Math.round(temp.getX1()));
    int y1 = Integer.parseInt(""+Math.round (temp.getY1()));
    int x2 = Integer.parseInt(""+Math.round (temp.getX2()));
    int y2 = Integer.parseInt(""+Math.round (temp.getY2()));

    g.drawLine(x1-focusX,y1-focusY,x2-focusX,y2-focusY);
  }
  if(drawGuy)
  {

    try
    {
      //draw the guy:
      URL url = this.getClass().getResource ("guy.png");
      Image img = Toolkit.getDefaultToolkit(). getImage(url);
      g.drawImage(img, guy.x-focusX, guy.y-focusY, this);
```

```
    }
    catch(Exception e){}
    //reset the focus to the guy's position
    focusX = guy.x-100;
    focusY = guy.y-100;
  }
}

//these are the mouse listener methods:
public void mouseClicked(MouseEvent e){}

//these are the mouse listener methods:
public void mouseEntered(MouseEvent e){}

//these are the mouse listener methods:
public void mouseExited(MouseEvent e){}

//these are the mouse listener methods:
public void mousePressed(MouseEvent e)
{
  //this code will run when the mouse is first pressed, getting the
  //starting point of the line.
  holder = new Point2D.Double(e.getX()+ focusX,e.getY()+focusY);
}

//these are the mouse listener methods:
public void mouseReleased(MouseEvent e)
{
  //this code will run when the mouse is released, completing the
  //line and adding it to the ArrayList.
  Point2D.Double end = new Point2D.Double(e.get X()+focusX, e.getY()+focusY);
  lines.add(new Line2D.Double(holder,end));

  //now, repinat the screen so the line is drawn:
  repaint();
}

//these are the key listener methods:
public void keyPressed(KeyEvent e){}

//these are the key listener methods:
public void keyReleased(KeyEvent e){}

//these are the key listener methods:
public void keyTyped(KeyEvent e)
{
  //if the user presses "p" OR "P," start the Thread
  if(e.getKeyChar()=='p' || e.getKeyChar()=='P' )
  {
    //init the Thread:
    move = new Move();
    move.start();
    move.action(true);
  }

  //if the user presses "q" OR "Q," stop the Thread
  if(e.getKeyChar()=='q' || e.getKeyChar()=='Q' )
  {
    move.action(false);
    drawGuy = false;
    focusX = 0;
    focusY = 0;
    move = null;
  }

  //if the user presses "a" OR "A," move the focus
```

```
  if(e.getKeyChar()=='a' || e.getKeyChar()=='A')
  {
    focusX-=100;
    repaint();
  }
  //if the user presses "s" OR "S," move the focus
  if(e.getKeyChar()=='s' || e.getKeyChar()=='S')
  {
    focusY+=100;
    repaint();
  }
  //if the user presses "w" OR "W," move the focus
  if(e.getKeyChar()=='w' || e.getKeyChar()=='W')
  {
    focusY-=100;
    repaint();
  }
  //if the user presses "d" OR "D," move the focus
  if(e.getKeyChar()=='d' || e.getKeyChar()=='D')
  {
    focusX+=100;
    repaint();
  }
}
//this is the thread that will make the character move
private class Move extends Thread
{
  //these variables will hold the player's speed and gravity
  double velocity;
  double gravity;
  //stops/starts the thread:
  boolean go = false;
  public void run()
  {
    if(go)
    {
      initGuy();
      velocity = 0;
      gravity = 1;
    }
    while(go)
    {
      try
      {
        //this will hold the line the guy is on (null if none)
        Line2D.Double lineTaken = null;
        //this will say whether the character is even on a line
        boolean onLine = false;
        //gravity needs to be reset when the guy first lands on
        //the line. This will hold that info. It holds the line #
        int firstOnLine = -1;
        //check if he is on a line:
        for(int i = lines.size()-1; i>=0; i--)
        {
          //get the line:
          Line2D.Double temp = (Line2D.Double) lines.get(i);
          if(temp.intersects(guy.x,guy.y,30,30))
```

```
        {
          lineTaken = temp;
          onLine = true;

          if(firstOnLine!=i)
          {
            firstOnLine = i;
            gravity = 0;
          }
          break;
        }
      }
      //if there is a line it is on ...
      if(onLine)
      {
        //now, get the new gravity by subtracting the y's and
        //dividing by 20
        double mGrav = (lineTaken.y2-lineTaken.y1)/50;

        //now, get the new velocity by subtracting the x's and
        //dividing by 20
        double mVel = (lineTaken.x2-lineTaken.x1)/100;

        //set the maximum values
        if(velocity<5)
          velocity+=mVel;
        if(gravity<2.5)
          gravity+=mGrav;
      }
      else
      {
        gravity+=.2;
      }

      //alter the guy's movements:
      guy.x += velocity;
      guy.y += gravity;

      //delay before repainting:
      Thread.sleep(75);
      //repaint:
      repaint();
    }
    catch(Exception e){ break; }
  }
}
public void action(boolean a)
{
  //stops the thread:
  go = a;
}
public void initGuy()
{
  /*
   * This code will set up the character's position
   */
  //get the first line
  Line2D.Double firstLine = (Line2D.Double) lines.get(0);

  //get the first "x" and "y" of that line:
  int x = Integer.parseInt(""+Math.round(firstLine.x1));
  int y = Integer.parseInt(""+Math.round(firstLine.y1));
```

```
      guy = new Rectangle(x+30,y-20,30,30);

      drawGuy = true;
    }
  }

  public static void main (String[ ] args)
  {

    //begin the program:
    new G2P3();
  }
}
```


Figures 15-2 through 15-6 depict the game play of Screen Skier.

With editing tools, you can make endless variations: add more track, undo the last piece of slope, or even bulldoze everything and start again.

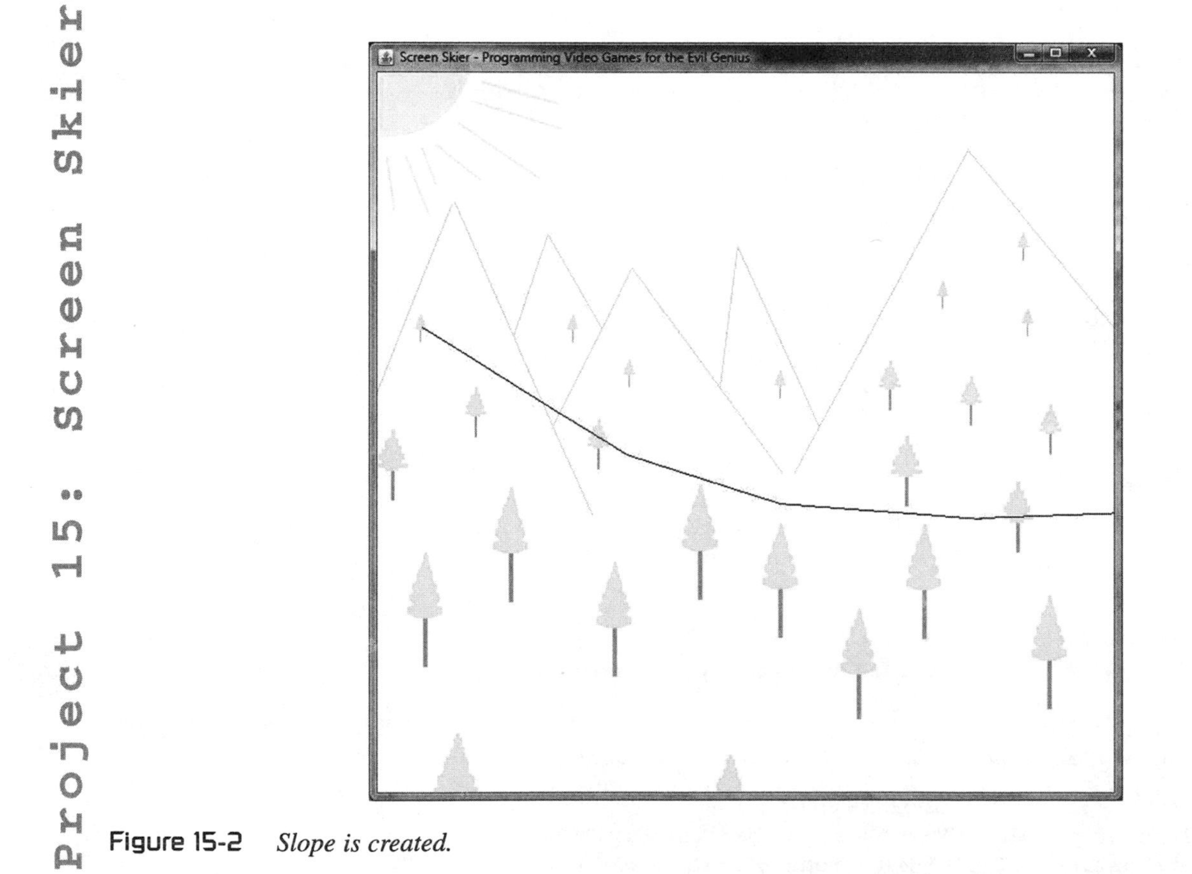

Figure 15-2 *Slope is created.*

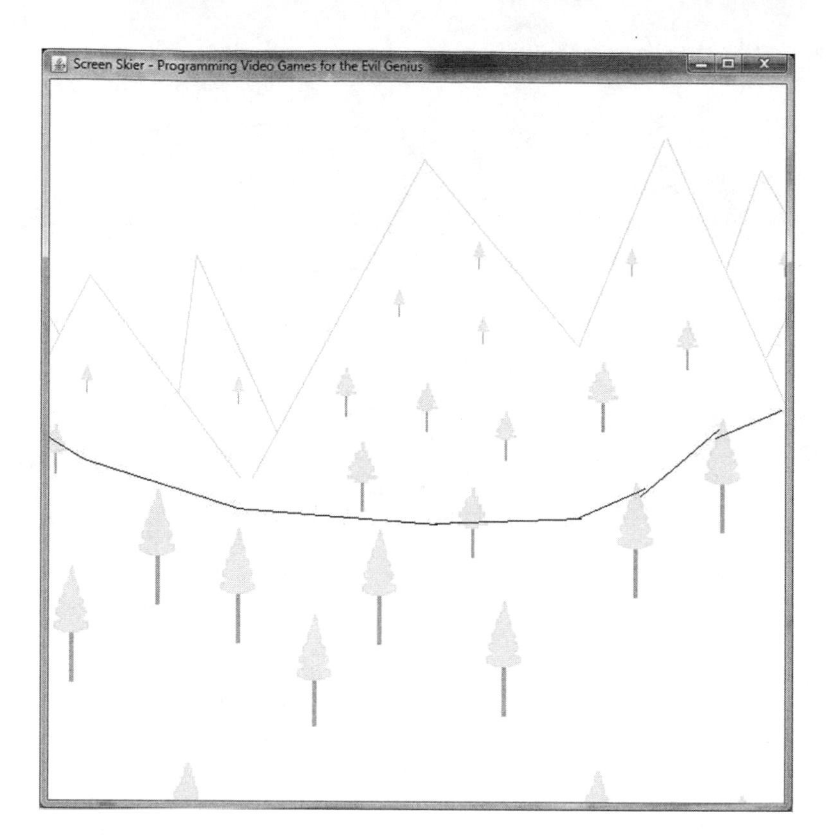

Figure 15-3 *Screen moves to expand the slope.*

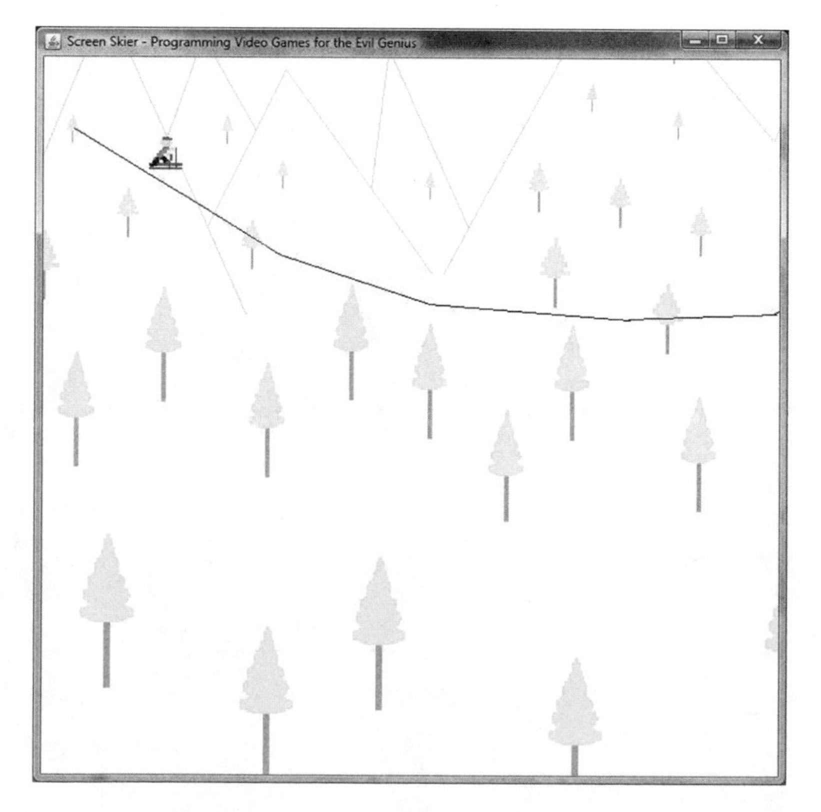

Figure 15-4 *Skier journeys down the slope.*

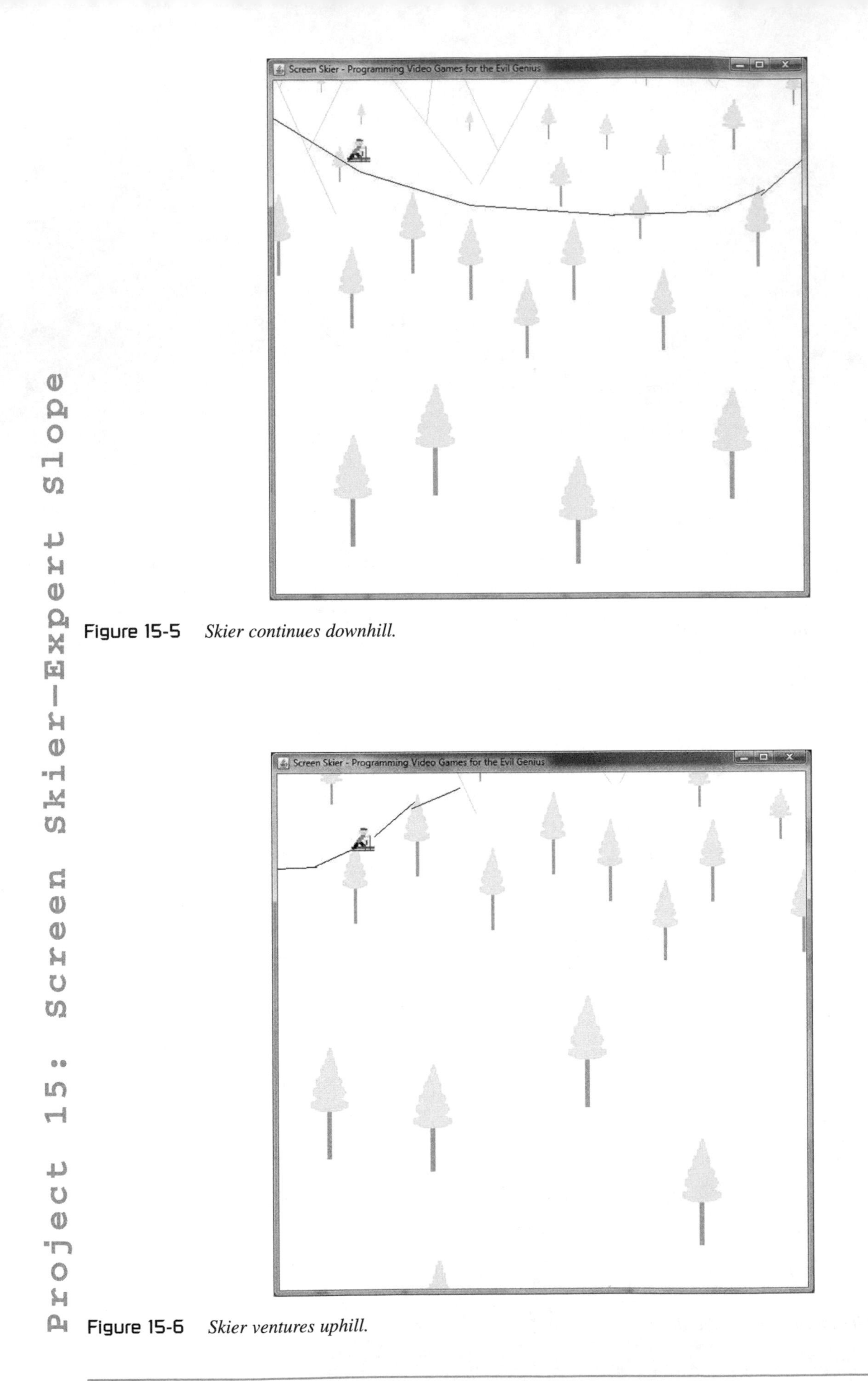

Figure 15-5 *Skier continues downhill.*

Figure 15-6 *Skier ventures uphill.*

Project 16: Screen Skier—Bulldozer

Project

Change the course: delete it, add to it ... make it extreme!

Making the game

To allow the player to delete his/her last line, create a new keyEvent for "z" (because "Crtl+z" is undo). When the user presses "z," execute the following code:

```
lines.remove(lines.size()-1);
```

"`lines.remove`" deletes the variable inside of the parentheses at the given element position. Because you are giving it the position of "`lines.size()-1`," the computer will remove the last line (remember `lines.size` returns the size of the ArrayList, not the last element!).

To let the user clear the entire track, create a keyEvent for "x." To clear the entire ArrayList, use the following code:

```
lines.clear();
```

Figure 16-1 *"confirmDialog" box.*

Warning: if the user accidentally presses "x," hours of hard work will be deleted! Prevent this distaster by using a JOptionPane, which will provide a safety net. Use the following code:

```
int answer = JOptionPane.showConfirmDialog
                (null, "Delete?");
```

A ConfirmDialog box is displayed in Figure 16-1.

The variable "`answer`" represents one of the three options. Use the following code inside an if-statement to check if the user pressed "OK." Ways to check for other responses can be found in the API.

```
answer==JOptionPane.OK_OPTION
```

```
//import everything:
import javax.swing.*;
import javax.swing.event.*;
import java.awt.*;
import java.awt.event.*;
import java.awt.geom.*;
import java.util.*;
import java.io.*;
import java.net.*;
import java.applet.AudioClip;
//the actual class:
public class G2P4 extends JFrame implements MouseListener, KeyListener
{
  //this ArrayList holds the lines:
  ArrayList lines = new ArrayList();

  //this will hold the first point of the line
  Point2D.Double holder;

  //these are the Threads:
  Move move;
```

```
Counter cnt;

//this will hold the character's information:
Rectangle guy = null;

//this will tell repaint whether or not to draw the guy:
boolean drawGuy = false;

//this will hold the seconds the skier has been skiing
int counter = 0;

//this will make the screen focus on the guy:
int focusX = 0;
int focusY = 0;

//if the guys hits the bottom, change the icon with this boolean
boolean alive = true;

//true if the sound has been played
boolean sndPlayed = false;

//the current time:
int count = 0;

//this holds the final time
int holdTime = 0;

//this says whether or not to use the "hold" value
boolean useHold = false;

//the constructor:
public G2P4()
{
  //set the title:
  super("Screen Skier — Programming Video Games for the Evil Genius");
  setSize(700,700);
  setVisible(true);
  setDefaultCloseOperation(JFrame.EXIT_ON_CLOSE);

  JOptionPane.showMessageDialog(null,
      "SCREEN SKIER\n\n" +
      "Create your own ski run! Simply\n" +
      "click and drag the mouse to create\n" +
      "a track. Press 'p' to start the\n" +
      "run and 'q' to end it. To create\n" +
      "a larger track, use the WASD keys\n" +
      "to move the focus of the screen.\n" +
      "All uphill slopes become ski lifts.\n" +
      "If you make them too steep, however,\n" +
      "they may collapse! Also, if you try\n" +
      "to even the track out from a large\n" +
      "decline, the skier may fall through\n" +
      "the track into the snow! Also, if you\n" +
      "made a mistake designing your track,\n" +
      "you can erase it by pressing 'x'.\n" +
      "You can also undo the last line by\n" +
      "pressing 'z'.\n\n\n" +
      "Your goal: Create a track that keeps the\n" +
      "skier alive for exactly 45 seconds.\n\n" +
      "Be Careful and Good Luck!");

  addMouseListener(this);
  addKeyListener(this);

  //set up the counter
  cnt = new Counter();
  cnt.go = false;
  cnt.start();
```

```
}
public void paint(Graphics g)
{
  super.paint(g);
  //draw the background:
  try
  {
    //this is the general, plain white background:
    g.setColor(Color.white);
    g.fillRect(-2000,-2000,5000,5000);
    //this is the image background
    URL url = this.getClass().getResource ("bg.png");
    Image img = Toolkit.getDefaultToolkit(). getImage(url);
    g.drawImage(img, 0-focusX, 0-focusY, this);
  }
  catch(Exception e){}
  //set the color to black (for the lines):
  g.setColor(Color.black);
  //this for loop will go thru every line in the ArrayList and draw them:
  for(int i = 0; i <lines.size(); i++)
  {
    //get the line in the ArrayList:
    Line2D.Double temp = (Line2D. Double) lines.get(i);
    //now get all the x's and y's
    int x1 = Integer.parseInt(""+Math.round (temp.getX1()));
    int y1 = Integer.parseInt(""+Math.round (temp.getY1()));
    int x2 = Integer.parseInt(""+Math.round (temp.getX2()));
    int y2 = Integer.parseInt(""+Math.round (temp.getY2()));
    g.drawLine(x1-focusX,y1-focusY,x2-focusX,y2-focusY);
  }
  if(drawGuy)
  {
    try
    {
      //draw the time:
      g.setFont(new Font("times new roman", Font.BOLD, 16));
      URL urlTime = this.getClass().getResource ("time.png");
      Image imgT = Toolkit.getDefaultToolkit().getImage (urlTime);
      g.drawImage(imgT, -35, 10, this);
      if(!useHold)
        g.drawString("Current Time: "+count, 50, 50);
      else
        g.drawString("Current Time: "+holdTime, 50, 50);
      //draw the guy:
      if(alive)
      {
        URL url = this.getClass().getResource ("guy.png");
        Image img = Toolkit.getDefaultToolkit().getImage(url);
        g.drawImage(img, guy.x-focusX, guy.y-focusY, this);
      }
      else
      {
        URL url = this.getClass().getResource ("guyDead.png");
        Image img = Toolkit.getDefaultToolkit().getImage(url);
        g.drawImage(img, guy.x-focusX, guy.y-focusY, this);
        //if the snd hasn't been played, play it!
        if(!sndPlayed)
```

```
        {
          //set up the hold time variable:
          holdTime = count;
          useHold = true;
          //play the sound
          URL snd = this.getClass().getResource("scream.wav");
          AudioClip scream = JApplet.newAudioClip(snd);
          scream.play();
          sndPlayed = true;

          //check for a win:
          checkWin();
        }
      }
    }
    catch(Exception e){}

    //reset the focus to the guy's position
    focusX = guy.x-100;
    focusY = guy.y-100;
  }
}

//check to see if the goal is accomplished
public void checkWin()
{
  if(holdTime==45)
  {
    JOptionPane.showMessageDialog(null,
        "Congrats!\n\n" +
        "MISSION ACCOMPLISHED!");
  }
}

//these are the mouse listener methods:
public void mouseClicked(MouseEvent e){}

//these are the mouse listener methods:
public void mouseEntered(MouseEvent e){}

//these are the mouse listener methods:
public void mouseExited(MouseEvent e){}

//these are the mouse listener methods:
public void mousePressed(MouseEvent e)
{
  //this code will run when the mouse is first pressed, getting the
  //starting point of the line.
  holder = new Point2D.Double(e.getX()+focusX,e.getY()+focusY);
}

//these are the mouse listener methods:
public void mouseReleased(MouseEvent e)
{
  //this code will run when the mouse is released, completing the
  //line and adding it to the ArrayList.
  Point2D.Double end = new Point2D.Double(e.getX()+focusX,e.getY()+focusY);
  lines.add(new Line2D.Double(holder,end));
  //now, repinat the screen so the line is drawn:
  repaint();
}

//these are the key listener methods:
public void keyPressed(KeyEvent e){}

//these are the key listener methods:
```

```
public void keyReleased(KeyEvent e){}

//these are the key listener methods:
public void keyTyped(KeyEvent e)
{
  //if the user presses "p" OR "P," start the Thread
  if(e.getKeyChar()=='p'  || e.getKeyChar()=='P' )
  {
    alive = true;
    count = 0;
    useHold = false;
    //init the Thread:
    move = new Move();
    cnt.go = true;
    move.start();
    move.action(true);
    sndPlayed = false;
  }

  //if the user presses "q" OR "Q," stop the Thread
  if(e.getKeyChar()=='q'  || e.getKeyChar()=='Q' )
  {
    move.action(false);
    drawGuy = false;
    focusX = 0;
    focusY = 0;
    move = null;
    cnt.go = false;
  }

  //if the user presses "a" OR "A," move the focus
  if(e.getKeyChar()=='a'  || e.getKeyChar()=='A' )
  {
    focusX-=100;
    repaint();
  }

  //if the user presses "s" OR "S," move the focus
  if(e.getKeyChar()=='s'  || e.getKeyChar()=='S' )
  {

    focusY+=100;
    repaint();
  }

  //if the user presses "w" OR "W," move the focus
  if(e.getKeyChar()=='w'  || e.getKeyChar()=='W' )
  {
    focusY-=100;
    repaint();
  }

  //if the user presses "d" OR "D," move the focus
  if(e.getKeyChar()=='d'  || e.getKeyChar()==' D' )
  {
    focusX+=100;
    repaint();
  }

  //if the user presses "z" OR "Z," undo the last line
  if(e.getKeyChar()=='z'  || e.getKeyChar()=='Z' )
  {
    lines.remove(lines.size()-1);
    repaint();
```

```
  }
  //if the user presses "x" OR "X," clear the track
  if(e.getKeyChar()=='x'  || e.getKeyChar()=='X')
  {
    //double check before deleting!
    int answer = JOptionPane.showConfirm Dialog(null,
       "Do you really want to delete your track?");
    if(answer==JOptionPane.OK_OPTION)
      lines.clear();
    repaint();
  }
}

private class Counter extends Thread
{
  public boolean go = true;
  public void run()
  {
    try
    {
      while(true)
      {
        if(go)
        {
          Thread.sleep(1000);
          count++;
        }
      }
    }
    catch(Exception e){}
  }
}
//this is the thread that will make the character move
private class Move extends Thread
{

  //these variables will hold the player's speed and gravity
  double velocity;
  double gravity;

  //stops/starts the thread:
  boolean go = false;

  public void run()
  {

    if(go)
    {
      initGuy();
      velocity = 0;
      gravity = 1;
    }
    while(go)
    {
      try
      {
        //this will hold the line the guy is on (null if none)
        Line2D.Double lineTaken = null;
        //this will say whether the character is even on a line
        boolean onLine = false;
        //gravity needs to be reset when the guy first lands on
        //the line. This will hold that info. It holds the line #
```

```
        int firstOnLine = -1;

        //check if he is on a line:
        for(int i = lines.size()-1; i>=0; i--)
        {
          //get the line:
          Line2D.Double temp = (Line2D.Double) lines.get(i);

          if(temp.intersects(guy.x,guy.y,30,30))
          {
            lineTaken = temp;
            onLine = true;

            if(firstOnLine!=i)
            {
              firstOnLine = i;
              gravity = 0;
            }

            break;
          }
        }
        //if there is a line it is on ...
        if(onLine)
        {
          //now, get the new gravity by subtracting the y's and
          //dividing by 20
          double mGrav = (lineTaken.y2-lineTaken.y1)/50;

          //now, get the new velocity by subtracting the x's and
          //dividing by 20
          double mVel = (lineTaken.x2-lineTaken.x1)/100;

          //set the maximum values
          if(velocity<5)
            velocity+=mVel;
          if(gravity<2.5)
            gravity+=mGrav;
        }
        else
        {
          gravity+=.2;
        }

        //alter the guy's movements:
        guy.x += velocity;
        guy.y += gravity;

        //check to see if the guy died:
        if(guy.y>1400)
        {
          alive = false;
        }

        //delay before repainting:
        Thread.sleep(75);
        //repaint:
        repaint();
      }
      catch(Exception e){ break; }
    }
  }

  public void action(boolean a)
  {
    //stops the thread:
```

```
      go = a;
    }
    public void initGuy()
    {
      /*
       * This code will set up the character's position
       */
      //get the first line
      Line2D.Double firstLine = (Line2D.Double) lines.get(0);
      //get the first "x" and "y" of that line:
      int x = Integer.parseInt(""+Math.round (firstLine.x1));
      int y = Integer.parseInt(""+Math.round(firstLine.y1));

      guy = new Rectangle(x+30,y-20,30,30);

      drawGuy = true;

    }
  }
  public static void main (String[ ] args)
  {
    //begin the program:
    new G2P4();
  }
}
```

Figures 16-2 through 16-5 demonstrate the program's ability to add or delete track.

Right now, Screen Skier is a fun drawing tool. In the next project, learn how to transform Screen Skier into a high pressure "time-trial" game.

Figure 16-2 *Pre-made slope.*

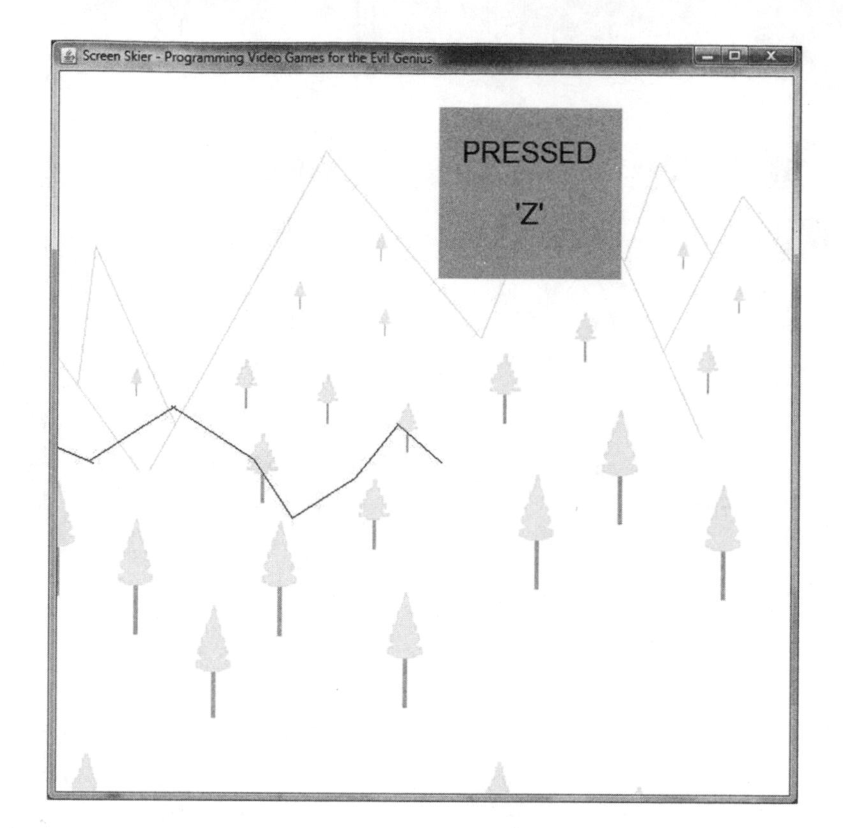

Figure 16-3 *"Undoing" the last part of the slope (by pressing "Z").*

Figure 16-4 *The slope is about to be cleared (by pressing "X").*

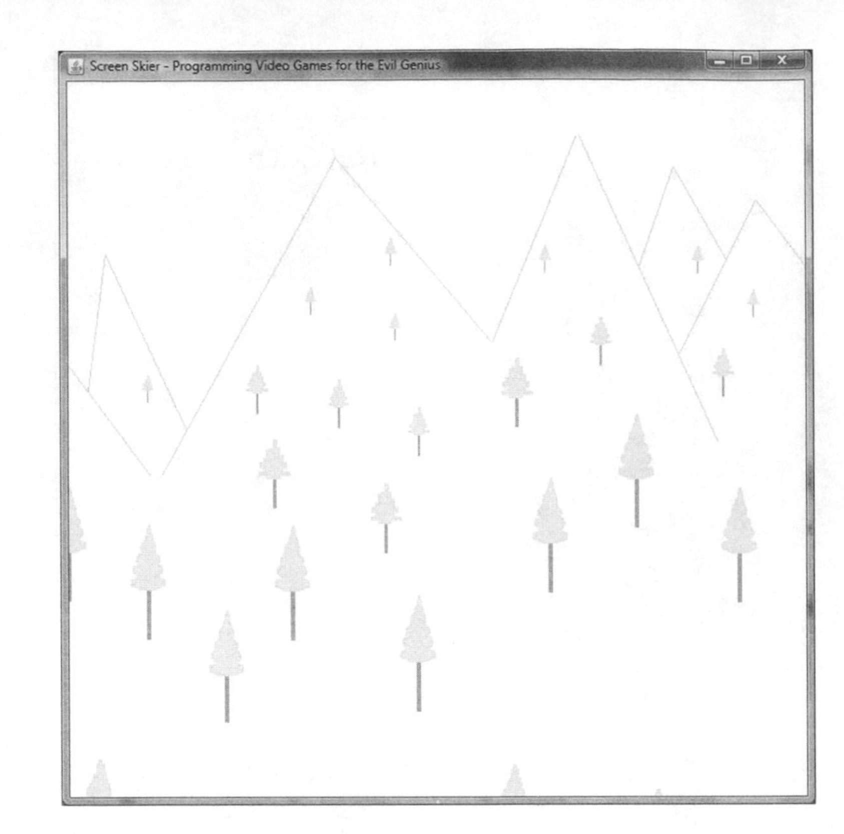

Figure 16-5 *Slope destruction (by clicking "Okay").*

Project 17: Screen Skier—Competition

Project

It all comes together right here—the program you created turns into an adrenaline pumping game. The goal? Make the skier complete his/her run in exactly 45 seconds. Also, add code for sound effects of screams when the skier falls off the mountain.

Making the game

First, create a counter that times the skier's run by adding a global variable called "`count.`" Then, initialize a new Thread called "`Counter.`" In the run method, insert a `while loop`. Inside the loop, delay for one second and increment "count" by one. Figure 17-1 demonstrates an innovative way to display the counter.

Because you want the timer to run only when the skier is in motion, go to the keyListener code that ends the skier's run. Insert code that will end the `while loop` of the counter by altering a boolean in the Thread. Also, be sure to make the boolean true in the keyListener code that makes the skier begin skiing. This stops the clock when the player passes the finish line ... or falls down the ravine.

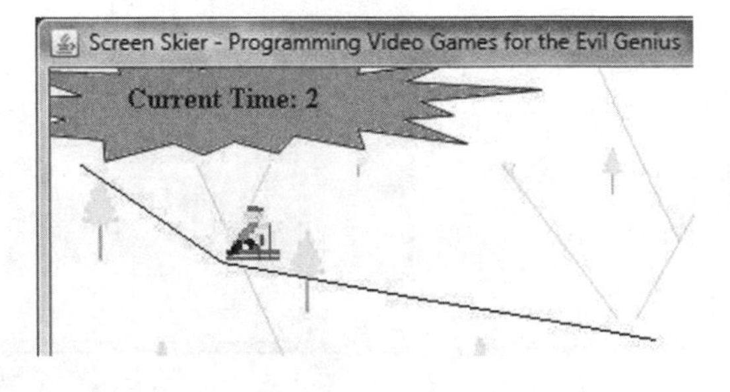

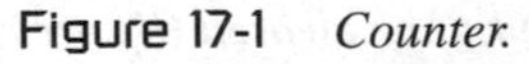
Figure 17-1 *Counter.*

To draw the text, use the method "g.drawString("text",x,y)"

You can also set the text before hand with the method "g.setFont(new Font())." To specify the font, look up Font's constructors in the API.

To check if the player has passed the finish line, create a boolean called "alive." In the Thread that controls movement, check to see if the skier's y coordinate is greater than 1400 (because 1400 is the bottom of the background image). If so, he/she has crashed. If you want, you can change the icon and add sound effects (record your own shrieks!).

When the player is flat on his/her face, call a method that checks the time against 45 seconds. Dead or alive, if the slope was completed in exactly 45 seconds, the player wins.

Be sure to stop the counter and display the final time at the end of the run. Do this by creating a variable called "holdTime."

```
//import everything:
import javax.swing.*;
import javax.swing.event.*;
import java.awt.*;
import java.awt.event.*;
import java.awt.geom.*;
import java.util.*;
import java.io.*;
import java.net.*;
import java.applet.AudioClip;
//the actual class:
public class G2P5 extends JFrame implements MouseListener, KeyListener
{
  //this ArrayList holds the lines:
  ArrayList lines = new ArrayList();
  //this will hold the first point of the line
  Point2D.Double holder;
  //these are the Threads:
  Move move;
  Counter cnt;
  //this will hold the character's information:
  Rectangle guy = null;
  //this will tell repaint whether or not to draw the guy:
  boolean drawGuy = false;
  //this will hold the seconds the skier has been skiing
  int counter = 0;
  //this will make the screen focus on the guy:
  int focusX = 0;
  int focusY = 0;
  //if the guys hits the bottom, change the icon with this boolean
  boolean alive = true;
  //true if the sound has been played
  boolean sndPlayed = false;
  //the current time:
  int count = 0;
  //this holds the final time
  int holdTime = 0;
  //this says whether or not to use the "hold" value
  boolean useHold = false;
```

```
//the constructor:
public G2P5()
{
  //set the title:
  super("Screen Skier - Programming Video Games for the Evil Genius");
  setSize(700,700);
  setVisible(true);
  setDefaultCloseOperation(JFrame.EXIT_ON_CLOSE);

  JOptionPane.showMessageDialog(null,
     "SCREEN SKIER\n\n" +
     "Create your own ski run! Simply\n" +
     "click and drag the mouse to create\n" +
     "a track. Press 'p' to start the\n" +
     "run and 'q' to end it. To create\n" +
     "a larger track, use the WASD keys\n" +
     "to move the focus of the screen.\n" +
     "All uphill slopes become ski lifts.\n" +
     "If you make them too steep, however,\n" +
     "they may collapse! Also, if you try\n" +
     "to even the track out from a large\n" +
     "decline, the skier may fall through\n" +
     "the track into the snow! Also, if you\n" +
     "made a mistake designing your track,\n" +
     "you can erase it by pressing 'x'.\n" +
     "You can also undo the last line by\n" +
     "pressing 'z'.\n\n\n" +
     "Your goal: Create a track that keeps the\n" +
     "skiier alive for exactly 45 seconds.\n\n" +
     "Be Careful and Good Luck!");

  addMouseListener(this);
  addKeyListener(this);

  //set up the counter
  cnt = new Counter();
  cnt.go = false;
  cnt.start();
}

public void paint(Graphics g)
{
  super.paint(g);

  //draw the background:
  try
  {
    //this is the general, plain white background:
    g.setColor(Color.white);
    g.fillRect(-2000,-2000,5000,5000);
    //this is the image background
    URL url = this.getClass().getResource ("bg.png");
    Image img = Toolkit.getDefaultToolkit(). getImage(url);
    g.drawImage(img, 0-focusX, 0-focusY, this);

  }
  catch(Exception e){}

  //set the color to black (for the lines):
  g.setColor(Color.black);

  //this for loop will go thru every line in the ArrayList and draw them:
  for(int i = 0; i <lines.size(); i++)
  {
```

```
    //get the line in the ArrayList:
    Line2D.Double temp = (Line2D.Double) lines.get(i);

    //now get all the x's and y's
    int x1 = Integer.parseInt(""+Math.round (temp.getX1()));
    int y1 = Integer.parseInt(""+Math.round (temp.getY1()));
    int x2 = Integer.parseInt(""+Math.round (temp.getX2()));
    int y2 = Integer.parseInt(""+Math.round (temp.getY2()));

    g.drawLine(x1-focusX,y1-focusY,x2-focusX,y2-focusY);
  }
  if(drawGuy)
  {
    try
    {
      //draw the time:
      g.setFont(new Font("times new roman", Font.BOLD, 16));
      URL urlTime = this.getClass().getResource ("time.png");
      Image imgT = Toolkit.getDefaultToolkit(). getImage(urlTime);
      g.drawImage(imgT, -35, 10, this);
      if(!useHold)
        g.drawString("Current Time: "+count, 50, 50);
      else
        g.drawString("Current Time: "+holdTime, 50, 50);
      //draw the guy:
      if(alive)
      {
        URL url = this.getClass().getResource ("guy.png");
        Image img = Toolkit.getDefaultToolkit(). getImage(url);
        g.drawImage(img, guy.x-focusX, guy.y-focusY, this);
      }
      else
      {
        URL url = this.getClass().getResource ("guyDead.png");
        Image img = Toolkit.getDefaultToolkit().getImage(url);
        g.drawImage(img, guy.x-focusX, guy.y-focusY, this);
        //if the snd hasn't been played, play it!
        if(!sndPlayed)
        {
          //set up the hold time variable:
          holdTime = count;
          useHold = true;
          //play the sound
          URL snd = this.getClass().getResource("scream.wav");
          AudioClip scream = JApplet.newAudioClip(snd);
          scream.play();
          sndPlayed = true;

          //check for a win:
          checkWin();
        }
      }
    }
    catch(Exception e){}

    //reset the focus to the guy's position
    focusX = guy.x-100;
    focusY = guy.y-100;
  }

}

//check to see if the goal is accomplished
public void checkWin()
```

```
{
  if(holdTime==45)
  {
    JOptionPane.showMessageDialog(null,
        "Congrats!\n\n" +
        "MISSION ACCOMPLISHED!");
  }
}

//these are the mouse listener methods:
public void mouseClicked(MouseEvent e){}

//these are the mouse listener methods:
public void mouseEntered(MouseEvent e){}

//these are the mouse listener methods:
public void mouseExited(MouseEvent e){}

//these are the mouse listener methods:
public void mousePressed(MouseEvent e)
{
  //this code will run when the mouse is first pressed, getting the
  //starting point of the line.
  holder = new Point2D.Double(e.getX()+ focusX,e.getY()+focusY);
}

//these are the mouse listener methods:
public void mouseReleased(MouseEvent e)
{
  //this code will run when the mouse is released, completing the
  //line and adding it to the ArrayList.
  Point2D.Double end = new Point2D.Double(e.getX()+focusX,e.getY()+focusY);
  lines.add(new Line2D.Double(holder,end));

  //now, repinat the screen so the line is drawn:
  repaint();
}
//these are the key listener methods:
public void keyPressed(KeyEvent e){}

//these are the key listener methods:
public void keyReleased(KeyEvent e){}

//these are the key listener methods:
public void keyTyped(KeyEvent e)
{
  //if the user presses "p" OR "P," start the Thread
  if(e.getKeyChar()=='p' || e.getKeyChar()=='P')
  {
    alive = true;
    count = 0;
    useHold = false;
    //init the Thread:
    move = new Move();
    cnt.go = true;
    move.start();
    move.action(true);
    sndPlayed = false;
  }

  //if the user presses "q" OR "Q," stop the Thread
  if(e.getKeyChar()=='q' || e.getKeyChar()=='Q')
  {
    move.action(false);
    drawGuy = false;
```

```
    focusX = 0;
    focusY = 0;
    move = null;
    cnt.go = false;
  }

  //if the user presses "a" OR "A," move the focus
  if(e.getKeyChar()=='a' || e.getKeyChar()=='A')
  {
    focusX-=100;
    repaint();
  }

  //if the user presses "s" OR "S," move the focus
  if(e.getKeyChar()=='s' || e.getKeyChar()=='S')
  {
    focusY+=100;
    repaint();
  }

  //if the user presses "w" OR "W," move the focus
  if(e.getKeyChar()=='w' || e.getKeyChar()=='W')
  {
    focusY-=100;
    repaint();
  }

  //if the user presses "d" OR "D," move the focus
  if(e.getKeyChar()=='d' || e.getKeyChar()=='D')
  {
    focusX+=100;
    repaint();
  }

  //if the user presses "z" OR "Z," undo the last line
  if(e.getKeyChar()=='z' || e.getKeyChar()=='Z')
  {
    lines.remove(lines.size()-1);
    repaint();
  }

  //if the user presses "x" OR "X," clear the track
  if(e.getKeyChar()==' x' || e.getKeyChar()==' X')
  {
    //double check before deleting!
    int answer = JOptionPane.showConfirmDialog(null,
        "Do you really want to delete your track?");
    if(answer==JOptionPane.OK_OPTION)
      lines.clear();
    repaint();
  }
}

private class Counter extends Thread
{
  public boolean go = true;
  public void run()
  {
    try
    {
      while(true)
      {
        if(go)
```

```
        {
          Thread.sleep(1000);
          count++;
        }
      }
    }
    catch(Exception e){}
  }
}

//this is the thread that will make the character move
private class Move extends Thread
{
  //these variables will hold the player's speed and gravity
  double velocity;
  double gravity;

  //stops/starts the thread:
  boolean go = false;

  public void run()
  {
    if(go)
    {
      initGuy();
      velocity = 0;
      gravity = 1;
    }
    while(go)
    {
      try
      {
        //this will hold the line the guy is on (null if none)
        Line2D.Double lineTaken = null;
        //this will say whether the character is even on a line
        boolean onLine = false;
        //gravity needs to be reset when the guy first lands on
        //the line. This will hold that info. It holds the line #
        int firstOnLine = -1;

        //check if he is on a line:
        for(int i = lines.size()-1; i>=0; i--)
        {
          //get the line:
          Line2D.Double temp = (Line2D.Double) lines.get(i);

          if(temp.intersects(guy.x,guy.y,30,30))
          {
            lineTaken = temp;
            onLine = true;

            if(firstOnLine!=i)
            {
              firstOnLine = i;
              gravity = 0;
            }
            break;
          }
        }
        //if there is a line it is on...
        if(onLine)
        {
          //now, get the new gravity by subtracting the y's and
```

```
          //dividing by 20
          double mGrav = (lineTaken.y2-line Taken.y1)/50;

          //now, get the new velocity by subtracting the x's and
          //dividing by 20
          double mVel = (lineTaken.x2-line Taken.x1)/100;

          //set the maximum values
          if(velocity<5)
            velocity+=mVel;
          if(gravity<2.5)
            gravity+=mGrav;
        }
        else
        {
          gravity+=.2;
        }

        //alter the guy's movements:
        guy.x += velocity;
        guy.y += gravity;
        //check to see if the guy died:
        if(guy.y>1400)
        {
          alive = false;
        }
        //delay before repainting:
        Thread.sleep(75);
        //repaint:
        repaint();
      }
      catch(Exception e){ break; }
    }
  }
  public void action(boolean a)
  {
    //stops the thread:
    go = a;
  }

  public void initGuy()
  {

    /*
     * This code will set up the character's position
     */
    //get the first line
    Line2D.Double firstLine = (Line2D.Double) lines.get(0);

    //get the first "x" and "y" of that line:
    int x = Integer.parseInt(""+Math.round (firstLine.x1));
    int y = Integer.parseInt(""+Math.round(firstLine.y1));

    guy = new Rectangle(x+30,y-20,30,30);

    drawGuy = true;
  }
}

public static void main (String[ ] args)
{
  //begin the program:
  new G2P5();
}
}
```

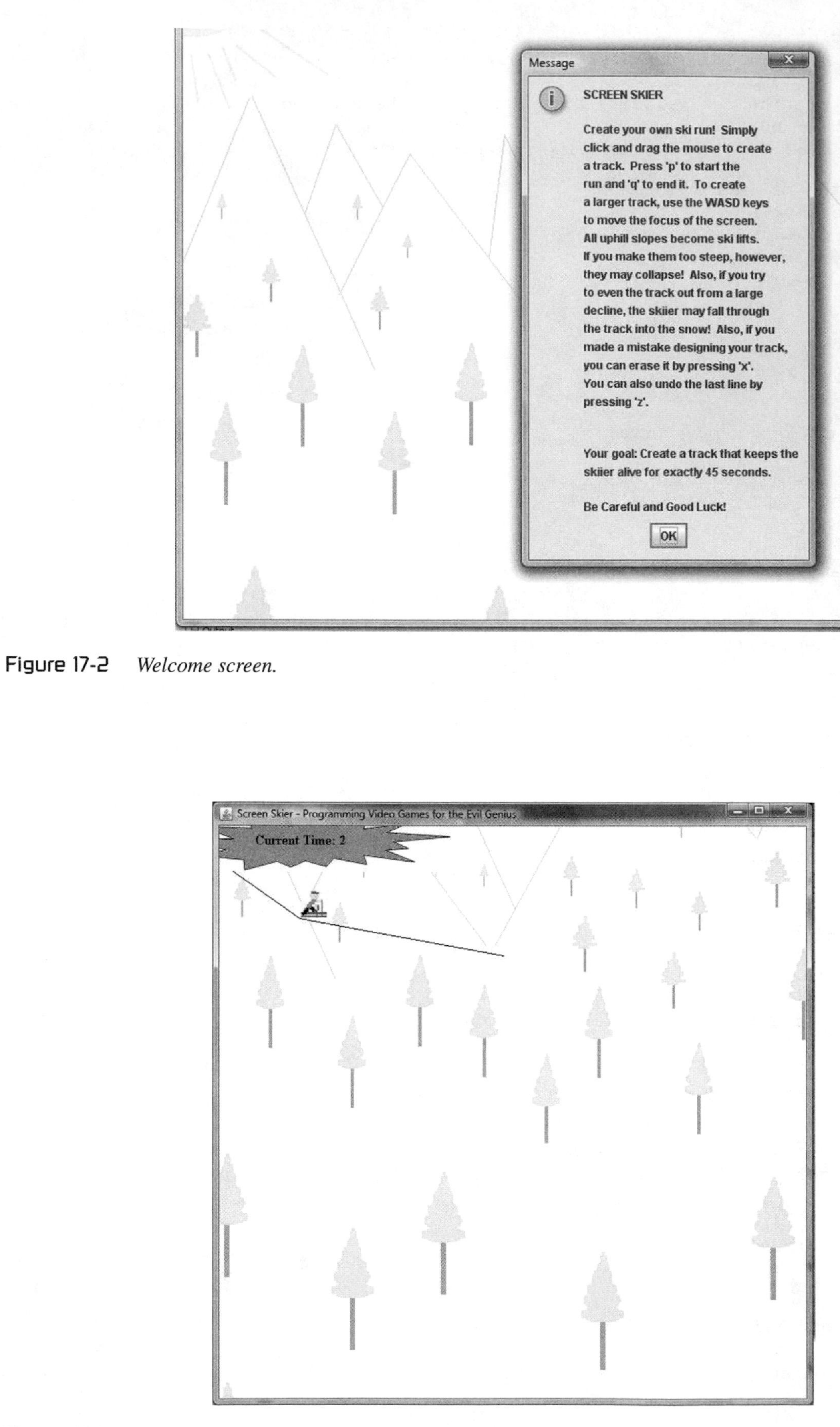

Figure 17-2 *Welcome screen.*

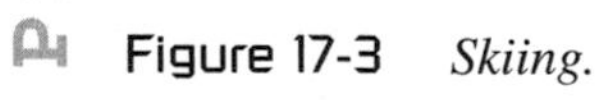

Figure 17-3 *Skiing.*

Figure 17-4 *End of the run.*

Figures 17-2 through 17-4 illustrate the game play of the final version of Screen Skier.

Customizing the game

Look up a class called MouseMotionListener in the API. It works just like MouseListener. Utilizing the new class, generate a line from the first point of the slope to the mouse's current position. This lets the player preview his/her line before actually drawing it.

Draw your own icon for the skier.

Create your own background graphic (blacken the background, change the trees to stars and ski between galaxies).

When the user presses a number, the coordinating line is deleted.

Turn the trees into obstacles. Run into them and you lose!

Change the time goal.

Add a second skier racing at a different speed—tracked by a second counter.

Make the lines automatically lock together (use the previous line's last point) to create a continuous slope.

Record affirmations to get you through the slope ("You can do it!" "Fantastic skiing!").

Create random avalanches in the middle of the run.

Intersect with an out of control snowboarder.

Add cheat codes ... let the player jump into a snowmobile.

Section Three

Board Games

Project 18: Whack-an Evil Genius—The Lab

Whack-an Evil Genius

An Evil Genius is hiding in his lab waiting to unleash havoc upon the world. Because he fears sunshine and fresh air, he only comes out briefly. When he does, your job is to whack him with a click of the mouse so he goes back to his underground lab. The problem? He keeps cloning himself.

Project

Begin by creating the laboratory.

New building blocks

Components, JButtons, JLabels

Components

A component is a device that can be added to any Java program to increase functionality. Some common components include buttons, textfields, and labels. Before you add components, however, you need a place to put them. That place is called a Container. To create a container, place the following code in your constructor after the JFrame is created.

```
Container <varName> = getContentPane();
```

`<varName>` is a variable; it can be named anything.

After this code, you must specify how you want the components added. There are many ways to orient them: you can place them in a grid, on the sides, one after another, etc. In this book, we will focus on placing the components one after another – a FlowLayout. This layout is the easiest to code and the easiest to understand. To set the Container to a FlowLayout, use the following code.

```
cont.setLayout(new FlowLayout());
```

Now, before you start creating components, there is just one more thing: the very last line of the constructor should now say:

```
setContentPane(cont);
```

"`cont`" can be replaced with the name of your Container. This line makes sure the JFrame displays every component you have added.

JButtons

A button is a clickable area. In Java, buttons are called JButtons.

A simple button is illustrated in Figure 18-1.

When creating JButtons, it is best to create them globally (outside of the constructor but inside the class). This way, it will be easier to modify their text anywhere in the program.

To create a JButton, use the following code:

```
JButton <varName> = new JButton("Text");
```

`<varName>` can be anything; it is a variable. `"Text"` is the text that is displayed on the inside of a

Figure 18-1 *JButton.*

JButton. If you wish to display an image instead of text, use the following code:

```
ImageIcon <imgName> = new ImageIcon("path
                         with extension");
JButton <varName> = new JButton(<imgName>);
```

<imgName> and <varName> are variable name; you can name them whatever you want. "path with extension" should be the path of the file from the location of the Java program. A button with an image is illustrated in Figure 18-2.

To add a JButton to the container (if the Container's name is "cont" and the JButton's name is "button"), simply use the following code:

```
cont.add(button);
```

JButtons have many methods for disabling, enabling, setting their icons, among other things. A full list is available in the API. For this game, however, there is one method you should know. It is the "setEnabled()" method. It takes in a boolean. For example, if you have a JButton named "button," the following code disables the button.

```
button.setEnabled(false);
```

Or, to enable the JButton, use the following code:

```
button.setEnabled(true);
```

JLabels

A JLabel is another type of component. It allows text to be displayed on the JFrame. A JLabel is illustrated in Figure 18-3.

To create a JLabel, use the following code:

```
JLabel <varName> = new JLabel("Text");
```

Like a JButton, JLabels can be added to the JFrame using the following code (if Container is called "cont"):

```
cont.add(<varName>);
```

Now that you know how to add JLabels, it is helpful to know how to position them. By using a method called "setBounds()," you can position the JLabel wherever you want. "setBounds()" takes in an x and y coordinate as well as the width and height of the label. For example, if you want to position a label at (50,100), simply use the following code:

```
cont.setLocation(50,100,<width>,<height>);
```

Figure 18-2 *JButton with an ImageIcon.*

Figure 18-3 *JLabel.*

JLabels can also hold an image like JButtons. The code is almost identical:

```
ImageIcon image = new ImageIcon("path with
                    extension");
JLabel label = new JLabel (image);
```

When using JLabels, it is often useful to update the text. To do this, use a method called "setText()." It takes in a String. For example (if your JLabel is named "label"):

label.setText("This is the new text.");

Making the game

Create a JFrame. Then, add a container. Next, create a 5×5 array of JButtons, use `for loops` to initialize the JButtons in the array, and add the buttons to the container. Make sure to add images of an Evil Genius to the buttons! And don't forget to disable all of the buttons. Lastly, create and add a JLabel that holds the score (for now, just make it say "SCORE: ").

```
import javax.swing.event.*;
import javax.swing.*;
import java.awt.*;
import java.awt.event.*;

public class Whack-an extends JFrame
{
  //this array holds the 25 "evil geniuses"
  JButton[ ][ ] spots = new JButton[ 5][ 5];
  //these are the two icons that pop up when the "evil genius"
  //comes out of his lab
  ImageIcon alive = new ImageIcon("alive.GIF");

  //the label:
  JLabel score = new JLabel("SCORE: ");

  //the constructor
  public Whack-an()
  {
    //create the JFrame
    super("Whack-an Evil Genius");
    setSize(350,325);
    setVisible(true);
    setDefaultCloseOperation(JFrame.EXIT_ON_CLOSE);
    //this holds the buttons and labels
    Container cont = getContentPane();
    cont.setLayout(new FlowLayout());

    //this prepares the buttons for displaying
    for(int i = 0; i < spots.length; i++)
    {
      for(int j = 0; j < spots[ 0].length; j++)
      {
        //create the JButton
        spots[ i][ j] = new JButton(alive);
        //add it to the JFrame
        cont.add(spots[ i][ j]);
        //make it disabled
        spots[ i][ j].setEnabled(false);
      }
    }
    //this makes the JLabel display the score
    score.setText("SCORE: ");
    cont.add(score);
```

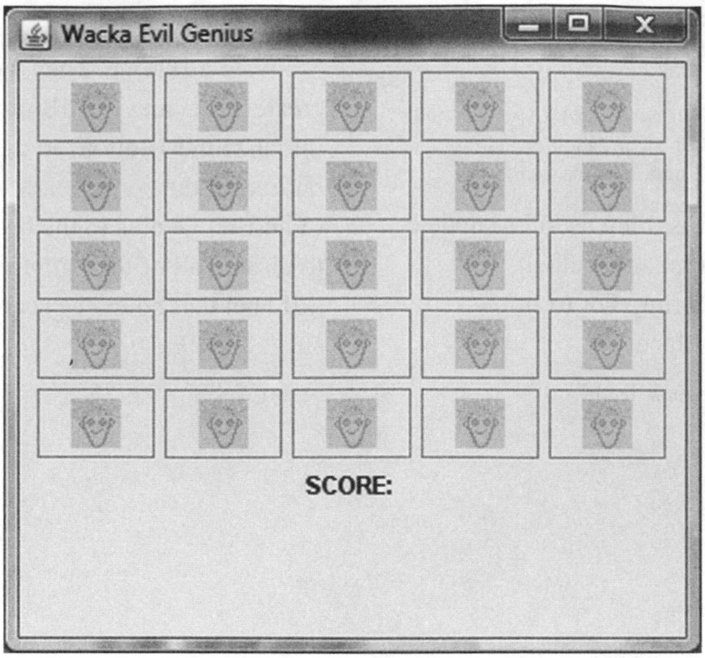

Figure 18-4 *The game so far.*

```
    setContentPane(cont);
  }
  public static void main (String[ ] args)
  {
    //start the game
    new Whack-an();
  }
}
```

The first part of Whack-an Evil Genius is illustrated in Figure 18-4.

In the next project, you will learn how to make Whack-an Evil Genius operational by determining if a JButton has already been pressed.

Project 19: Whack-an Evil Genius—Quick! Get 'em!

Project

The game takes form. The face of the Evil Genius appears. But have no fear, you will be able to whack him away by adding user interaction.

New building blocks

ActionListener

ActionListener

An ActionListener does exactly as it sounds: it listens for actions. In this case, actions are the player clicking buttons. An ActionListener must be added to every button you want to listen to. In the case of Whack-an Evil Genius, you will want each and every button to have an ActionListener. To add an ActionListener to a JButton called "`button`," for example, use the following code:

```
button.addActionListener(this);
```

If you are using ActionListeners, you must use "`implements ActionListener`." If you are already implementing "`KeyListener`," you can implement both by separating them with a comma.

Remember how the KeyListener had so many mandatory methods? Well, luckily, ActionListener only has one mandatory method. When a button is pressed, Java goes to that method. This is how that method should be created:

```
public void actionPerformed (ActionEvent event)
{
}
```

To find out which button has been pressed, simply put the following condition inside of an if-statement:

```
event.getSource()==<button name>
```

Remember how I suggested that you make your buttons global? This is why. You need to refer to the name of the button to see which one has been clicked.

Making the game

First, add ActionListeners to every button in the array. Then, implement ActionListener and add the mandatory method.

Next, create a new Thread. In its infinite `while loop`, make the program wait a random amount of time (between 0 and 1500 milliseconds). Then, have a random Evil Genius pop up (by enabling it). Wait 1000 milliseconds, and then disable the button.

Now, go back to the actionPerformed method. When a button is pressed, increment the counter that keeps track of the score, and then pause both threads for a quarter of a second (this gives the player time to catch his or her breath).

You can then display the score in the JLabel.

```
import javax.swing.event.*;
import javax.swing.*;
import java.awt.*;
import java.awt.event.*;

public class Whack-an extends JFrame implements ActionListener
{
  //this array holds the 25 "evil geniuses"
  JButton[ ][ ] spots = new JButton[ 5][ 5];

  //this displays the game's status
  JLabel score = new JLabel();

  //the score:
  double hits = 0;

  //these are the two icons that pop up when the "evil genius"
  //comes out of his lab
  ImageIcon alive = new ImageIcon("alive.GIF");

  //this is the thread:
  T runner = null;

  //the constructor
  public Whack-an()
  {
```

```
    //create the JFrame
    super("Whack-an Evil Genius");
    setSize(350,325);
    setVisible(true);
    setDefaultCloseOperation(JFrame.EXIT_ON_CLOSE);

    //this holds the buttons and labels
    Container cont = getContentPane();
    cont.setLayout(new FlowLayout());

    //this prepares the buttons for displaying
    for(int i = 0; i < spots.length; i++)
    {
      for(int j = 0; j < spots[0].length; j++)
      {
        //create the JButton
        spots[i][j] = new JButton(alive);
        //add it to the JFrame
        cont.add(spots[i][j]);
        //make it disabled
        spots[i][j].setEnabled(false);
        //make it respond to clicks
        spots[i][j].addActionListener(this);
      }
    }
    cont.add(score);

    setContentPane(cont);
    //this starts the Thread
    runner = new T();
    runner.start();
    setContentPane(cont);
  }
  //the thread:
  private class T extends Thread
  {
    public void run()
    {
      //an infinite loop
      while(true)
      {
        //create a random time delay between 0 and 1.5 seconds
        int timeDelay = (int)(Math.random()*1500);
        try
        {
          //pause for the random time
          Thread.sleep(timeDelay);
        }
        catch(Exception e)
        {}

        //make a random genius pop up
        int genius = (int)(Math.random()*5);
        int genius2 = (int)(Math.random()*5);
        //make the genius come out by enabling the button
        spots[genius][genius2].setEnabled(true);

        try
        {
          //pause to let the user try to catch
          //the evil genius
          Thread.sleep(1000);
        }
```

```
      catch(Exception e)
      {}
      //make the genius disappear
      spots[genius][genius2].setEnabled(false);
      //display the stats
      score.setText("SCORE: "+hits);
    }
  }
}
//this checks to see if the button was clicked
public void actionPerformed(ActionEvent e)
{
  //increase the score
  hits++;
  //pause the game for 1/2 a second
  try
  {
    runner.sleep(500);
    Thread.sleep(500);
  }
  catch(Exception ex)
  {}
}
public static void main (String[ ] args)
{
  //start the game
  new Whack-an();
}
}
```

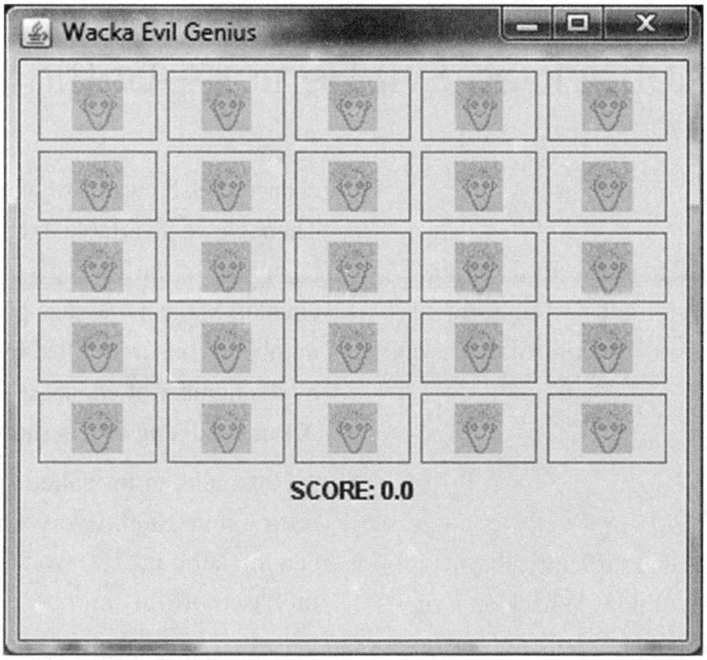

Figure 19-1 *Evil Geniuses are hiding ... .*

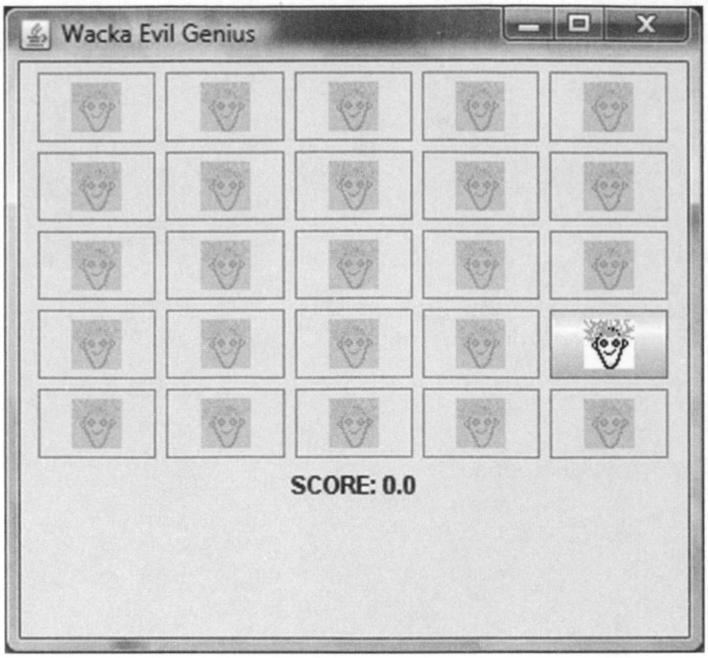

Figure 19-2 *Quick! Catch him!*

Figures 19-1 and 19-2 illustrate the game play of Whack-an Evil Genius.

Move on to increasing the difficulty levels as the player keeps on trying to whack the Evil Genius.

Warning: this guy can get pretty slippery as he learns from his mistakes ... After all, he is an Evil Genius!!!

Project 20: Whack-an Evil Genius—Getting Smarter ...

Project

In this project, tension builds as the Evil Genius appears and disappears faster and faster. Torment your players as you learn how to control the levels of difficulty.

Making the game

When games do not increase in difficulty, they quickly become boring. Right now, Whack-an Evil Genius is set at an intermediate level. To solve this problem, start by asking the player how many chances he/she wants to whack away the Evil Genius. Then, create a variable to count the attempts. At the beginning of the Thread's `while loop`, increment the counter. Make a message pop up displaying the number of hits multiplied by 10,000 and then divided by the number of chances. This ends the game. A sample dialog box is shown in Figure 20-1.

Next, add an int called "maxDelay." Start by setting it at 1000. Every time an Evil Genius is caught, subtract 100 from it. Then, create a random int based off this number:

```
int delay = (int)(Math.random()*maxDelay);
```

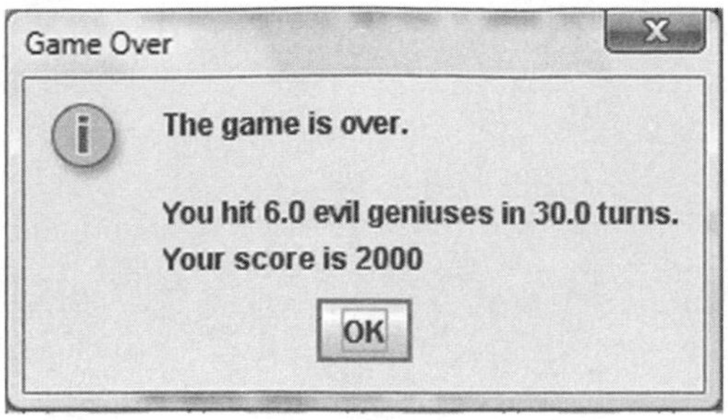

Use this new number as the delay time (how long the Evil Genius makes an appearance). This way, the better you do, the more difficult the game becomes.

Game Over

The game is over.

You hit 6.0 evil geniuses in 30.0 turns.
Your score is 2000

OK

Figure 20-1 *Score is displayed; game over.*

```
import javax.swing.event.*;
import javax.swing.*;
import java.awt.*;
import java.awt.event.*;

public class Whack-an extends JFrame implements ActionListener
{
  //this array holds the 25 "evil geniuses"
  JButton[ ][ ] spots = new JButton[ 5][ 5] ;

  //this displays the game's status
  JLabel score = new JLabel();

  //these are the variables that keep track of the score
  int maxDelay = 1000;
  double hits = 0;
  double turns = 0;
  double maxTurn = 0;

  //these are the two icons that pop up when the "evil genius"
  //comes out of his lab
  ImageIcon alive = new ImageIcon("alive.GIF");

  //this is the thread:
  T runner = null;

  //the constructor
  public Whack-an()
  {
    //create the JFrame
    super("Whack-an Evil Genius");
    setSize(350,325);
    setVisible(true);
    setDefaultCloseOperation(JFrame.EXIT_ON_CLOSE);

    //this holds and inputs the number
    //of rounds the user will play
    maxTurn = Double.parseDouble
        (JOptionPane.showInputDialog("How many chances do you\n"+"want to whack an evil
             genius?"));
    //this holds the buttons and labels
    Container cont = getContentPane();
    cont.setLayout(new FlowLayout());

    //this prepares the buttons for displaying
    for(int i = 0; i < spots.length; i++)
    {
      for(int j = 0; j < spots[ 0] .length; j++)
```

```
      {
        //create the JButton
        spots[ i][ j] = new JButton(alive);
        //add it to the JFrame
        cont.add(spots[ i][ j] );
        //make it disabled
        spots[ i][ j] .setEnabled(false);
        //make it respond to clicks
        spots[ i][ j] .addActionListener(this);
      }
    }
    //this makes the JLabel display the score
    score.setText("Turn "+turns+"/"+maxTurn+". Current Score:
          "+((int)((hits/maxTurn)*100)));
    cont.add(score);
    setContentPane(cont);
    //this starts the Thread
    runner = new T();
    runner.start();
  }
  //the thread:
  private class T extends Thread
  {
    public void run()
    {
      //an infinite loop
      while(true)
      {
        //check for game over
        if(turns>=maxTurn)
        {
          //if the game is over, display the score
          JOptionPane.showMessageDialog(null,
            "The game is over.\n\n"+"You hit "+
            hits+" evil geniuses in "+turns+" turns.\n"+"Your score is "+
              ((int)(((hits*10000)/turns))),
            "Game Over",
            JOptionPane.INFORMATION_MESSAGE);
          break;
        }
        //count the number of turns
        turns++;
        //create a random time delay between 0 and 1.5 seconds
        int timeDelay = (int)(Math.random()*1500);
        try
        {
          //pause for the random time
          Thread.sleep(timeDelay);
        }
        catch(Exception e)
        {}
        //make a random genius pop up
        int genius = (int)(Math.random()*5);
        int genius2 = (int)(Math.random()*5);
        //make the genius come out by enabling the button
        spots[ genius][ genius2] .setEnabled(true);
        try
        {
          //pause to let the user try to catch
          //the evil genius
```

```
        Thread.sleep(maxDelay);
      }
      catch(Exception e)
      { }
      //make the genius disappear
      spots[ genius][ genius2] .setEnabled(false);
      //display the stats
    score.setText("Turn "+turns+"/"+maxTurn+
        ". Current Score: "+
        ((int)(((hits*10000)/maxTurn))));
      }
    }
  }
  //this checks to see if the button was clicked
  public void actionPerformed(ActionEvent e)
  {
    //if an evil genius was caught,
    //decrease the time each genius is shown,
    //making the game harder
    maxDelay-=100;
    //increase the score
    hits++;
    //pause the game for 1/2 a second
    try
    {
      runner.sleep(500);
      Thread.sleep(500);
    }
    catch(Exception ex)
    { }
  }
  public static void main (String[ ] args)
  {
      //start the game
      new Whack-an();
  }
}
```

Figures 20-2 through 20-4 illustrate the game play of Whack-an Evil Genius.

Ready for special effects? In the next project, explore lots of ways to add realistic and/or wild features to your game for enhanced play value.

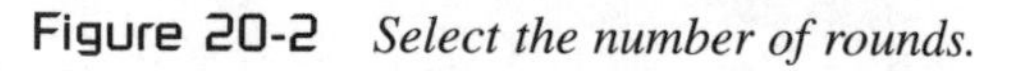

Figure 20-2 *Select the number of rounds.*

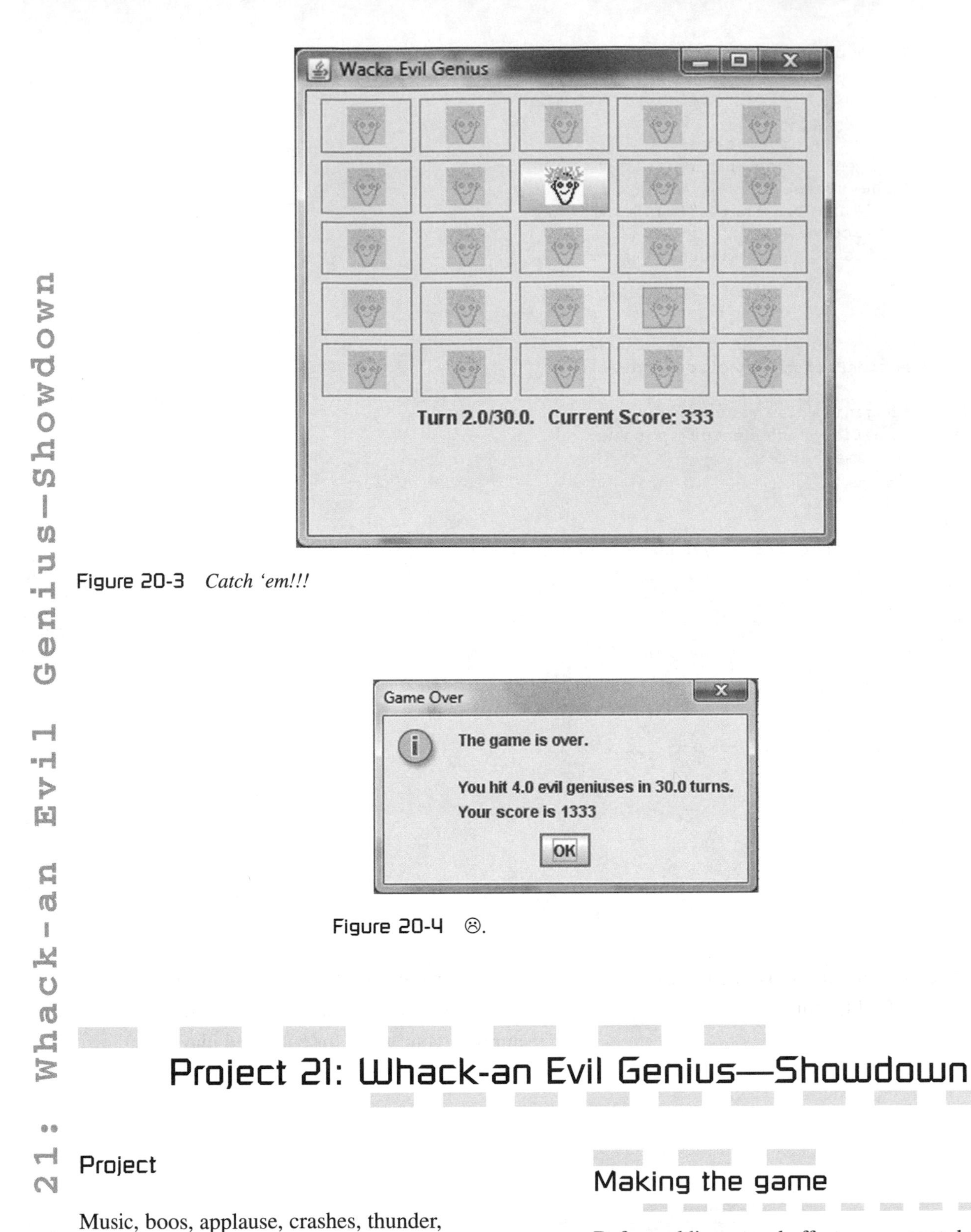

Figure 20-3 *Catch 'em!!!*

Figure 20-4 ☹.

Project 21: Whack-an Evil Genius—Showdown

Project

Music, boos, applause, crashes, thunder, explosions, or any other sounds can be added for more intense game play. But it doesn't stop there. To give the game even more dimensional quality, you'll also learn how to add images.

Making the game

Before adding sound effects, you must decide where to put them and what they should be. For example, when the player whacks an Evil Genius away, he could say something scientific ... "$E=MC^2$," for example. When an Evil Genius

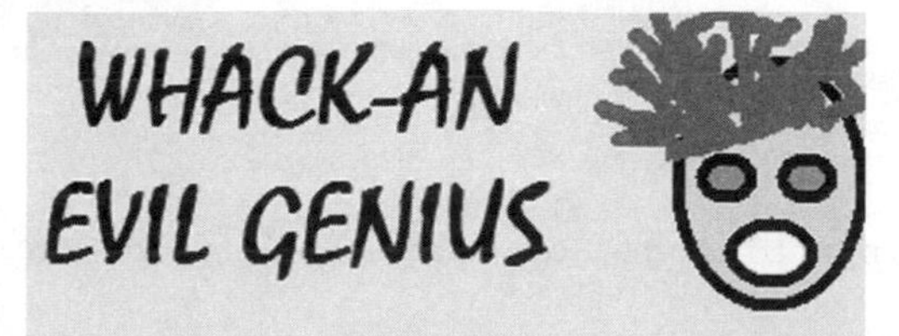

Figure 21-1 *Title image.*

disappears, the sound of a door slamming shut could be played.

Now, after retrieving the sound file, program the sounds globally. Next, in the actionPerformed method, play the scientific sound. In the Thread, right before you disable the button, play the sound of the door slamming shut.

Lastly, create an amazing title image like the one shown in Figure 21-1. Attach it to a JLabel and add the JLabel to the container before creating any of the buttons.

Remember, all of these sounds and images can be downloaded for free from www.books.mcgraw-hill.com/authors/cinnamon

```
import javax.swing.event.*;
import javax.swing.*;
import java.awt.*;
import java.awt.event.*;
import java.io.*;
import java.net.*;
import java.applet.AudioClip;

public class Whack-an extends JFrame implements ActionListener
{
  //this array holds the 25 "evil geniuses"
  JButton[ ][ ] spots = new JButton[ 5][ 5];

  //this displays the game's status
  JLabel score = new JLabel();

  //these are the variables that keep track of the score
  int maxDelay = 1000;
  double hits = 0;
  double turns = 0;
  double maxTurn = 0;

  //these are the two icons that pop up when the "evil genius"
  //comes out of his lab
  ImageIcon alive = new ImageIcon("alive.GIF");

  //this is the thread:
  T runner = null;

  URL sci = null, dr = null;
  AudioClip scientific = null, door = null;

  //the constructor
  public Whack-an()
  {
    //create the JFrame
    super("Whack-an Evil Genius");
    setSize(350,475);
    setVisible(true);
    setDefaultCloseOperation(JFrame.EXIT_ON_CLOSE);

    try
    {
      sci = this.getClass().getResource("sci.wav");
      dr = this.getClass().getResource ("door.wav");
      scientific = JApplet.newAudioClip(sci);
      door = JApplet.newAudioClip(dr);
```

```
 }
 catch(Exception e){ }

 //this holds and inputs the number
 //of rounds the user will play
 maxTurn = Double.parseDouble(JOptionPane.showInputDialog("How many chances do you\n"+
          "want to whack an evil genius?"));
 //this holds the buttons and labels
 Container cont = getContentPane();
 cont.setLayout(new FlowLayout());

 //this JLabel will be the title image.
 //notice how you can create a JLabel and
 //add an image all on one line ...
 JLabel title = new JLabel(new ImageIcon("title.PNG"));
 //now add it:
 cont.add(title);

 //this prepares the buttons for displaying
 for(int i = 0; i < spots.length; i++)
 {
   for(int j = 0; j < spots[0].length; j++)
   {
     //create the JButton
     spots[i][j] = new JButton(alive);
     //add it to the JFrame
     cont.add(spots[i][j]);
     //make it disabled
     spots[i][j].setEnabled(false);
     //make it respond to clicks
     spots[i][j].addActionListener(this);
   }
 }

 //this makes the JLabel display the score
 score.setText("Turn "+turns+"/"+maxTurn+". Current Score: "+((int)
         ((hits/maxTurn)*100)));
 cont.add(score);
 setContentPane(cont);
 //this starts the Thread
 runner = new T();
 runner.start();
}

//the thread:
public class T extends Thread
{
  public void run()
  {
  //an infinite loop
  while(true)
  {
    //check for game over
    if(turns>=maxTurn)
    {
      //if the game is over, display the score
      JOptionPane.showMessageDialog(null,
        "The game is over.\n\n"+"You hit "+
        hits+" evil geniuses in "+turns+" turns.\n"+"Your score is
           "+((int)(((hits*10000)/turns))),
        "Game Over",
        JOptionPane.INFORMATION_MESSAGE);
      break;
    }
```

```
      //count the number of turns
      turns++;
      //create a random time delay between 0 and 1.5 seconds
      int timeDelay = (int)(Math.random()*1500);
      try
      {
        //pause for the random time
        Thread.sleep(timeDelay);
      }
      catch(Exception e)
      {}
      //make a random genius pop up
      int genius = (int)(Math.random()*5);
      int genius2 = (int)(Math.random()*5);
      //make the genius come out by enabling the button
      spots[genius][genius2].setEnabled(true);
      try
      {
        door.play();
        //pause to let the user try to catch
        //the evil genius
        Thread.sleep(maxDelay);
      }
      catch(Exception e)
      {}
      //make the genius disappear
      spots[genius][genius2].setEnabled(false);
      //display the stats
    score.setText("Turn "+turns+"/"+maxTurn+". Current Score: "+((int)(((hits*10000)/
          maxTurn))));
      }
    }
  }
  //this checks to see if the button was clicked
  public void actionPerformed(ActionEvent e)
  {
    //if an evil genius was caught,
    //decrease the time each genius is shown,
    //making the game harder
    maxDelay-=100;
    scientific.play();
    //increase the score
    hits++;
    //pause the game for 1/2 a second
    try
    {
      runner.sleep(500);
      Thread.sleep(500);
    }
    catch(Exception ex)
    {}
  }
  public static void main (String[] args)
  {
    //start the game
    new Whack-an();
  }
}
```

Figures 21-2 and 22-3 illustrate the addictive game play of Whack-an Evil Genius.

Customizing the game

Change the dimensions of the board: you can make it anything from a simple 2×2 button setup to a super-complicated 100×100 buttons setup!

Change the icon of the Evil Genius: use a picture of yourself ... or your "favorite" coworker.

Add background music.

Create your own title image.

Develop different levels (when you capture 10 Evil Geniuses, move on to the next, faster-paced level).

Create a high score list using file writing and reading.

Every few rounds, have your Evil Genius morph into a different icon. Create some "good-guy icons." If you accidentally whack them, you automatically lose!

Input
How many chances do you want to whack an evil genius?
30
OK
Cancel

Figure 21-2 *30 whacks.*

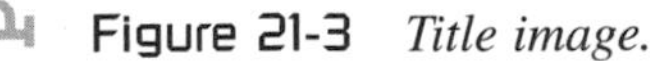

Figure 21-3 *Title image.*

Project 22: Tic-Tac-Toe Boxing—The Ring

Tic-Tac-Toe Boxing

The classic game of Tic-Tac-Toe is transformed into a boxing match in which each "fighter" is represented by a different boxing glove image. By adding punching sounds, groans, or taunts, you will be able to challenge you opponent (or the computer) to win, lose, or draw.

Project

Start by creating the ring so you can touch gloves and come out fighting!!!

New building blocks

NullLayout

NullLayout

Instead of using a FlowLayout, which simply places components one after another, you can use a NullLayout which allows you to specify the location of the component in pixel coordinates. To use a NullLayout, change the "`setLayout`" line to the following:

```
container.setLayout(null);
```

You can still add components the same way you did with the FlowLayout. To specify the pixel location of the component, use the following code only after you added the component to the Container:

```
<component>.setBounds(<x>, <y>, <width>,
<height>);
```

`<component>` is the variable that represents the component. The rest of the parameters are self-explanatory.

Before you create original games and add components based on pixel location, you need to know one more thing: a NullLayout is not the most reliable layout. There is always a chance that the screen will not refresh and your change will not be displayed. No worries, however! There is an easy fix to this problem so your changes will always register. After you add everything you need to the container, type the following:

```
container.repaint();
```

Making the game

First, create an array of JButtons nine elements long to hold the game board. Add the buttons from the array to the container inside a loop. Remember—you will be adding the components based on pixel location. So, figure out a simple algorithm that spaces the pieces of the ring correctly. Here is a suggestion: each of the images used with the game are 100 by 100 pixels, so the boxes should be 100 pixels apart. In addition, create two variables set to zero outside of the loop. At each iteration, increment one of the variables by one. Then, multiply this value by 100 to get the correct "x" spacing. Add an if statement inside of your `for loop`. Make it check to see if the counter is a multiple of 3 (because the board is three pieces long). If so, drop the other pieces down one line by incrementing the second variable by one. Okay ... it sounds confusing, but once you see the code, as shown below, it will all make sense.

```
int newLine = 0;
```

```
int lineCount = 0;
for(int i = 0; i < spots.length; i++)
{
  //initialize it with a blank image
  spots[ i] = new JButton(blank);
  //this checks whether to use a new row
  if(i==3 || i ==6)
  {
    newLine++;
    lineCount = 0;
  }
  //set the position of the button
  spots[ i] .setBounds(lineCount*100,newLine*
            100,100,100);
  //add it to the container
  container.add(spots[ i] );
  //and now add the action listener:
  spots[ i] .addActionListener(this);

  lineCount++;
}
```

Now that the board is complete, create a global variable called "turn." Everytime the "actionPerformed" method executes, "turn" should be incremented by one. Then, you will be able to check to see whose turn it is based on "turn" being even or odd.

In "actionPerformed," if "turn" is odd, it is player O's turn. If it is even, player X is up. Depending on whose turn it is, set the icons to the correct images. After setting the icon, remember to remove the ActionListener so a player can not steal his/her opponent's position!

```
//import everything:
import javax.swing.*;
import javax.swing.event.*;
import java.awt.*;
import java.awt.event.*;
import java.io.*;
import java.net.*;

//the class with the JFrame and the ActionListner:
public class TicTacToe extends JFrame implements ActionListener
{
  //this is an array of the buttons (spots). We use an array, not
  //an arrayList because the number of buttons is constant; it does
  //not change.
  JButton spots[ ] = new JButton[ 9] ;

  //this will keep track of turns: even is player 1; odd is player2
  //we will use mod (%) to differentiate the values:
  int turn = 1;

  //these images represent the sides
  ImageIcon red = new ImageIcon("red.png");
  ImageIcon blue = new ImageIcon("blue.png");
  ImageIcon blank = new ImageIcon("blank.png");

  //the constructor
  public TicTacToe()
  {
    super("TicTacToe: Boxing Style");
    setSize(330,350);
    setVisible(true);
    setDefaultCloseOperation(JFrame.EXIT_ON_CLOSE);

    //this will hold the buttons:
    Container container = getContentPane();
    //this will tell the computer how to display the buttons:
    container.setLayout(null);
int newLine = 0;
    int lineCount = 0;
    for(int i = 0; i < spots.length; i++)
```

```
    {
      //initialize it with a blank image
      spots[ i] = new JButton(blank);
      //this checks whether to use a new row
      if(i==3 || i ==6)
      {
        newLine++;
        lineCount = 0;
      }
      //set the position of the button
      spots[ i].setBounds(lineCount*100,newLine*100,100,100);
      //add it to the container
      container.add(spots[ i]);
      //and now add the action listener:
      spots[ i].addActionListener(this);
      lineCount++;
    }
  }
  //the mandatory method:
  public void actionPerformed(ActionEvent e)
  {
    //this will check the button pressed:
    for(int i = 0; i < spots.length; i++)
    {
      if(e.getSource()==spots[ i])
      {
        //check which turn:
        if(turn%2==0)
        {
          //if even, player 1's turn (X)
          spots[ i].setIcon(red);
        }
        else
        {
          //if odd, player 2's turn (O)
          spots[ i].setIcon(blue);
        }
        //disable the btn so it can't be re-pressed
        spots[ i].removeActionListener(this);
      }
    }
    turn++;
    //before letting the other player go, check whhether a player won:
    checkWin();
  }
  public void checkWin()
  {
    //first, we will use to go through three iterations. This allows us to
    //check for both horizontal and vertical wins without using too
    //much code:
    for(int i = 0; i < 3; i++)
    {
      //this checks for a vertical X win
      if(spots[ i].getText()=="X" &&
          spots[ i+3].getText()=="X" &&
          spots[ i+6].getText()=="X")
        JOptionPane.showMessageDialog(null,"X Wins");
      //this checks for a vertical O win
      if(spots[ i].getText()=="O" &&
          spots[ i+3].getText()=="O" &&
```

```
          spots[i+6].getText()=="O")
        JOptionPane.showMessageDialog(null,"O Wins");
      //this checks for a vertical X win
      if(spots[i*3].getText()=="X" &&
          spots[(i*3)+1].getText()=="X" &&
          spots[(i*3)+2].getText()=="X")
        JOptionPane.showMessageDialog(null,"X Wins");
      //this checks for a vertical X win
      if(spots[i*3].getText()=="O" &&
        spots[(i*3)+1].getText()=="O" &&
          spots[(i*3)+2].getText()=="O")
        JOptionPane.showMessageDialog(null,"O Wins");
    }
    //now, this loop will check for diagnol wins
    for(int i = 0; i <= 2; i+=2)
    {
      //this will check for diagnol X wins
      if(spots[i].getText()=="X" &&
          spots[4].getText()=="X" &&
          spots[8-i].getText()=="X")
        JOptionPane.showMessageDialog(null,"X Wins");
      //this will check for diagnol O wins
      if(spots[i].getText()=="O" &&
          spots[4].getText()=="O" &&
          spots[8-i].getText()=="O")
        JOptionPane.showMessageDialog(null,"O Wins");
    }
  }
  //starter (main) method:
  public static void main(String[] args) {
    TicTacToe ttt = new TicTacToe();
  }
}
```

Figures 22-1 through 22-4 illustrate the exciting game play of Boxing Tic-Tac-Toe.

Fair fighting requires an impartial judge. In the next project, you are going to program the computer to determine whether Player One is the winner, Player Two is a winner, or if the boxing match ends in a draw.

Figure 22-1 *Game start up.*

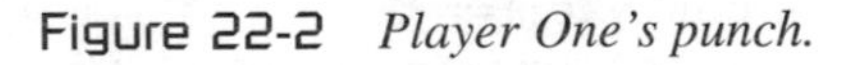

Figure 22-2 *Player One's punch.*

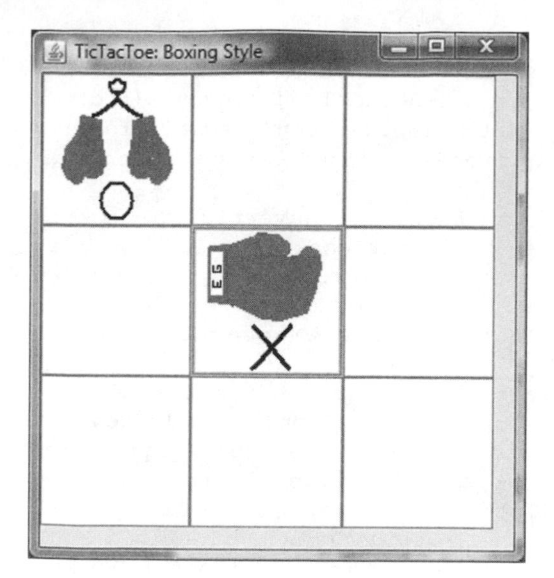

Figure 22-3 *Player Two's counter-punch.*

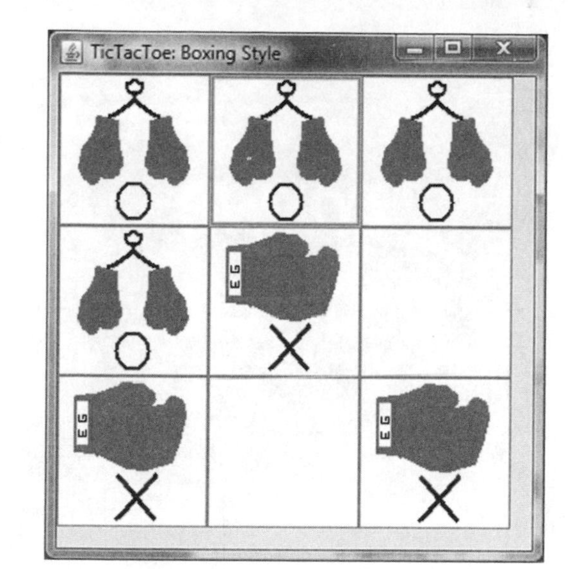

Figure 22-4 *Player Two ("O") wins ... but the computer cannot yet detect the result.*

Project 23: Tic-Tac-Toe Boxing—Fight!!!

Project

Welcome to training camp. It is here where you will enable your two fighters to square off in the ring. Add a title image to hype the fans, and program your computer to determine who's the champ and who's the chump.

Making the game

First, start with the easier of the two objectives: adding the title image. To do this, create a title image in paint. Next, create a JLabel that will display that image. Add this JLabel above the buttons (at location 0,0). Don't forget to change the loop that adds the buttons so they will not be covered by the title image. Do this by simply adding the height of the image to the y value of the buttons. You can also add a set number to the x value of the buttons so they will be centered, as shown in Figure 23-1.

Next, you will teach the computer how to check for knockouts. Create a method named `checkWin`. At the end of the `actionPerformed` method, call `checkWin`. Now that you have created the methods, there are two techniques you can use to check for wins. The first is a heck of a lot of if-statements ... not fun! The second technique requires much less code. Because of the nature of Tic-Tac-Toe (being a square), it is easier to use if-statements inside of `for loops`. This cuts down nearly 66% of the code. As illustrated in Figure 23-2, you can see that a single loop can check six different cases.

And now the difficult part: writing that pesky `for loop`. To do this, create a `for loop` that counts from 0 to 2 (three iterations). Inside the loop, add four different if-statements. One will check for a vertical X win; one will check for a vertical O win; one for a horizontal X win; and the last one, for a horizontal O win. Each loop determines if the icon of the buttons in the appropriate position are all the same color. The actual code in the if-statements is shown below:

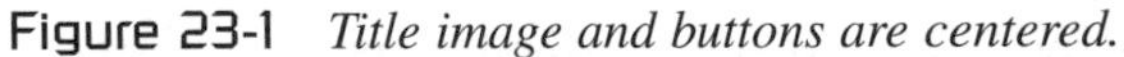

Figure 23-1 *Title image and buttons are centered.*

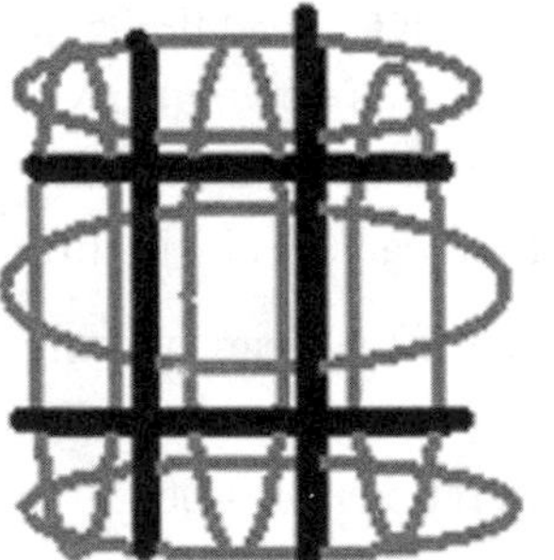

Figure 23-2 *Single for loop checks six different possible wins.*

```
//this checks for a vertical X win
if(spots[ i] .getIcon().equals(red) &&
    spots[ i+3] .getIcon().equals(red) &&
    spots[ i+6] .getIcon().equals(red))
  JOptionPane.showMessageDialog(null,"X
          Wins");
//this checks for a vertical O win
if(spots[ i] .getIcon().equals(blue) &&
    spots[ i+3] .getIcon().equals(blue) &&
    spots[ i+6] .getIcon().equals(blue))
  JOptionPane.showMessageDialog(null,"O
          Wins");
//this checks for a horizontal X win
if(spots[ i*3] .getIcon().equals(red) &&
    spots[ (i*3)+1] .getIcon().equals(red) &&
    spots[ (i*3)+2] .getIcon().equals(red))
  JOptionPane.showMessageDialog(null,"X
          Wins");
//this checks for a horizontal O win
if(spots[ i*3] .getIcon().equals(blue) &&
  spots[ (i*3)+1] .getIcon().equals(blue) &&
    spots[ (i*3)+2] .getIcon().equals(blue))
  JOptionPane.showMessageDialog(null,"O
          Wins");
```

Almost there! You only need one more thing: a second loop to check for diagonal wins. This can be accomplished the same way as checking for horizontal and vertical wins, except the loop should only iterate two times. The entire loop is shown below:

```
//now, this loop will check for diagnol wins
for(int i = 0; i <= 2; i+=2)
{
  //this will check for diagnol X wins
  if(spots[ i] .getIcon().equals(red) &&
     spots[ 4] .getIcon().equals(red) &&
     spots[ 8-i] .getIcon().equals(red))
   JOptionPane.showMessageDialog(null,
          "X Wins");
  //this will check for diagnol O wins
  if(spots[ i] .getIcon().equals(blue) &&
     spots[ 4] .getIcon().equals(blue) &&
     spots[ 8-i] .getIcon().equals(blue))
   JOptionPane.showMessageDialog(null,
          "O Wins");
}
```

Congrats! Now you have all the code necessary to check for wins. The full code is below.

```
//import everything:
import javax.swing.*;
import javax.swing.event.*;
import java.awt.*;
import java.awt.event.*;
import java.io.*;
import java.net.*;

//the class with the JFrame and the
```

ActionListner:

```
public class TicTacToe extends JFrame implements ActionListener
{
  //this is an array of the buttons (spots). We use an array, not
  //an arrayList because the number of buttons is constant; it does
  //not change.
  JButton spots[ ] = new JButton[ 9] ;
  //this will keep track of turns: even is player 1; odd is player2
  //we will use mod (%) to differentiate the values:
  int turn = 1;
  //these images represent the sides
  ImageIcon red = new ImageIcon("red.png");
  ImageIcon blue = new ImageIcon("blue.png");
  ImageIcon blank = new ImageIcon("blank.png");
  //the title image:
  JLabel title = new JLabel(new ImageIcon("title.png"));
  //the constructor
  public TicTacToe()
  {
    super("TicTacToe: Boxing Style");
    setSize(350,625);
    setVisible(true);
    setDefaultCloseOperation(JFrame.EXIT_ON_CLOSE);
    //this will hold the buttons:
    Container container = getContentPane();
    //this will tell the computer how to display the buttons:
    container.setLayout(null);
    container.add(title);
    title.setBounds(0,0,350,288);
    int newLine = 0;
    int lineCount - 0;
    for(int i = 0; i < spots.length; i++)
    {
      //initialize it with a blank image
      spots[ i] = new JButton(blank);
      //this checks whether to use a new row
      if(i==3 || i ==6)
      {
        newLine++;
        lineCount = 0;
      }
      //set the position of the button
      spots[ i] .setBounds(15+(lineCount*100),288+(newLine*100),100,100);
      //add it to the container
      container.add(spots[ i] );

      //and now add the action listener:
      spots[ i] .addActionListener(this);

      lineCount++;
    }
    //remember to refresh the screen!
    container.repaint();
  }

  //the mandatory method:
  public void actionPerformed(ActionEvent e)
  {
    //this will check the button pressed:
    for(int i = 0; i < spots.length; i++)
    {
      if(e.getSource()==spots[ i] )
      {
```

```
        //check which turn:
        if(turn%2==0)
        {
          //if even, player 1's turn (X)
          spots[i].setIcon(red);
        }
        else
        {
          //if odd, player 2's turn (O)
          spots[i].setIcon(blue);
        }
        //disable the btn so it can't be re-pressed
        spots[i].removeActionListener(this);
      }
    }
    turn++;
    //before letting the other player go, check whether a player won:
    checkWin();
  }
  public void checkWin()
  {
    //first, we will use to go through three iterations. This allows us to
    //check for both horizontal and vertical wins without using too
    //much code:
    for(int i = 0; i < 3; i++)
    {
      //this checks for a vertical X win
      if(spots[i].getIcon().equals(red) &&
          spots[i+3].getIcon().equals(red) &&
          spots[i+6].getIcon().equals(red))
        JOptionPane.showMessageDialog(null,"X Wins");
      //this checks for a vertical O win
      if(spots[i].getIcon().equals(blue) &&
          spots[i+3].getIcon().equals(blue) &&
          spots[i+6].getIcon().equals(blue))
        JOptionPane.showMessageDialog(null,"O Wins");
      //this checks for a horizontal X win
      if(spots[i*3].getIcon().equals(red) &&
          spots[(i*3)+1].getIcon().equals(red) &&
          spots[(i*3)+2].getIcon().equals(red))
        JOptionPane.showMessageDialog(null,"X Wins");
      //this checks for a horizontal O win
      if(spots[i*3].getIcon().equals(blue) &&
        spots[(i*3)+1].getIcon().equals(blue) &&
          spots[(i*3)+2].getIcon().equals(blue))
        JOptionPane.showMessageDialog(null,"O Wins");
    }
    //now, this loop will check for diagnol wins
    for(int i = 0; i <= 2; i+=2)
    {
      //this will check for diagnol X wins
      if(spots[i].getIcon().equals(red) &&
          spots[4].getIcon().equals(red) &&
          spots[8-i].getIcon().equals(red))
        JOptionPane.showMessageDialog(null,"X Wins");
      //this will check for diagnol O wins
      if(spots[i].getIcon().equals(blue) &&
          spots[4].getIcon().equals(blue) &&
```

```
        spots[ 8-i] .getIcon().equals(blue))
      JOptionPane.showMessageDialog(null,"O Wins");
    }
  }
  //starter (main) method:
  public static void main(String[ ] args) {
    TicTacToe ttt = new TicTacToe();
  }
}
```

Figures 23-3 through 23-5 show Tic-Tac-Toe Boxing's new title image and ability to check for knockouts.

Ali, Tyson, Foreman—in the next project, program the ultimate boxer by adding artificial intelligence.

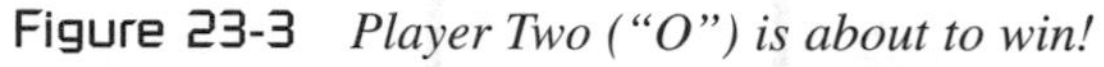

Figure 23-3 *Player Two ("O") is about to win!*

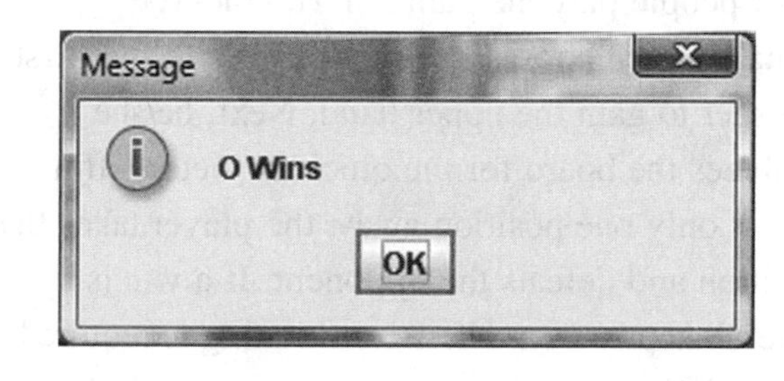

Figure 23-4 *Computer recognizes Player Two ("O") as winner.*

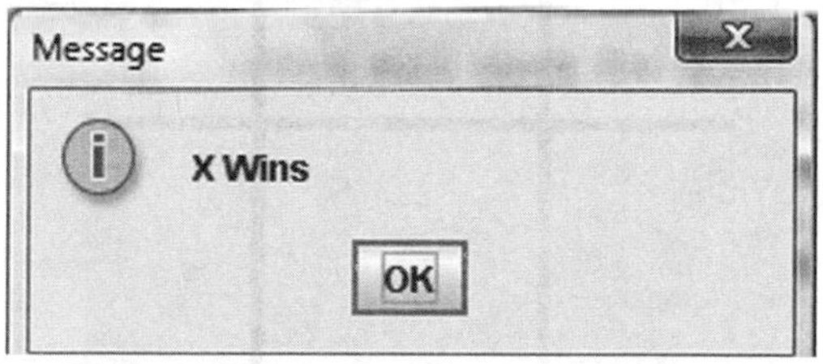

Figure 23-5 *Player One ("X") wins next round.*

Project

Forget fancy footwork, quick jabs, and thunderous punches. Artificial Intelligence is the most powerful force in the boxing ring. This project shows you how to use AI to create an opponent that gives you and your friends the ultimate challenge.

Making the game

Although you do not have to learn new code or syntax, there is a lot of math involved in creating Artificial Intelligence (AI). Not a problem. Thinking the game play through before actually programming the AI makes this process easier.

Whenever you create AI, you must analyze how most people play the game of Tic-Tac-Toe. Usually, one chooses a corner or middle spot first in order to gain the upper hand. Next, he/she analyzes the board for the quickest victory. If a win is only one position away, the player takes that position and defeats the opponent. If a win is several steps away, defensive thinking is required to block the opponent.

Once you have taken into consideration the human thought process for securing a victory, you can start programming the AI! First, instead of switching icons in the `actionPerformed` method, simply make it display the X icon. Then, at the end of `actionPerformed`, call a method named `ai`.

In the method `ai`, first check the turn counter. If it is the computer's first turn, choose either the center or top left square. Next, type "`return`" so the method ends and the fighter can make his/her move. If it is not the computer's first turn, check for two O's in a row. If two are found, act offensively and grab that spot, as illustrated in Figure 24-1. The computer knocks out the player!

If two O's in a row are not found, play defensively and check for two X's in a row. Block that third spot so the other contender cannot win, as illustrated in Figure 24-2.

To do this, create a method that takes in an `Icon`. It will check for two in a row of that Icon using `for loops`. If one is found, take the spot and type "`return`," which ends the method so the computer cannot move twice in a row.

That's it! Begin Round 1! The completed code is below.

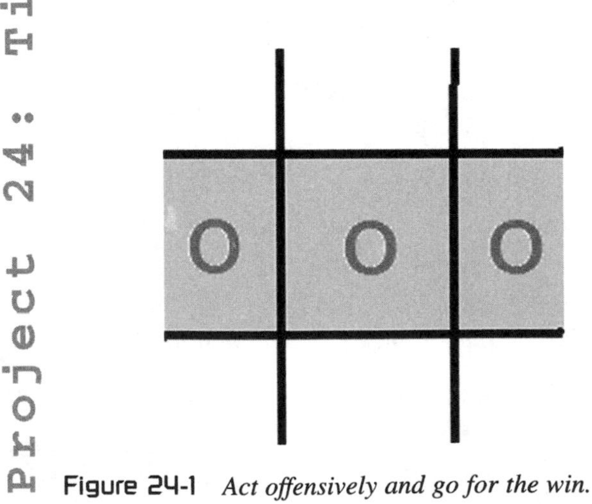

Figure 24-1 *Act offensively and go for the win.*

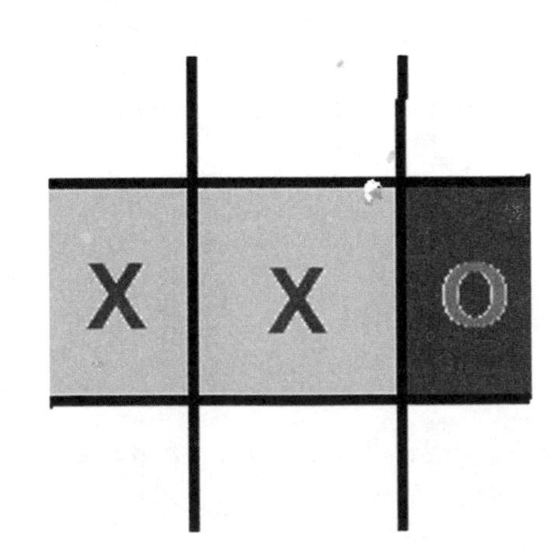

Figure 24-2 *Play defensively and prevent the X's from winning.*

```
//import everything:
import javax.swing.*;
import javax.swing.event.*;
import java.awt.*;
import java.awt.event.*;
import java.io.*;
import java.net.*;

//the class with the JFrame and the ActionListner:
public class TicTacToe extends JFrame implements ActionListener
{
  //this is an array of the buttons (spots). We use an array, not
  //an arrayList because the number of buttons is constant; it does
  //not change.
  JButton spots[ ] = new JButton[ 9] ;
  //this will keep track of turns: even is player 1; odd is player2
  //we will use mod (%) to differentiate the values:
  int turn = 1;
  //these images represent the sides
  ImageIcon red = new ImageIcon("red.png");
  ImageIcon blue = new ImageIcon("blue.png");
  ImageIcon blank = new ImageIcon("blank.png");
  //the title image:
  JLabel title = new JLabel(new ImageIcon("title.png"));
  //the constructor
  public TicTacToe()
  {
    super("TicTacToe: Boxing Style");
    setSize(350,625);
    setVisible(true);
    setDefaultCloseOperation(JFrame.EXIT_ON_CLOSE);
    //this will hold the buttons:
    Container container = getContentPane();
    //this will tell the computer how to display the buttons:
    container.setLayout(null);
    container.add(title);
    title.setBounds(0,0,350,288);
    int newLine = 0;
    int lineCount = 0;
    for(int i = 0; i < spots.length; i++)
    {
      //initialize it with a blank image
      spots[ i] = new JButton(blank);
      //this checks whether to use a new row
      if(i==3 || i ==6)
      {
        newLine++;
        lineCount = 0;
      }
      //set the position of the button
      spots[ i] .setBounds(15+(lineCount*100),288+(newLine*100),100,100);
      //add it to the container
      container.add(spots[ i] );

      //and now add the action listener:
      spots[ i] .addActionListener(this);

      lineCount++;
    }
    //remember to refresh the screen!
    container.repaint();
  }
```

```
//the mandatory method:
public void actionPerformed(ActionEvent e)
{
  //this will check the button pressed:
  for(int i = 0; i < spots.length; i++)
  {
    if(e.getSource()==spots[i])
    {
      //set the button to X
      spots[i].setIcon(red);
      //disable the btn so it can't be re-pressed
      spots[i].removeActionListener(this);
    }
  }
  turn++;

  //before letting the other player go, check whether a player won:
  checkWin();

  //this method lets the computer select its turn
  ai();
}

public void checkWin()
{
  //first, we will use to go through three iterations. This allows us to
  //check for both horizontal and vertical wins without using too
  //much code:
  for(int i = 0; i < 3; i++)
  {
    //this checks for a vertical X win
    if(spots[i].getIcon().equals(red) &&
       spots[i+3].getIcon().equals(red) &&
       spots[i+6].getIcon().equals(red))
    {
      JOptionPane.showMessageDialog(null,"You Win!");
      return;
    }

    //this checks for a vertical O win
    if(spots[i].getIcon().equals(blue) &&
       spots[i+3].getIcon().equals(blue) &&
       spots[i+6].getIcon().equals(blue))
    {
      JOptionPane.showMessageDialog(null,"YOU LOSE!");
      return;
    }

    //this checks for a horizontal X win
    if(spots[i*3].getIcon().equals(red) &&
       spots[(i*3)+1].getIcon().equals(red) &&
       spots[(i*3)+2].getIcon().equals(red))
    {
      JOptionPane.showMessageDialog(null,"You Win!");
      return;
    }

    //this checks for a horizontal O win
    if(spots[i*3].getIcon().equals(blue) &&
       spots[(i*3)+1].getIcon().equals(blue) &&
       spots[(i*3)+2].getIcon().equals(blue))
    {
```

```
      JOptionPane.showMessageDialog(null,"YOU LOSE!");
      return;
    }
  }
  //now, this loop will check for diagnol wins
  for(int i = 0; i <= 2; i+=2)
  {
    //this will check for diagnol X wins
    if(spots[i].getIcon().equals(red) &&
       spots[4].getIcon().equals(red) &&
       spots[8-i].getIcon().equals(red))
    {
      JOptionPane.showMessageDialog(null,"You Win!");
      return;
    }
    //this will check for diagnol O wins
    if(spots[i].getIcon().equals(blue) &&
       spots[4].getIcon().equals(blue) &&
       spots[8-i].getIcon().equals(blue))
    {
      JOptionPane.showMessageDialog(null,"YOU LOSE!");
      return;
    }
  }
}
public void ai()
{
  boolean movedYet;
  //if this is the computer's first turn, then try to go in the top left
  //if already taken, take the middle
  if(turn == 2)
  {
  //if the top left is taken, take the middle
  if(spots[0].getIcon().equals(red))
  {
    spots[4].setIcon(blue);
    spots[4].removeActionListener(this);
    movedYet = true;
  }
  //else, take the top left
  else
  {
    spots[0].setIcon(blue);
    spots[0].removeActionListener(this);
    movedYet = true;
  }
}
//if this is not the first turn, then check for 2 out of 3 spots
//taken. If there are none, go to a random location
else
{
  //callin this method checks for two in a row of the first String passed in.
  //It then takes the 3rd spot with the 2nd String passed in:
  movedYet = twoInARow(blue);
  //if the computer didn't take an offensive spot, take a defensive
  //one.
  if(!movedYet)
```

```
    {
      movedYet = twoInARow(red);
      //if there is no defensive move, take the next open one.
      if(!movedYet)
      {
        //this loop finds the first untaken spot:
        for(int i = 0; i < spots.length; i++)
        {
          //if empty, take it!
          if(spots[i].getIcon().equals(blank))
          {
            spots[i].setIcon(blue);
            spots[i].removeActionListener(this);
            movedYet = true;
            break;
          }
        }
      }
    }
  }

  turn++;
  System.out.println(turn);
  checkWin();

  if(!movedYet)
  {
    //if no spot was taken, it must be a cat's game:
    JOptionPane.showMessageDialog(null,"DRAW!!!");
  }
}
public boolean twoInARow(Icon a)
{

    for(int i = 0; i < 3; i++)
    {
      //this checks for 2 in a row from the top
      if(spots[i].getIcon().equals(a) &&
          spots[i+3].getIcon().equals(a) &&
          spots[i+6].getIcon().equals(blank))
      {
        spots[i+6].setIcon(blue);
        spots[i+6].removeActionListener(this);
        return true;
      }

      //this checks (from the top and bottom)
      //for a taken spot, then a gap, then a taken one:
      if(spots[i].getIcon().equals(a) &&
          spots[i+6].getIcon().equals(a) &&
          spots[i+3].getIcon().equals(blank))
      {
        spots[i+3].setIcon(blue);
        spots[i+3].removeActionListener(this);
        return true;
      }

      //this checks for 2 in a row from the bottom
      if(spots[i+6].getIcon().equals(a) &&
        spots[i+3].getIcon().equals(a) &&
          spots[i].getIcon().equals(blank))
      {
        spots[i].setIcon(blue);
```

```
    spots[ i] .removeActionListener(this);
    return true;
  }

  //this checks for 2 in a row from the left
  if(spots[ i*3] .getIcon().equals(a) &&
    spots[ (i*3)+1] .getIcon().equals(a) &&
      spots[ (i*3)+2] .getIcon().equals(blank))
  {
    spots[ (i*3)+2] .setIcon(blue);
    spots[ (i*3)+2] .removeActionListener(this);
    return true;
  }

  //this checks (from the left and right)
  //for a taken spot, then a gap, then a taken one
  if(spots[ i*3] .getIcon().equals(a) &&
    spots[ (i*3)+2] .getIcon().equals(a) &&
      spots[ (i*3)+1] .getIcon().equals(blank))
  {
    spots[ (i*3)+1] .setIcon(blue);
    spots[ (i*3)+1] .removeActionListener(this);
    return true;
  }
  //this checks for 2 in a row from the right
  if(spots[ (i*3)+2] .getIcon().equals(a) &&
    spots[ (i*3)+1] .getIcon().equals(a) &&
      spots[ i*3] .getIcon().equals(blank))
  {
    spots[ i*3] .setIcon(blue);
    spots[ i*3] .removeActionListener(this);
    return true;
  }
  //now we will check for a diagnol 2 in a row:
  for(int j = 0; j <= 2; j+=2)
  {
    //this will check for diagnol X wins
    if(spots[ j] .getIcon()==a &&
      spots[ 4] .getIcon()==a &&
      spots[ 8-j] .getIcon().equals(blank))
    {
      spots[ 8-j] .setIcon(blue);
      spots[ 8-j] .removeActionListener(this);
      return true;
    }

    //this checks (from a diagnol)
    //for a taken spot, then a gap, then a taken one
    if(spots[ j] .getIcon()==a &&
      spots[ 8-j] .getIcon()==a &&
      spots[ 4] .getIcon().equals(blank))
    {
      spots[ 4] .setIcon(blue);
      spots[ 4] .removeActionListener(this);
      return true;
    }

    if(spots[ 8-j] .getIcon()==a &&
      spots[ 4] .getIcon()==a &&
      spots[ j] .getIcon().equals(blank))
    {
      spots[ j] .setIcon(blue);
```

```
            spots[ j] .removeActionListener(this);
            return true;
         }
      }
    }
    return false;
  }
  //starter (main) method:
  public static void main(String[ ] args) {
    TicTacToe ttt = new TicTacToe();
  }
}
```


Figures 24-3 through 24-6 demonstrate Tic-Tac-Toe Boxing's incredible new AI!

Blood and gore. That's what fighting fans pay to see. Next up, you'll add the sounds and graphics to make Tic-Tac-Toe Boxing worth the price of admission.

Figure 24-3 *Boxing begins!*

Figure 24-4 *Player takes top left, computer takes middle.*

Figure 24-5 *Player takes top middle, but is blocked by computer.*

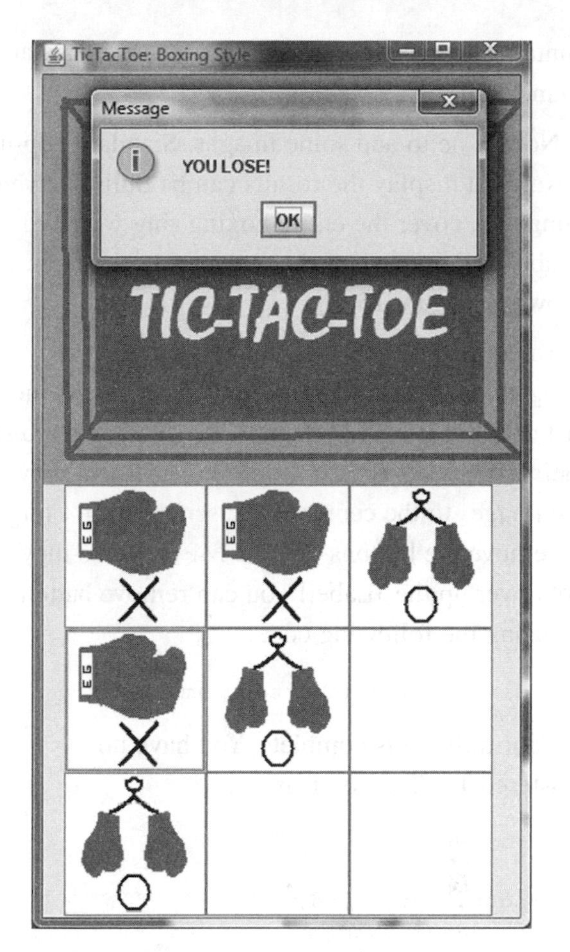

Figure 24-6 *Player loses ☹ (but don't get too cocky, computer, a human programmed your brain!).*

Project 25: Tic-Tac-Toe Boxing—Championship

Project

Want to make the frenzied crowd scream for their favorite boxer or boo the opponent? Want to have great displays showing a "Win," "Lose," or "Draw?" Here you add audio clips and graphic images to make the boxing match come to life.

Making the game

Begin by adding the sound effects. Declare three `AudioClips` globally, but initialize them inside of the constructor. The first `AudioClip` should contain a punching sound. Play the punching sound in the beginning of the `actionPerformed` method. This way, every time a user throws a punch, you hear a corresponding POW! The other two `AudioClips` should contain cheering and booing sounds. If the player wins, sound the cheers. If he/she loses, it is time to release the boos!

Instead of having to repeat code each time a sound is played, create a new method called `win`. It has two parameters: a boolean and a String. If the boolean is false, the game is a draw. If it is true, check whether the second parameter, a String, contains "win" or "lose." The correct message and sounds will be displayed.

Now, time to add some images. Standard pop-up boxes that display the results can be dull. To liven things up, cover the entire boxing ring with a giant JLabel that announces the outcome of the fight, as shown in Figure 25-1.

To do this, create three JLabels to hold the images: one called `win`, one called `lose`, and the last called `draw`. Add them to the container and set their bounds off screen. In the `win` method, move the images to the center of the screen. Don't forget to remove the buttons using a `for loop` so they do not cover up the JLabel! You can remove buttons by using the following code:

```
container.remove(<button name>);
```

Your training is complete. You have now mastered Tic-Tac-Toe Boxing.

Figure 25-1 *JLabel displays outcome of fight.*

```
//import everything:
import javax.swing.*;
import javax.swing.event.*;
import java.awt.*;
import java.awt.event.*;
import java.io.*;
import java.applet.AudioClip;
import java.net.*;

//the class with the JFrame and the ActionListner:
public class TicTacToe extends JFrame implements ActionListener
{
  //this is an array of the buttons (spots). We use an array, not
  //an arrayList because the number of buttons is constant; it does
  //not change.
  JButton spots[ ] = new JButton[ 9] ;

  //this will keep track of turns
  int turn = 1;

  //this is a JLabel: it will display the text
  JLabel lbl = new JLabel(new ImageIcon("title.png"));

  //this JLabel will display whose turn it is
  JLabel turnLbl = new JLabel("X' s Turn");

  ImageIcon red = new ImageIcon("red.png");
  ImageIcon blue = new ImageIcon("blue.png");
  ImageIcon blank = new ImageIcon("blank.png");
  ImageIcon loseImg = new ImageIcon("lose.png");
  ImageIcon winImg = new ImageIcon("win.png");
  ImageIcon drawImg = new ImageIcon("draw.png");

  JLabel lose = new JLabel(loseImg);
  JLabel win = new JLabel(winImg);
```

```
JLabel draw = new JLabel(drawImg);

URL hitURL = null, cheerURL = null, booURL = null;
AudioClip hit = null, cheer = null, boo = null;

Container container;

//the constructor
public TicTacToe()
{
  super("TicTacToe: Boxing Style");
  setSize(450,700);
  setVisible(true);
  setDefaultCloseOperation(JFrame.EXIT_ON_CLOSE);

  //this will hold the buttons:

  container = getContentPane();

  //this will tell the computer how to display the buttons:
  container.setLayout(null);

  //we will add the winning, losing, and draw labels, butt
  //move them off the screen.
  container.add(lose);
  lose.setBounds(-500,-500,331,438);
  container.add(win);
  lose.setBounds(-500,-500,331,438);
  container.add(draw);
  lose.setBounds(-500,-500,331,438);
  //add the lbl:
  container.add(lbl);
  lbl.setBounds(20,0,400,288);

  try
  {

    hitURL = this.getClass().getResource("hit.wav");
    booURL = this.getClass().getResource("boo.wav");
    cheerURL = this.getClass().getResource("cheer.wav");
    hit = JApplet.newAudioClip(hitURL);
    boo = JApplet.newAudioClip(booURL);
    cheer = JApplet.newAudioClip(cheerURL);

  }

  catch(Exception e){}

  //before we can add the buttons to the container, we must
  //initialize them. Use a for loop to do it:
  int newLine = 0;
  int lineCount = 0;
  for(int i = 0; i < spots.length; i++)

  {

    //initialize it with a blank image
    spots[i] = new JButton(blank);
    //this checks whether to use a new row
    if(i==3 || i ==6)
    {
      newLine++;
      lineCount = 0;
    }

    //set the position of the button
    spots[i].setBounds(75+(lineCount*100),300+(newLine*100),100,100);

    //add it to the container
```

```
      container.add(spots[ i] );

      //and now add the action listener:
      spots[ i] .addActionListener(this);

      lineCount++;

    }

    //add the JLabel that describes the turn

    container.add(turnLbl);

    turnLbl.setBounds(200,630,100,30);

    container.setComponentZOrder(lose,0);

    container.setComponentZOrder(win,0);

    container.setComponentZOrder(draw,0);

    //make sure everything is displayed:

    container.repaint();

  }

  public void reset()
  {
    for(int i = 0; i < spots.length; i++)
    {
      spots[ i] .setIcon(blank);
      spots[ i] .addActionListener(this);
    }
    turn = 1;
  }

  //the mandatory method:
  public void actionPerformed(ActionEvent e)
  {
    hit.play();
    try
    {
      Thread.sleep(600);
      }
    catch(Exception excep){ }

    //this will check the button pressed:

    for(int i = 0; i < spots.length; i++)

    {

      if(e.getSource()==spots[ i] )
      {

        //set the button to X
        spots[ i] .setIcon(red);

        //now, change the JLabel that describes the player's turn
        turnLbl.setText("X's (Red's) Turn");

        //disable the btn so it can't be re-pressed
        spots[ i] .removeActionListener(this);

      }
    }

    turn++;

    //before letting the other player go, check whether a player won:

    checkWin();
```

```
  //this method lets the computer select its turn

  ai();

}

public void ai()
{

  boolean movedYet;

  //if this is the computer's first turn, then try to go in the top left

  //if already taken, take the middle

  if(turn == 2)

  {

    //if the top left is taken, take the middle
    if(spots[0].getIcon().equals(red))
    {
      spots[4].setIcon(blue);
      spots[4].removeActionListener(this);
      movedYet = true;
    }
    //else, take the top left
    else
    {
      spots[0].setIcon(blue);
      spots[0].removeActionListener(this);
      movedYet = true;
    }
}
//if this is not the first turn, then check for 2 out of 3 spots
//taken. If there are none, go to a random location
else
{
  //callin this method checks for two in a row of the first String passed in.
  //It then takes the 3rd spot with the 2nd String passed in:
  movedYet = twoInARow(blue);

  //if the computer didn't take an offensive spot, take a defensive
  //one.
  if(!movedYet)
  {

    movedYet = twoInARow(red);
    //if there is no defensive move, take the next open one.
    if(!movedYet)
    {
      //this loop finds the first untaken spot:
      for(int i = 0; i < spots.length; i++)
      {
        //if empty, take it!
        if(spots[i].getIcon().equals(blank))
        {
          spots[i].setIcon(blue);
          spots[i].removeActionListener
            (this);
          movedYet = true;
          break;
        }
      }
    }
  }
```

```
    }

    turn++;
    System.out.println(turn);
    checkWin();

    if(!movedYet)
    {
      //if no spot was taken, it must be a cat's game:
      win(false,"");
    }
  }

  public void win(boolean notADraw,String result)
  {
    //remove the jbuttons:
    for(int i = 0; i < spots.length; i++)
    {
      container.remove(spots[ i] );
    }

    //checks to see if is a win or a draw
    if(notADraw)
    {
      if(result.equals("win"))
      {
        //the player won:
        cheer.play();
        try
        {
          Thread.sleep(600);
        }
        catch(Exception excep){ }

        win.setBounds(50,50,331,438);
        }
        else
        {
        //the computer won:
        boo.play();
        try
        {
          Thread.sleep(600);
        }
        catch(Exception excep){ }
        lose.setBounds(50,50,331,438);;
        reset();
        return;
      }
    }
    else
    {
      //a draw:
      draw.setBounds(50,50,331,438);
    }
  }
  public boolean twoInARow(Icon a)
  {
    for(int i = 0; i < 3; i++)
    {
        //this checks for 2 in a row from the top
        if(spots[ i] .getIcon().equals(a) &&
            spots[ i+3] .getIcon().equals(a) &&
```

```
    spots[i+6].getIcon().equals(blank))
  {
    spots[i+6].setIcon(blue);
    spots[i+6].removeActionListener(this);
    return true;
  }
  //this checks (from the top and bottom)
  //for a taken spot, then a gap, then a taken one:
  if(spots[i].getIcon().equals(a) &&
     spots[i+6].getIcon().equals(a) &&
     spots[i+3].getIcon().equals(blank))
  {
    spots[i+3].setIcon(blue);
    spots[i+3].removeActionListener(this);
    return true;
  }

  //this checks for 2 in a row from the bottom
  if(spots[i+6].getIcon().equals(a) &&
     spots[i+3].getIcon().equals(a) &&
     spots[i].getIcon().equals(blank))
  {
    spots[i].setIcon(blue);
    spots[i].removeActionListener(this);
    return true;
  }
  //this checks for 2 in a row from the left
  if(spots[i*3].getIcon().equals(a) &&
     spots[(i*3)+1].getIcon().equals(a) &&
     spots[(i*3)+2].getIcon().equals(blank))
  {
    spots[(i*3)+2].setIcon(blue);
    spots[(i*3)+2].removeActionListener(this);
    return true;
  }
  //this checks (from the left and right)
  //for a taken spot, then a gap, then a taken one
  if(spots[i*3].getIcon().equals(a) &&
    spots[(i*3)+2].getIcon().equals(a) &&
    spots[(i*3)+1].getIcon().equals(blank))
  {
    spots[(i*3)+1].setIcon(blue);
    spots[(i*3)+1].removeActionListener(this);
    return true;
  }
  //this checks for 2 in a row from the right
  if(spots[(i*3)+2].getIcon().equals(a) &&
    spots[(i*3)+1].getIcon().equals(a) &&
    spots[i*3].getIcon().equals(blank))
  {
    spots[i*3].setIcon(blue);
    spots[i*3].removeActionListener(this);
    return true;
  }

  //now we will check for a diagnol 2 in a row:
  for(int j = 0; j <= 2; j+=2)
  {
    //this will check for diagnol X wins
    if(spots[j].getIcon()==a &&
      spots[4].getIcon()==a &&
```

```
        spots[ 8-j] .getIcon().equals(blank))
      {
        spots[ 8-j] .setIcon(blue);
        spots[ 8-j] .removeActionListener(this);
        return true;
      }
      //this checks (from a diagnol)
      //for a taken spot, then a gap, then a taken one
      if(spots[ j] .getIcon()==a &&
        spots[ 8-j] .getIcon()==a &&
        spots[ 4] .getIcon().equals(blank))
      {
        spots[ 4] .setIcon(blue);
        spots[ 4] .removeActionListener(this);
        return true;
      }
      if(spots[ 8-j] .getIcon()==a &&
        spots[ 4] .getIcon()==a &&
        spots[ j] .getIcon().equals(blank))
      {
        spots[ j] .setIcon(blue);
        spots[ j] .removeActionListener(this);
        return true;
      }
    }
  }
  return false;
}
public void checkWin()
{
  //first, we will use to go through three iterations. This allows us to
  //check for both horizontal and vertical wins without using too
  //much code:
  for(int i = 0; i < 3; i++)
  {
  //this checks for a vertical X win
  if(spots[ i] .getIcon().equals(red) &&
      spots[ i+3] .getIcon().equals(red) &&
      spots[ i+6] .getIcon().equals(red))
   {
     win(true,"win");
     reset();
     return;
   }
   //this checks for a vertical O win
   if(spots[ i] .getIcon().equals(blue) &&
      spots[ i+3] .getIcon().equals(blue) &&
      spots[ i+6] .getIcon().equals(blue))
      {
     win(true,"lose");
     reset();
     return;
   }
   //this checks for a horizontal X win
   if(spots[ i*3] .getIcon().equals(red) &&
      spots[ (i*3)+1] .getIcon().equals(red) &&
      spots[ (i*3)+2] .getIcon().equals(red))
   {
     win(true,"win");
     reset();
```

```
        return;
      }
        //this checks for a horizontal O win
        if(spots[ i*3] .getIcon().equals(blue) &&
            spots[ (i*3)+1] .getIcon().equals(blue) &&
            spots[ (i*3)+2] .getIcon().equals(blue))
        {
          win(true,"lose");
          reset();
          return;
        }
    }
    //now, this loop will check for diagnol wins
    for(int i = 0; i <= 2; i+=2)
    {
      //this will check for diagnol X wins
      if(spots[ i] .getIcon().equals(red) &&
            spots[ 4] .getIcon().equals(red) &&
            spots[ 8-i] .getIcon().equals(red))
        {
          win(true,"win");
          reset();
          return;

      }
    //this will check for diagnol O wins
    if(spots[ i] .getIcon().equals(blue) &&
          spots[ 4] .getIcon().equals(blue) &&
          spots[ 8-i] .getIcon().equals(blue))
          {
        win(true,"lose");
        reset();
        return;
        }
    }
  }
  //starter (main) method:
  public static void main(String[ ] args) {
    TicTacToe ttt = new TicTacToe();
    }
  }
```

Figures 25-2 through 25-5 shows Tic-Tac-Toe Boxing's attention-grabbing GUI!

Customizing the game

Customize the images: you can turn the gloves into full boxing figures, growling faces ... anything!

Customize the sounds: punch pillows, jump on empty plastic bottles, blow up balloons and pop them to simulate body blows; record your doorbell ringing to signal the next round.

Expand the 3 by 3 boxing ring into a 4 by 4 space requiring more complicated moves.

If you really want a challenge, modify the boxing ring into a three dimensional space with multiple fighters.

Allow the user to select various difficulty levels.

Play ten rounds against the computer: the catch—the computer progresses in intelligence after each round.

Figure 25-2 *Although you cannot hear it, trust me, a punching sound just played.*

Figure 25-3 *Player loses.*

Figure 25-4 *Match ends in draw!*

Figure 25-5 *KNOCK OUT! Player wins!*

Section Four

Shoot 'em up Games

Project 26: Snake Pit—The Arena

Snake Pit

Two snakes. One mouse. Venom flying everywhere! Snake Pit is a classic "shoot 'em up" game in which you must prevent your opponent from capturing prey as you avoid being hit by his/her poisonous venom.

Project

Create the snake pit.

New building blocks

Component Color, JLabel Font

Component color

When using a Container, you can easily set the color of the background of any component, such as a JFrame, by using the command:

```
cont.setBackground(Color.<color>);
```

<color> can be any you choose. The option to select a color pops up after you type the period. Figure 26-1 illustrates a new background color.

You can also set the color of the text of a JLabel by using the command:

```
lbl.setForeground(Color.<color>);
```

Making the game

To begin, create a JFrame that is 800 by 600 pixels. Next, set the background color of the JFrame to black.

Now that you have the arena set, add the two snakes. Create two ImageIcons per snake—one of the basic snake (Figure 26-2); the other, of the snake being hit by venom (Figure 26-3).

Initialize a JLabel for the two snakes and assign the snake's basic image to it (you will use the second ImageIcon later when you create venom). Use the setBounds method to center each snake on the far left and far right sides of the pit. Then, use a JLabel to add the image of a mouse to the center of the screen. Now, create two JLabels to represent the score of each snake. Remember to set the JLabels' fonts to white so the scores can be seen against the dark background of the arena.

Figure 26-1 *JFrame with black background.*

```
import javax.swing.*;
import javax.swing.event.*;
import java.awt.*;
import java.awt.event.*;
import java.util.*;
public class SnakePit extends JFrame
{
  //these variable will help to determine when the
  //ball leaves the screen
  int width = 800;
  int height = 600;
  //the two scores and the JLabels to show them:
```

Figure 26-2 *Basic snake.*

Figure 26-3 *Snake struck by venom.*

```
int scoreLeft = 0;
int scoreRight = 0;
JLabel left = new JLabel("Score: "+scoreLeft);
JLabel right = new JLabel("Score: "+scoreRight);
//SNAKE ICONS:
ImageIcon snakeLeftImg = new ImageIcon("snakeLeft.PNG");
ImageIcon snakeRightImg = new ImageIcon("snakeRight.PNG");

ImageIcon snakeLeftHit = new ImageIcon("snakeLeftHit.PNG");
ImageIcon snakeRightHit = new ImageIcon("snakeRightHit.PNG");

//these are the snake JLabels
JLabel snakeLeft = new JLabel(snakeLeftImg);
JLabel snakeRight = new JLabel(snakeRightImg);
JLabel food = new JLabel(new ImageIcon("food.PNG"));

JLabel venomRight = new JLabel(new ImageIcon("venomRight.png"));
JLabel venomLeft = new JLabel(new ImageIcon("venomLeft.png"));

//global container:
Container cont;

public SnakePit()
{
  //create the JFrame...
  super("Snake Pit");
  setVisible(true);
  setSize(width,height);
  setDefaultCloseOperation(JFrame.EXIT_ON_CLOSE);

  //this is the container:
  cont = getContentPane();
```

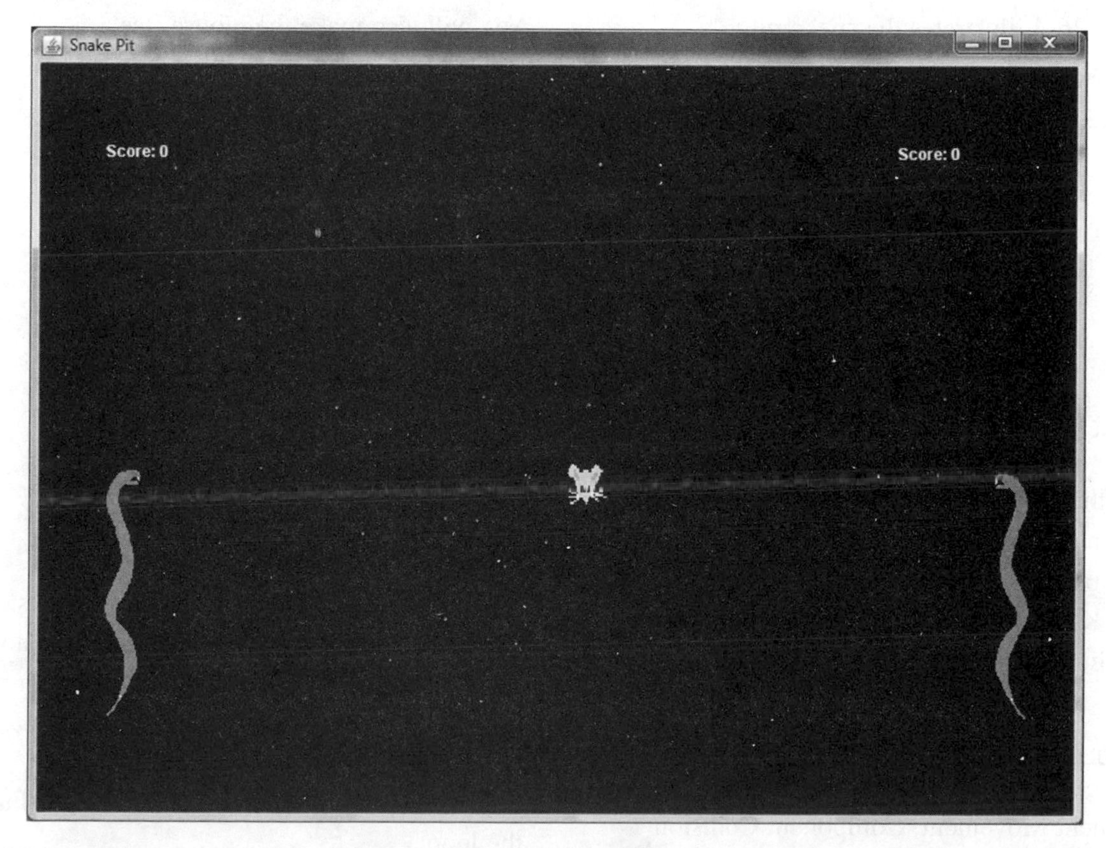

Figure 26-4 *Snake Pit Arena.*

```
    cont.setLayout(null);
    cont.setBackground(Color.BLACK);

    //add the score labels:
    cont.add(left);
    left.setBounds(50,50,100,30);
    left.setForeground(Color.WHITE);
    cont.add(right);
    right.setBounds(width-150,50,100,30);
    right.setForeground(Color.WHITE);

    //set the layout to null
    cont.setLayout(null);

    //add the two snakes and set their bounds to the correct positions
    cont.add(snakeLeft);
    snakeLeft.setBounds(50,height/2,30,200);
    cont.add(snakeRight);
    snakeRight.setBounds(width-80,height/2,30,200);
    //now, add the food (the mouse), and put it in the center
    cont.add(food);
    food.setBounds(width/2,height/2,30,30);
    }
  public static void main (String[ ] args)
  {
    new SnakePit();
  }
  }
```

Figure 26-4 illustrates the snake pit.

Slither forward, snakes! Go on to bring the snakes to life by adding a KeyListener. You will also make the mouse flee for its life.

Project 27: Snake Pit—Snake Bait

Project

One of the key features of Snake Pit is making the snakes move so they can sink their fangs into their bait—a bouncing mouse. You accomplish this by using a KeyListener. Careful, those snakes are getting hungry!

New building blocks

Component Movement, Component Collision Detection

Component movement

Moving components is easy to do in Java. Remember that setBounds method? Use the same method to move components. All you do is create a Thread and an infinite loop to set the bounds of the component during each iteration. For example, if you want a component to move to the right, use the following code, which should be inserted into the loop:

component.setBounds(component.getX()+1, component.getY(), <width>, <height>);

The `getX()` and `getY()` methods return the component's current X and Y positions, respectively.

Component collision detection

Collision detection with components can be written in one simple condition. For this example, `c1` and `c2` are the names of the two components.

c1.getBounds().intersects(c2.getBounds)

The above code checks for the situation illustrated in Figure 27-1.

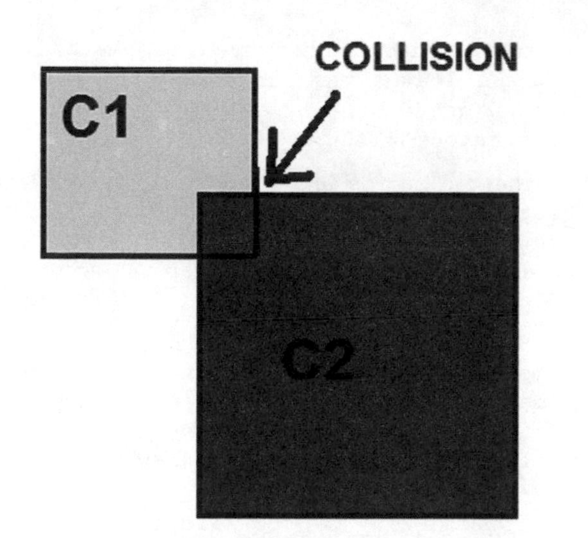

Figure 27-1 *Collision detection.*

Making the game

Create a Thread and add a KeyListener to it. Then, create two global variables that represent the velocity of the food in the x and y directions. Set each variable to 1. Within the thread's infinite loop, use setBounds to move the food by the velocity variables, as shown below:

food.setBounds(food.getX()+velX, food.getY()+velY, 30, 30);

Now, use your new collision detection skills to make the mouse bounce around the arena. If the mouse tries to exit in the y direction (vertically), make the mouse ricochet by multiplying the y velocity by −1. If the mouse tries to exit the arena in the x direction (horizontally), add one point to the opposing snake's score (Note: scoring is based upon preventing your opponent from capturing prey and, as you'll soon see, striking the opposing snake with venom). If one of the snakes is able to capture the mouse, make the mouse bounce by multiplying the x velocity by −1.

Now, use the thread's KeyListener to add user movement. When the "w" key is pressed, use the setBounds method to move the left snake up. When "s" is pressed, move the left snake down. The right snake will be controlled by AI, which you will program in the next project.

```
import javax.swing.*;
import javax.swing.event.*;
import java.awt.*;
import java.awt.event.*;
import java.util.*;

public class SnakePit extends JFrame
{
  //these two arraylists will hold the labels that represent the bullets
  ArrayList bulletsLeft = new ArrayList();
  ArrayList bulletsRight = new ArrayList();

  //this array will hold both arraylists
  ArrayList[ ] bullets = {bulletsLeft, bulletsRight};
```

```
//these variable will help to determine when the
//ball leaves the screen
int width = 800;
int height = 600;

//the two scores and the JLabels to show them:
int scoreLeft = 0;
int scoreRight = 0;
JLabel left = new JLabel("Score: "+scoreLeft);
JLabel right = new JLabel("Score: "+scoreRight);

//SNAKE ICONS:
ImageIcon snakeLeftImg = new ImageIcon("snakeLeft.PNG");
ImageIcon snakeRightImg = new ImageIcon("snakeRight.PNG");
ImageIcon snakeLeftHit = new ImageIcon("snakeLeftHit.PNG");
ImageIcon snakeRightHit = new ImageIcon("snakeRightHit.PNG");

//these are the snake JLabels
JLabel snakeLeft = new JLabel(snakeLeftImg);
JLabel snakeRight = new JLabel(snakeRightImg);
JLabel food = new JLabel(new ImageIcon("food.PNG"));

JLabel venomRight = new JLabel(new ImageIcon("venomRight.png"));
JLabel venomLeft = new JLabel(new ImageIcon("venomLeft.png"));

//these variables represent how much the food moves
//in the x and y directions:
int foodX = 1;
int foodY = 1;

//global container:
Container cont;

public SnakePit()
{
  //create the JFrame...
  super("Snake Pit");
  setVisible(true);
  setSize(width,height);
  setDefaultCloseOperation(JFrame.EXIT_ON_CLOSE);

  //this is the container:
  cont = getContentPane();
  cont.setLayout(null);
  cont.setBackground(Color.black);

  //add the score labels:
  cont.add(left);
  left.setBounds(50,50,100,30);
  left.setForeground(Color.WHITE);
  cont.add(right);
  right.setBounds(width-150,50,100,30);
  right.setForeground(Color.WHITE);

  //add the two snakes and set their bounds to the correct positions
  cont.add(snakeLeft);
  snakeLeft.setBounds(50,height/2,30,200);
  cont.add(snakeRight);
  snakeRight.setBounds(width-80,height/2,30,200);
  //now, add the food (the mouse), and put it in the center
  cont.add(food);
  food.setBounds(width/2,height/2,30,30);

  //the thread for the snakes
  SnakeThread st = new SnakeThread();
  st.start();
```

```
  }

//the thread:
public class SnakeThread extends Thread implements KeyListener
{
  public void run()
  {
    addKeyListener(this);
    while(true)
    {
      try
      {
        //if the food hits the left or right sides...
        if(food.getX()<0 || food.getX()>width)
        {
          //then reset the ball and add a point to the correct player
          food.setBounds(width/2,height/2,30,30);
          foodX = 1;
          foodY = 1;
          if(food.getX()<0)
          {
          scoreLeft++;
          }
          else
          {
          scoreRight++;
          }
          left.setText("Score: "+scoreLeft);
          right.setText("Score: "+scoreRight);
        }
        //if the food goes too high or low, make it bounce
        else if(food.getY()>height-30 || food.getY()<0)
        {
          foodY *= -1;
        }
        //if the paddle is hit
        else if((food.getX()<80 && food.getY()>snakeLeft.getY() &&
            food.getY()<snakeLeft.getY()+200) || (food.getX()>width-80 &&
            food.getY()>snakeRight.getY() && food.getY()<snakeRight.getY()+200))
        {
          foodX *= -1;
        }

        //move the food
        food.setBounds(food.getX()-foodX, food.getY()-foodY,30,30);

        //the refresh delay
        Thread.sleep(4);
      }
      catch(Exception e){}
    }
  }

  //you must also implement this method from KeyListener
  public void keyPressed(KeyEvent event)
  {
    if(event.getKeyChar()=='w')
    {
      snakeLeft.setBounds(snakeLeft.getX(),snakeLeft.getY()-10,30,200);
    }
    if(event.getKeyChar()=='s')
    {
```

```
        snakeLeft.setBounds(snakeLeft.getX(),snakeLeft.getY()+10,30,200);
      }
    }

    //you must also implement this method from KeyListener
    public void keyReleased(KeyEvent event){ }

    //you must also implement this method from KeyListener
    public void keyTyped(KeyEvent event){ }
    }
    public static void main (String[ ] args)
    {
    new SnakePit();
    }
  }
```

Figures 27-2 through 27-4 display primary actions in the snake pit.

You're all alone and suddenly overcome with the urge to play Snake Pit. What do you do? Create an opponent with AI. And to make things even more interesting, you'll add code that allows the two snakes to spit deadly venom at one another while mouse hunting.

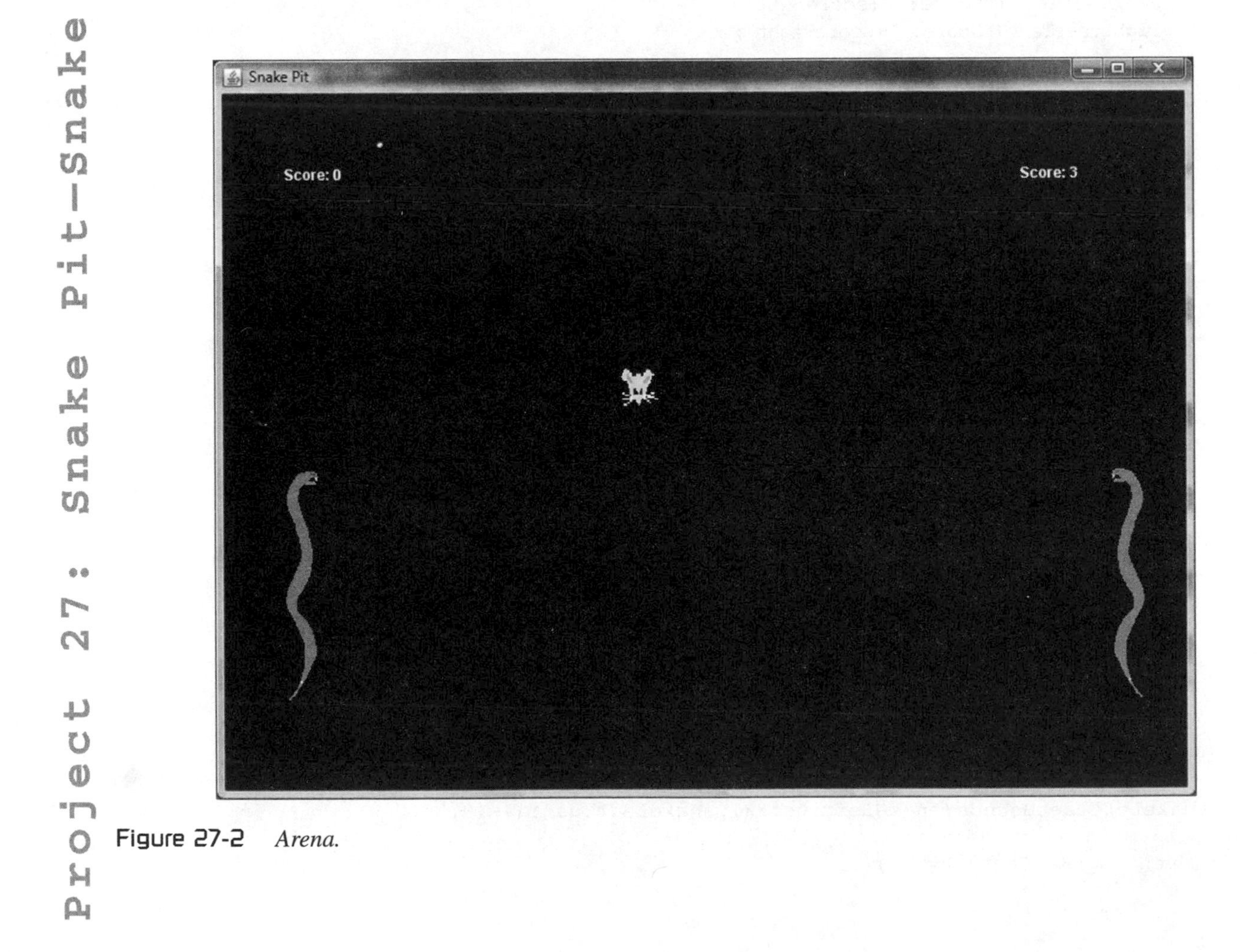

Figure 27-2 *Arena.*

Figure 27-3 *Snake is hungry.*

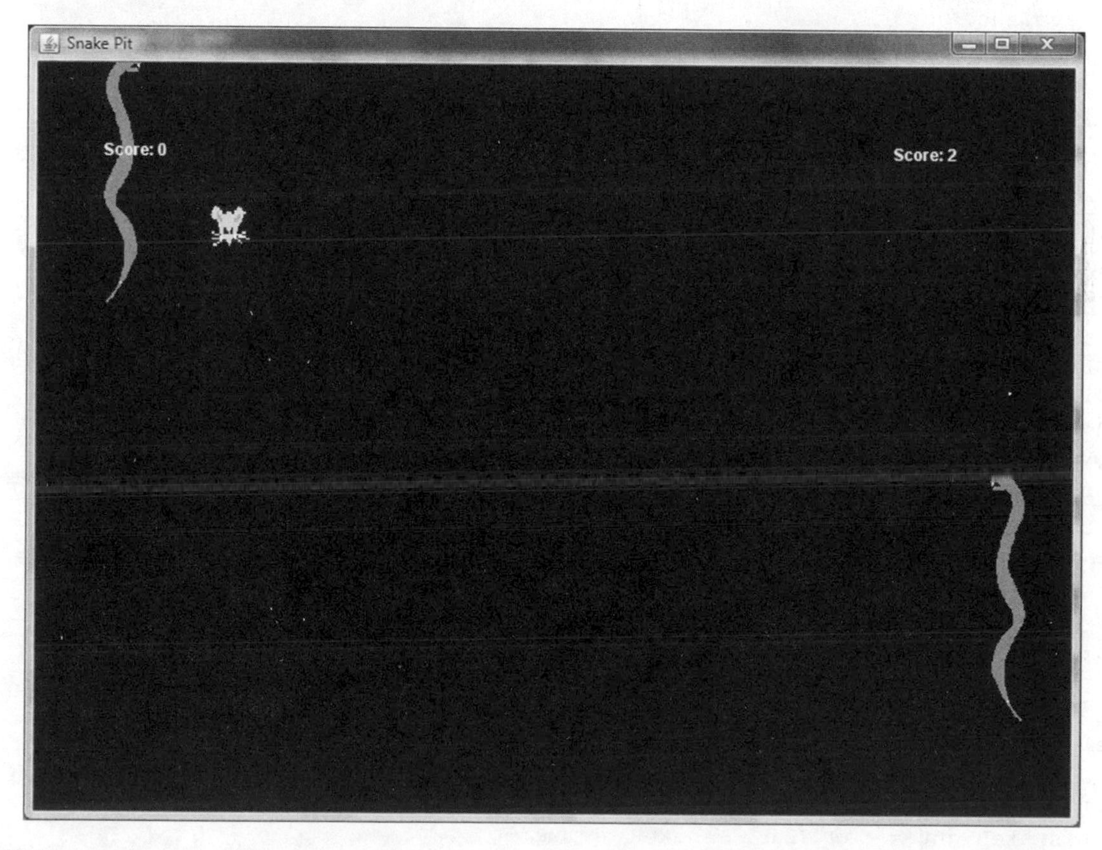

Figure 27-4 *One mouse gone, next mouse up!*

Project 28: Snake Pit—Snake Bite!

Project

With AI, turn your computer into an adversary that cannot be defeated. Also, you will implant venom glands into the snakes that allows them to spit out the deadly poison!

Making the game

First, you will add AI to the snake on the right side of the arena. The AI is very simple; in fact, it can be programmed with only two if-statements. The first if-statement should check to see if the mouse is higher than the right snake. If so, move the right snake up using the setBounds method. The second if-statement should check if the mouse is lower than the right snake. If so, move the right snake down. Both of these if statements belong in the infinite loop inside of the thread.

Now, it's time to add venom glands to the snake. To do this, create two ArrayLists: each one should hold the JLabels that represent the venom from each snake. In addition, to cut down on the repetition of your code, add both of the ArrayLists to an array. When the "q" key is pressed, add a new JLabel to the container and to the ArrayList. Next, create another thread. This one should control the AI's firing of venom. In the infinite loop, go through both ArrayLists and move the JLabels in the appropriate direction.

```
import javax.swing.*;
import javax.swing.event.*;
import java.awt.*;
import java.awt.event.*;
import java.util.*;
public class SnakePit extends JFrame
{
  //these two arraylists will hold the labels that represent the bullets
  ArrayList bulletsLeft = new ArrayList();
  ArrayList bulletsRight = new ArrayList();
  //this array will hold both arraylists
  ArrayList[ ] bullets = {bulletsLeft, bulletsRight};
  //these variable will help to determine when the
  //ball leaves the screen
  int width = 800;
  int height = 600;
  //the two scores and the JLabels to show them:
  int scoreLeft = 0;
  int scoreRight = 0;
  JLabel left = new JLabel("Score: "+scoreLeft);
  JLabel right = new JLabel("Score: "+scoreRight);
  //SNAKE ICONS:
  ImageIcon snakeLeftImg = new ImageIcon("snakeLeft.PNG");
  ImageIcon snakeRightImg = new ImageIcon("snakeRight.PNG");
  ImageIcon snakeLeftHit = new ImageIcon("snakeLeftHit.PNG");
  ImageIcon snakeRightHit = new ImageIcon("snakeRightHit.PNG");
  //these are the snake JLabels
  JLabel snakeLeft = new JLabel(snakeLeftImg);
  JLabel snakeRight = new JLabel(snakeRightImg);
  JLabel food = new JLabel(new ImageIcon("food.PNG"));
```

```
JLabel venomRight = new JLabel(new ImageIcon("venomRight.png"));
JLabel venomLeft = new JLabel(new ImageIcon("venomLeft.png"));

//these variables represent how much the food moves
//in the x and y directions:
int foodX = 1;
int foodY = 1;
//global container:
Container cont;

public SnakePit()
{
  //create the JFrame...
  super("Snake Pit");
  setVisible(true);
  setSize(width,height);
  setDefaultCloseOperation(JFrame.EXIT_ON_CLOSE);

  //this is the container:
  cont = getContentPane();
  cont.setLayout(null);
  cont.setBackground(Color.BLACK);

  //add the score labels:
  cont.add(left);
  left.setFont(new Font("Times New Roman", Font.BOLD, 20));
  left.setBounds(50,50,100,30);
  left.setForeground(Color.WHITE);
  cont.add(right);
  right.setFont(new Font("Times New Roman", Font.BOLD, 20));
  right.setBounds(width-150,50,100,30);
  right.setForeground(Color.WHITE);

  //add the two snakes and set their bounds to the correct positions
  cont.add(snakeLeft);
  snakeLeft.setBounds(50,height/2,30,200);
  cont.add(snakeRight);
  snakeRight.setBounds(width-80,height/2,30,200);
  //now, add the food (the mouse), and put it in the center
  cont.add(food);
  food.setBounds(width/2,height/2,30,30);

  //the thread for the snakes
  SnakeThread st = new SnakeThread();
  st.start();

  //this thread will control the intervals of the enemy's attack
  Attack attack = new Attack();
  attack.start();
  }
public class Attack extends Thread
{
  public void run()
  {
    while(true)
    {
      try
      {
        int interval = (int)(Math.random()*2000);
        Thread.sleep(interval);
        bulletsRight.add(venomRight);
        cont.add(venomRight);
        cont.setComponentZOrder(venomRight,0);
        venomRight.setBounds(snakeRight.getX()-30,snakeRight.getY()+30,20,10);
```

```
      }
      catch(Exception e){ }
    }
    }
  }
//the thread:
public class SnakeThread extends Thread implements KeyListener
{
  public void run()
  {
    addKeyListener(this);
    while(true)
    {
      try
      {
        //move the bullets
        //this for loop goes through the two arraylists
        for(int i = 0; i <bullets.length; i++)
        {
          int distance = 0;
          if(i==0)
            distance = 1;
          else
            distance = -1;
          for(int j = 0; j <bullets[ i] .size(); j++)
          {
            JLabel temp = ((JLabel)bullets [ i] .get(j));
          ((JLabel)bullets[ i] .get(j)).setBounds(
            temp.getX()+distance,temp.getY(),20,10);
          }
        }
        //if the food hits the left or right sides...
        if(food.getX()<0 || food.getX()>width)
        {
          //then reset the ball and add a point to the correct player
          food.setBounds(width/2,height/2,30,30);
          foodX = 1;
          foodY = 1;
          if(food.getX()<0)
          {
            scoreLeft++;
          }
          else
          {
            scoreRight++;
          }
          left.setText("Score: "+scoreLeft);
          right.setText("Score: "+scoreRight);
        }
        //if the food goes too high or low, make it bounce
        else if(food.getY()>height-30 || food.getY()<0)
        {
          foodY *= -1;
        }
        //if the paddle is hit
        else if((food.getX()<80 && food.getY()>snakeLeft.getY()
            &&food.getY()<snakeLeft.getY()+200) ||(food.getX()>width-80
            &&food.getY()>snakeRight.getY() &&food.getY()<snakeRight.getY()+200))
        {
          foodX *= -1;
```

```
            }
            //move the food
            food.setBounds(food.getX()-foodX, food.getY()-foodY,30,30);
            //this code controls the AI. If the food is higher
            //than the computer's paddle, move up. If not, move down.
            if(food.getY()>snakeRight.getY()+200)
            {
              snakeRight.setBounds(snakeRight.getX(),snakeRight.getY()+4,30,200);
            }
            if(food.getY()<snakeRight.getY())
            {
              snakeRight.setBounds(snakeRight.getX(),snakeRight.getY()-4,30,200);
            }

            //the refresh delay
            Thread.sleep(4);
          }
          catch(Exception e){ }
        }
      }

      //you must also implement this method from KeyListener
      public void keyPressed(KeyEvent event)
      {
        if(event.getKeyChar()=='w')
        {
          snakeLeft.setBounds(snakeLeft.getX(),snakeLeft.getY()-10,30,200);
        }
        if(event.getKeyChar()=='s')
        {
          snakeLeft.setBounds(snakeLeft.getX(),snakeLeft.getY()+10,30,200);
        }
        if(event.getKeyChar()=='q')
        {
          bulletsLeft.add(venomLeft);
          cont.add(venomLeft);
          cont.setComponentZOrder(venomLeft,0);
          venomLeft.setBounds(snakeLeft.getX()+30,snakeLeft.getY()+30,20,10);
        }
      }

      //you must also implement this method from KeyListener
      public void keyReleased(KeyEvent event){ }

        //you must also implement this method from KeyListener
        public void keyTyped(KeyEvent event){ }
        }
      public static void main (String[ ] args)
      {
        new SnakePit();
      }
    }
```

Figures 28-1 and 28-2 display the venom battle.

Chaos is about to unfold! In the next project, you will track collisions between the snakes and their venom. When one of these nasty creatures gets hit, not only will the score change, but the injured snake will wither in pain.

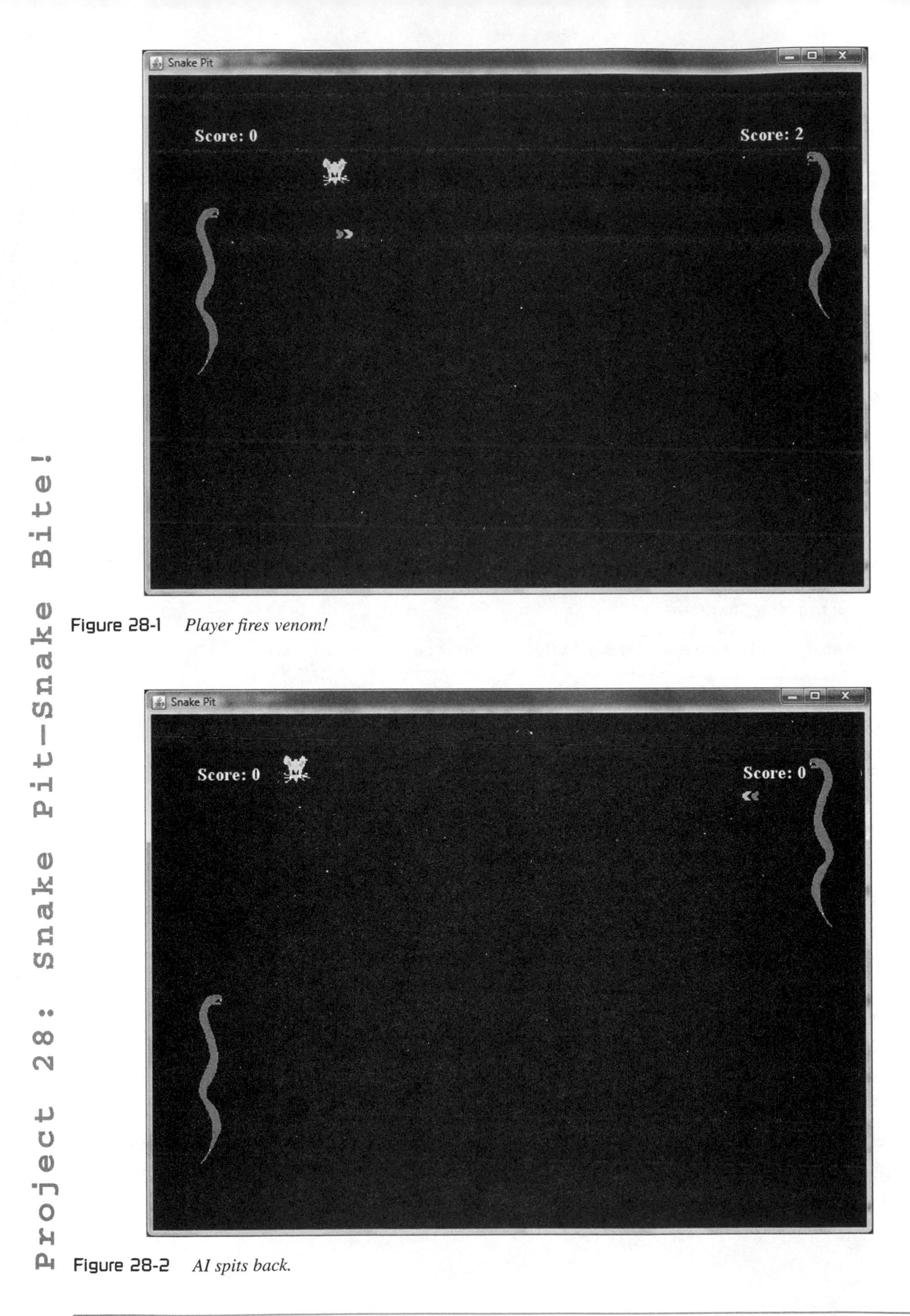

Figure 28-1 *Player fires venom!*

Figure 28-2 *AI spits back.*

Project 29: Snake Pit—King Cobra

Project

Stand back! You're about to make the snakes recoil when stricken and create a scoring system to keep track of venom hits. And there's more – you'll add a background image to the pit.

New building blocks

Font

Font

You can easily change the font of JLabels. Simply use the following code:

```
lbl.setFont(new Font("<font name>",
  <font type>, <size>));
```

`<font name>` is the name of the font, such as "Times New Roman" or "Arial."

`<font type>` give you options: bold, underlined, etc. To select the type, enter "Font."

When you type the period, options pop-up. Choose one of the uppercase choices, such as "BOLD" or "NORMAL."

Making the game

Add if-statements in the loops that move the venom to check for a collision between the venom and the snakes. If there is a collision, change the score accordingly.

Remember when you created that ImageIcon of the snakes being attacked? Finally, it is time to use it! Set the JLabel's icon to that image for 100 milliseconds. Then, add a background image by creating another ImageIcon and a JLabel. You can also change the font of the score JLabels to make them contrast against the background. You can even add a small delay to the game before the mouse starts moving so the player has a moment to prepare for the attack.

```
import javax.swing.*;
import javax.swing.event.*;
import java.awt.*;
import java.awt.event.*;
import java.util.*;

public class SnakePit extends JFrame
{
  //these two arraylists will hold the labels that represent the bullets
  ArrayList bulletsLeft = new ArrayList();
  ArrayList bulletsRight = new ArrayList();

  //this array will hold both arraylists
  ArrayList[ ] bullets = {bulletsLeft, bulletsRight};

  //these variable will help to determine when the
  //ball leaves the screen
  int width = 800;
  int height = 600;

  //the two scores and the JLabels to show them:
  int scoreLeft = 0;
  int scoreRight = 0;
  JLabel left = new JLabel("Score: "+scoreLeft);
  JLabel right = new JLabel("Score: "+scoreRight);
```

```
//SNAKE ICONS:
ImageIcon snakeLeftImg = new ImageIcon("snakeLeft.PNG");
ImageIcon snakeRightImg = new ImageIcon("snakeRight.PNG");
ImageIcon snakeLeftHit = new ImageIcon("snakeLeftHit.PNG");
ImageIcon snakeRightHit = new ImageIcon("snake RightHit.PNG");

//these are the snake JLabels
JLabel snakeLeft = new JLabel(snakeLeftImg);
JLabel snakeRight = new JLabel(snakeRightImg);
JLabel food = new JLabel(new ImageIcon("food.PNG"));
JLabel venomRight = new JLabel(new ImageIcon("venomRight.png"));
JLabel venomLeft = new JLabel(new ImageIcon ("venomLeft.png"));

//these variables represent how much the food moves
//in the x and y directions:
int foodX = 1;
int foodY = 1;

//global container:
Container cont;

public SnakePit()
{
  //create the JFrame...
  super("Snake Pit");
  setVisible(true);
  setSize(width,height);
  setDefaultCloseOperation(JFrame.EXIT_ON_CLOSE);

  //this is the container:
  cont = getContentPane();
  cont.setLayout(null);
  //add the score labels:
  cont.add(left);
  left.setFont(new Font("Times New Roman", Font.BOLD, 20));
  left.setBounds(50,50,100,30);
  left.setForeground(Color.WHITE);
  cont.add(right);
  right.setFont(new Font("Times New Roman", Font.BOLD, 20));
  right.setBounds(width-150,50,100,30);
  right.setForeground(Color.WHITE);

  //set the layout to null
  cont.setLayout(null);

  //add the two snakes and set their bounds to the correct positions
  cont.add(snakeLeft);
  snakeLeft.setBounds(50,height/2,30,200);
  cont.add(snakeRight);
  snakeRight.setBounds(width-80,height/2,30,200);
  //now, add the food (the mouse), and put it in the center
  cont.add(food);
  food.setBounds(width/2,height/2,30,30);

  //the thread for the snakes
  SnakeThread st = new SnakeThread();
  st.start();

  //this thread will control the intervals of the enemy's attack
  Attack attack = new Attack();
  attack.start();

  //set the background of the container with an image
  JLabel background = new JLabel(new ImageIcon("background.PNG"));
  cont.add(background);
  background.setBounds(0,0,800,600);
```

```
  }
public class Attack extends Thread
{
  public void run()
  {
    while(true)
    {
      try
      {
        int interval = (int)(Math.random()*2000);
        Thread.sleep(interval);
        bulletsRight.add(venomRight);
        cont.add(venomRight);
        cont.setComponentZOrder(venomRight,0);
        venomRight.setBounds(snakeRight.getX()-30, snakeRight.getY()+30,20,10);
      }
      catch(Exception e){ }
    }
  }
}
//the thread:
public class SnakeThread extends Thread implements KeyListener
{
  public void run()
  {
    try
    {
      Thread.sleep(2000);
    }
    catch(Exception e){ }
    addKeyListener(this);
    while(true)
    {
      try
      {
        //move the bullets
        //this for loop goes through the two arraylists
        for(int i = 0; i <bullets.length; i++)
        {
          int distance = 0;
          if(i==0)
          distance = 1;
          else
          distance = -1;
          for(int j = 0; j <bullets[ i] .size(); j++)
          {
            JLabel temp = ((JLabel)bullets[ i] .get(j));
            //check for collisions between the snakes and bullets
          if(i==0)
          {
            if(snakeRight.getBounds().intersects(temp.getX(),temp.getY(),20,10))
          {
            snakeRight.setIcon(snakeRightHit);
            scoreLeft++;
            left.setText("Score: "+scoreLeft);
            bullets[ i] .remove(j);
            Thread.sleep(100);
            snakeRight.setIcon(snakeRightImg);
          }
        }
```

```
      else
      {
        if(snakeLeft.getBounds().intersects(temp.getX(),temp.getY(),20,10))
        {
            snakeLeft.setIcon(snakeLeftHit);
            scoreRight++;
            right.setText("Score: "+scoreRight);
            bullets[ i] .remove(j);
            Thread.sleep(100);
            snakeLeft.setIcon(snakeLeftImg);
            }
        }
        ((JLabel)bullets[ i] .get(j)).setBounds(temp.getX()+distance,temp.getY(),20,10);
      }
    }

    //if the food hits the left or right sides...
    if(food.getX()<0 || food.getX()>width)
    {
      //then reset the ball and add a point to the correct player
      food.setBounds(width/2,height/2,30,30);
      foodX = 1;
      foodY = 1;
      if(food.getX()<0)
      {
        scoreLeft++;
      }
      else
      {
        scoreRight++;
      }
      left.setText("Score: "+scoreLeft);
      right.setText("Score: "+scoreRight);
    }
    //if the food goes too high or low, make it bounce
    else if(food.getY()>height-30 || food.getY()<0)
    {
      foodY *= -1;
    }

    //if the paddle is hit
    else if((food.getX()<80 && food.getY()>snakeLeft.getY()
          &&food.getY()<snakeLeft.getY()+200) ||(food.getX()>width-80
          &&food.getY()>snakeRight.getY() &&food.getY()<snakeRight.getY()+200))
    {
      foodX *= -1;
    }

    //move the food
    food.setBounds(food.getX()-foodX, food.getY()-foodY,30,30);
    //this code controls the AI. If the food is higher
    //than the computer's paddle, move up. If not, move down.
    if(food.getY()>snakeRight.getY()+200)
    {
      snakeRight.setBounds(snakeRight.getX(),snakeRight.getY()+4,30,200);
    }
    if(food.getY()<snakeRight.getY())
    {
      snakeRight.setBounds(snakeRight.getX(),snakeRight.getY()-4,30,200);
    }
      //the refresh delay
      Thread.sleep(4);
```

```
      }
      catch(Exception e){ }
    }
  }

  //you must also implement this method from KeyListener
  public void keyPressed(KeyEvent event)
  {
    if(event.getKeyChar()=='w')
    {
      snakeLeft.setBounds(snakeLeft.getX(),snakeLeft.getY()-10,30,200);
    }
    if(event.getKeyChar()=='s')
    {
      snakeLeft.setBounds(snakeLeft.getX(),snakeLeft.getY()+10,30,200);
    }
    if(event.getKeyChar()=='q')
    {
      bulletsLeft.add(venomLeft);
      cont.add(venomLeft);
      cont.setComponentZOrder(venomLeft,0);
      venomLeft.setBounds(snakeLeft.getX()+30,snakeLeft.getY()+30,20,10);
    }
  }

  //you must also implement this method from KeyListener
  public void keyReleased(KeyEvent event){ }

  //you must also implement this method from KeyListener
  public void keyTyped(KeyEvent event){ }

    }
  public static void main (String[ ] args)

  {
    new SnakePit();
  }
}
```

Figures 29-1 through 29-3 illustrate the vicious snake fight.

Customizing the game

Vary the speed of the snakes and mouse: faster, slower, or randomly changing.

Give the snake the ability to spit venom like a machine gun.

Add two more snakes to the top and bottom of the arena.

Add another mouse—double food!

Substitute a mongoose for the mouse—if it collides with the snake, the snake dies!

Create power-ups: freeze your opponent for several seconds.

Dual-wielding: shoot several venomous shots at once.

Change the snakes into scorpions and have them shoot stingers instead of venom.

Figure 29-1 *Creative background image.*

Figure 29-2 *Snake recoils from venom spit.*

Figure 29-3 *Player is winning!*

Project 30: Space Destroyers—The Landscape

Space Destroyers

Planet Earth is being invaded. You will design and command a military spaceship to destroy the aliens. Your arsenal of weapons includes plasma machine guns, health packs, and reflector beams. Save our planet!

Project

Create the landscape and the spacecraft.

New building blocks

MouseMotionListener

MouseMotionListener

The MouseMotionListener does exactly what is says: it listens for the mouse's movements. Just like the ActionListener and the KeyListener, the MouseMotionListener has two mandatory methods:

```
public void mouseMoved(MouseEvent event)
```

and

```
public void mouseDragged(MouseEvent event)
```

To get the X position of the cursor, use:

```
event.getX()
```

To get the Y position, use:

```
event.getY()
```

Making the game

Start making the game by creating a JFrame that is 500 by 700 pixels. Set the background color to black. For the military spaceship, create two ImageIcons: one of the basic spaceship; the other of the spaceship damaged. Then, create a JLabel and set the ImageIcon of the basic spaceship to it.

Add a MouseMotionListener to the class. In mouseMoved, use the spaceship's setBounds method to move the spaceship to the location of the cursor.

```
import javax.swing.*;
import javax.swing.event.*;
import java.awt.*;
import java.awt.event.*;
import java.util.*;
public class SpaceDestroyers extends JFrame implements MouseMotionListener
{
  //this holds the components
  Container cont;

  //these are the ship's images
  ImageIcon shipImg = new ImageIcon("ship.PNG");
  ImageIcon shipHit = new ImageIcon("shipHit.PNG");

  //this is the player's ship
  JLabel ship = new JLabel(shipImg);
```

Figure 30-1 *Spaceship ready for action!*

```
public SpaceDestroyers()
{
  super("Space Destroyers");
  setVisible(true);
  setDefaultCloseOperation(JFrame.EXIT_ON_CLOSE);
  setSize(500,700);

  cont = getContentPane();
  cont.setLayout(null);

  //set the background color
    cont.setBackground(Color.BLACK);
    cont.add(ship);
    ship.setBounds(225, 550,50,50);

    addMouseMotionListener(this);

    setContentPane(cont);
    }
  public void mouseMoved(MouseEvent event)
  {
    ship.setBounds(event.getX()-25,event.getY()-40,50,50);
  }
  public void mouseDragged(MouseEvent event){ }
  public static void main (String[ ] args)
  {
  new SpaceDestroyers();
  }
}
```

Figure 30-2 *Spaceship follows cursor.*

Figures 30-1 and 30-2 illustrate the spaceship and its initial movement.

Move on to add weapons to your spaceship. But what's the point of weapons if there is no target? That's why you will also be creating those threatening aliens.

Project 31: Space Destroyers—Lasers

Project

Who better than an Evil Genius to create evil aliens? But first, you will add lasers to your spaceship to give mankind a fighting chance.

Making the game

Create an ImageIcon and JLabel for the ship's lasers. Go on and make an ArrayList to store all of the lasers shot by the spaceship (you have already practiced this technique in Project 28, the Snake Pit game). Once the ArrayList is complete, add a KeyListener to determine when to fire the lasers. When the space bar is pressed, add the laser to both the content pane and the ArrayList. After creating a thread and an infinite loop, use a `for loop` and the setBounds method to shoot the lasers upward.

Now it's time to create the aliens. Make two ImageIcons: one of a basic alien (see Figure 31-1) and one of a damaged alien (see Figure 31-2).

Add the basic alien image to a JLabel. Next, create two variables: `level` and `numOfEnemies`. The number of enemies should correspond to the square of each level (e.g. the second level should have 2^2, 4 enemies; the third, 3^2, 9 enemies). In the constructor, use a loop to initialize the enemies. In the infinite while loop, move the aliens downward. Now, make an ArrayList to hold the aliens. Create an if-statement to check the number of enemies still viable. If it is 0, increment `level` by 1 and re-add the aliens to the screen and the ArrayList. To make life easier, you can also create a method that populates the ArrayList. Also, don't forget to have the aliens reappear on the top of the screen after disappearing offscreen!

Important: check for collisions between the ship's lasers and the aliens in the infinite loop. If a collision occurs, remove the aliens and the laser. And remember to set the alien's icon to the damaged image before removing it.

Figure 31-1 *Basic alien.*

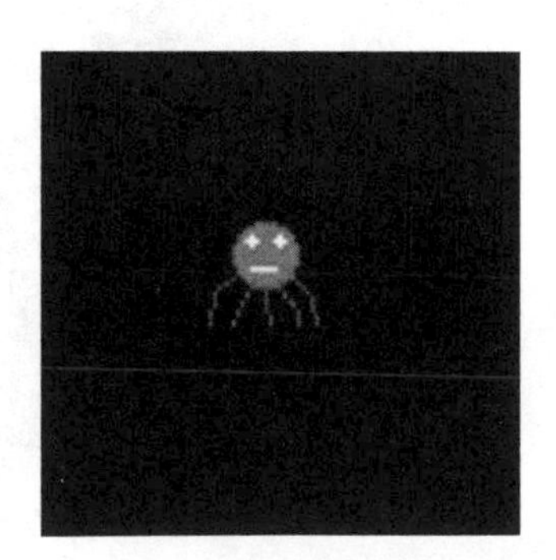

Figure 31-2 *Damaged alien.*

```
import javax.swing.*;
import javax.swing.event.*;
import java.awt.*;
import java.awt.event.*;
import java.util.*;

public class SpaceDestroyers extends JFrame implements KeyListener,MouseMotionListener
{
  //this holds the components
  Container cont;

  //this is the current level:
  int currentLevel = 1;

  //this is the number of enemies:
  int numOfEnemies = 1;

  //this is the bullet's image:
  ImageIcon shipBullet = new ImageIcon ("shipBullet.PNG");

  //this holds the player's bullets
  ArrayList playerBullets = new ArrayList();

  //this holds the enemies
  ArrayList enemies = new ArrayList();

  //these are the ship's images
  ImageIcon shipImg = new ImageIcon("ship.PNG");
  ImageIcon shipHit = new ImageIcon("shipHit.PNG");

  //these are the images of the enemies
  ImageIcon enemyImg = new ImageIcon ("enemy.PNG");
  ImageIcon enemyHit = new ImageIcon ("enemyHit.PNG");

  //this is the player's ship
  JLabel ship = new JLabel(shipImg);

  public SpaceDestroyers()
  {
    super("Space Destroyers");
    setVisible(true);
    setDefaultCloseOperation(JFrame.EXIT_ON_CLOSE);
    setSize(500,700);

    cont = getContentPane();
    cont.setLayout(null);

    //set the background color
    cont.setBackground(Color.BLACK);

    cont.add(ship);
    ship.setBounds(225, 550,50,50);

    addKeyListener(this);
    addMouseMotionListener(this);
    populateEnemies();

    Play play= new Play();
    play.start();

    setContentPane(cont);
    }
  public void populateEnemies()
  {
    for(int i = 0; i <= numOfEnemies; i++)
    {
      JLabel tempEnemy = new JLabel(enemyImg);
      int randLocation = (int)(Math.random()*500);
      enemies.add(tempEnemy);
```

```
      cont.add((JLabel)(enemies.get(i)));
      tempEnemy.setBounds(randLocation,10,30,30);
      cont.setComponentZOrder(((JLabel)(enemies.get(i))),0);
    }
  }
  public class Play extends Thread
  {
    public void run()
    {
      while(true)
      {
        try
        {
          for(int i = 0; i <enemies.size(); i++)
          {
            JLabel tempEnemy = (JLabel)(enemies.get(i));
            int distance = (int)(Math.random()*2);
            tempEnemy.setBounds(tempEnemy.getX(),tempEnemy.getY()+distance,30,30);
            if(tempEnemy.getBounds().intersects(ship.getBounds()))
            {
              cont.remove(tempEnemy);
            }
            if(tempEnemy.getY()>550)tempEnemy.setBounds(tempEnemy.getX(), 10, 30, 30);
          }

          //chack if the player's bullets hit the aliens
          boolean breakAll = false;
          for(int i = 0; i <playerBullets.size(); i++)
          {
            JLabel temp = ((JLabel)(playerBullets.get(i)));
            temp.setBounds(temp.getX(),temp.getY()-8,10,20);

            if(temp.getY()<0)
            {
            cont.remove(temp);
            playerBullets.remove(i);
            i--;
            }
            for(int j = 0; j < enemies.size(); j++)
            {
              JLabel tempEnemy = (JLabel)(enemies.get(j));
              if(temp.getBounds().intersects(tempEnemy.getBounds()))
              {
              tempEnemy.setIcon(enemyHit);
              Thread.sleep(100);
              enemies.remove(j);
              cont.remove(tempEnemy);
              numOfEnemies—;
              if(numOfEnemies<=0)
              {
                currentLevel++;
                numOfEnemies = currentLevel * currentLevel;
                populateEnemies();
                breakAll = true;
                break;
              }
            }
          }
          if(breakAll)
          break;
        }
```

```
        cont.repaint();
        Tread.sleep(10);
      }
      catch(Exception e){ }
    }
  }
}

public void mouseMoved(MouseEvent event)
{
  ship.setBounds(event.getX()-25,event.getY()-40,50,50);
}

public void mouseDragged(MouseEvent event){ }

public void keyPressed(KeyEvent event)
{
    if(event.getKeyChar()==' ')
    {
      JLabel tempBullet = new JLabel(shipBullet);
      tempBullet.setBounds(ship.getX()+20,ship.getY()-20,10,20);
      playerBullets.add(tempBullet);
      cont.add((JLabel)(playerBullets.get(playerBullets.size()-1)));
      cont.setComponentZOrder((JLabel)(playerBullets.get(playerBullets.size()-1)),0);
    }
  }
```

Figure 31-3 *Aliens are coming!*

Figure 31-4 *Aliens defeated. Second wave arrives.*

```
  public void keyReleased(KeyEvent event) { }
  public void keyTyped(KeyEvent event) { }
  public static void main (String[ ] args)
  {
    new SpaceDestroyers();
  }
}
```

Figures 31-3 through 35-6 show the player defending the planet from the aliens.

The aliens are now coming at you in droves. But it won't be an epic battle unless they can return fire. Go on to the next project to learn how.

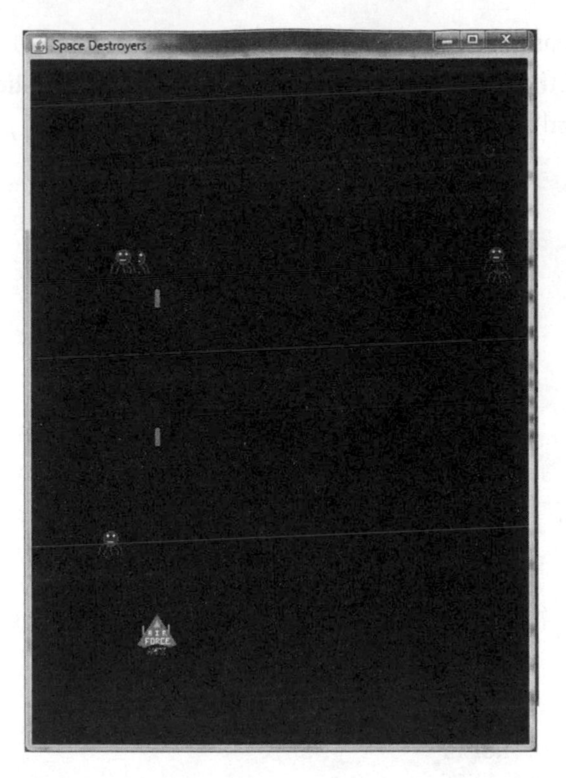

Figure 31-5 *Open fire!!!*

Figure 31-6 *Uh oh ... aliens spawn!*

Project 32: Space Destroyers—Retaliation

Project

The fight intensifies! Give the aliens firepower! Create power-ups that appear every few rounds.

Making the game

To arm the aliens, start by creating a random variable between 0 and 2500. If the value of that variable is less than or equal to 1, the alien fires its laser. Create an ArrayList and use the same code that triggers the spaceship to fire. In the infinite loop, check for collisions between the alien lasers and the spaceship. If they collide, change the icon of the basic spaceship to the damaged spaceship icon.

Next, create two variables. One represents the score and the other represents the health of the spaceship. When the spaceship is attacked, subtract from its health. Add to the score when an alien is destroyed.

Now for the power-ups. There are two types, reflector beams and health packs. Reflector beams randomize the aliens' particles, making them temporarily disappear. Health packs add energy to the spaceship. Every 3 rounds, a power-up appears. Every 5 rounds, a health pack appears. At the appropriate round, add the power-up to the top of the screen and make it descend so the spaceship can capture it.

If the ship collects a health pack, remove the health pack and increase the spaceship's health.

If the ship collides with a reflector beam, remove the reflector beam icon and add a new JLabel that covers the entire width of the screen. Make this beam move upward using the setBounds method. If the beam collides with an alien, remove the alien and add points to the player's score.

```
import javax.swing.*;
import javax.swing.event.*;
import java.awt.*;
import java.awt.event.*;
import java.util.*;
public class SpaceDestroyers extends JFrame implements KeyListener,MouseMotionListener
{
  //this holds the components
  Container cont;

  //this is the current level:
  int currentLevel = 1;

  //this is the number of enemies:
  int numOfEnemies = 1;

  //this is the bullet's image:
  ImageIcon shipBullet = new ImageIcon("shipBullet.PNG");
  ImageIcon enemyBullet = new ImageIcon("enemyBullet.PNG");

  //this holds the player's bullets
  ArrayList playerBullets = new ArrayList();

  //this holds the enemies
  ArrayList enemies = new ArrayList();

  //this holds the bullets of the enemies
  ArrayList enemyBullets = new ArrayList();

  //these are the ship's images
  ImageIcon shipImg = new ImageIcon("ship.PNG");
  ImageIcon shipHit = new ImageIcon("shipHit.PNG");

  //these are the images of the enemies
  ImageIcon enemyImg = new ImageIcon("enemy.PNG");
  ImageIcon enemyHit = new ImageIcon("enemyHit.PNG");

  //this is the powerup
  JLabel powerup = new JLabel(new ImageIcon("powerup.PNG"));
  //this is the player's ship
  JLabel ship = new JLabel(shipImg);

  //the attack from a powerup
  JLabel powerAttack = new JLabel(new ImageIcon("powerAttack.PNG"));

  //this boolean keeps track of whether the power attack was used
  boolean useAttack = false;

  //the health
  JLabel healthPack = new JLabel(new ImageIcon("health.PNG"));

  //the score:
  int score = 0;

  //health
  int health = 500;

  //the final score:
  int finalScore = 0;
```

```
public SpaceDestroyers()
{
  super("Space Destroyers");
  setVisible(true);
  setDefaultCloseOperation(JFrame.EXIT_ON_CLOSE);
  setSize(500,700);

  cont = getContentPane();
  cont.setLayout(null);

  //set the background color
  cont.setBackground(Color.BLACK);

  cont.add(ship);
  ship.setBounds(225, 550,50,50);

  addKeyListener(this);
  addMouseMotionListener(this);

  populateEnemies();

  Play play= new Play();
  play.start();

  setContentPane(cont);
}
public void title()
{
  try
  {
    JLabel title = new JLabel(new ImageIcon("title.PNG"));
    cont.add(title);
    title.setBounds(-60,-300,600,200);
    do
    {
      title.setBounds(title.getX(),title.getY()+1,600,200);
      Thread.sleep(3);
    }
    while(title.getY()<700);
  }
  catch(Exception e){ }
}
public void populateEnemies()
{
  for(int i = 0; i <= numOfEnemies; i++)
  {
    JLabel tempEnemy = new JLabel(enemyImg);
    int randLocation = (int)(Math.random()*500);
    enemies.add(tempEnemy);
    cont.add((JLabel)(enemies.get(i)));
    tempEnemy.setBounds(randLocation,10,30,30);
    cont.setComponentZOrder(((JLabel)(enemies.get(i))),0);
  }
}
public class Play extends Thread
{
  public void run()
  {
    while(true)
    {
      try
      {
        for(int i = 0; i < enemies.size(); i++)
        {
```

```
            JLabel tempEnemy = (JLabel)(enemies.get(i));
            int distance = (int)(Math.random()*2);
            tempEnemy.setBounds(tempEnemy.getX(), tempEnemy.getY()+distance,30,30);
            //check if the power attack hit the aliens
            if(useAttack)
            {
              if(powerAttack.getBounds().intersects(tempEnemy.getBounds()))
              {
              cont.remove(tempEnemy);
              i- -;
              numOfEnemies- -;
              score+=15;
            }
          }
          if(tempEnemy.getBounds().intersects(ship.getBounds()))
          {
            health- -;
            cont.remove(tempEnemy);
          }
          if(tempEnemy.getY()>550)tempEnemy.setBounds(tempEnemy.getX(), 10, 30, 30);
          int fire = (int)(Math.random()*2500);
          if(fire<=1)
          {
        JLabel tempBullet = new JLabel(enemyBullet);
        tempBullet.setBounds(tempEnemy.getX()+5,tempEnemy.getY()+30,10,20);
        enemyBullets.add(tempBullet);
        cont.add((JLabel)(enemyBullets.get(enemyBullets.size()-1)));
        cont.setComponentZOrder((JLabel)(enemyBullets.get(enemyBullets.size()-1)),0);
      }
    }
    //chack if the player's bullets hit the aliens
    boolean breakAll = false;
    for(int i = 0; i <playerBullets.size(); i++)
    {
      JLabel temp = ((JLabel)(playerBullets.get(i)));
        temp.setBounds(temp.getX(),temp.getY()-8,10,20);
      if(temp.getY()<0)
      {
        cont.remove(temp);
        playerBullets.remove(i);
        i--;
      }
      for(int j = 0; j < enemies.size(); j++)
      {
        JLabel tempEnemy = (JLabel)(enemies.get(j));
        if(temp.getBounds().intersects(tempEnemy.getBounds()))
        {
          score+=1000;
          tempEnemy.setIcon(enemyHit);
          Thread.sleep(100);
          enemies.remove(j);
          cont.remove(tempEnemy);
            numOfEnemies–;
          if(numOfEnemies<=0)
          {
            currentLevel++;

            if(currentLevel%3==0)
            {
              cont.add(powerup);
              int randLoc = (int)(Math.random()*450);
```

```
          powerup.setBounds(randLoc,0,30,30);
        }
        if(currentLevel%5==0)
        {
          cont.add(healthPack);
          int randLoc = (int)(Math.random()*450);
          healthPack.setBounds(randLoc,0,30,30);
        }
        numOfEnemies = currentLevel * currentLevel;
        populateEnemies();
        breakAll = true;
        break;
      }
    }
  }
  if(breakAll)break;
}
//move the power attack
if(useAttack)
{
  powerAttack.setBounds(0,powerAttack.getY()-1,500,10);
  if(powerAttack.getY()<0)
  {
    cont.remove(powerAttack);
    useAttack = false;
    currentLevel++;
    numOfEnemies = currentLevel * currentLevel;
    populateEnemies();
  }
}
//if it is every third round, allow the
//powerup to be moved
if(currentLevel%3==0)
{
  powerup.setBounds(powerup.getX(),powerup.getY()+1,30,30);
  if(powerup.getBounds().intersects(ship.getBounds()))
  {
    useAttack = true;
    cont.add(powerAttack);
    powerAttack.setBounds(0,ship.getY(),500,10);
    cont.remove(powerup);
    powerup.setBounds(-200,-200,30,30);
  }
}
if(currentLevel%5==0)
{
  healthPack.setBounds(healthPack.getX(),healthPack.getY()+1,30,30);
  if(healthPack.getBounds().intersects(ship.getBounds()))
  {
    health+=50;
    score+=100;
    cont.remove(healthPack);
    healthPack.setBounds(-100,-100,30,30);
  }
}
  //check if the aliens' bullets hit the player
  breakAll = false;
  for(int i = 0; i <enemyBullets.size(); i++)
```

```
    {
      JLabel temp = ((JLabel)(enemyBullets. get(i)));
          temp.setBounds(temp.getX(),temp.getY()+2,10,20);
      if(temp.getY()>550)
      {
        cont.remove(temp);
        enemyBullets.remove(i);
        i—;
      }
      if(ship.getBounds().intersects(temp.getBounds()))
      {
        ship.setIcon(shipHit);
        Thread.sleep(100);
        ship.setIcon(shipImg);
        score-=100;
        health-=50;
        cont.remove(temp);
        enemyBullets.remove(i);
            numOfEnemies—;
            if(numOfEnemies<=0)
            {
              currentLevel++;
              numOfEnemies = currentLevel * currentLevel;
              populateEnemies();
              breakAll = true;
              break;
            }
          }
          if(breakAll)
          break;
        }
        cont.repaint();
        Thread.sleep(10);
      }
      catch(Exception e){ }
    }
  }
}
public void mouseMoved(MouseEvent event)
{
  ship.setBounds(event.getX()-25,event.getY()-40,50,50);
}
public void mouseDragged(MouseEvent event){ }
public void keyPressed(KeyEvent event)
{
  if(event.getKeyChar()==' ')
  {
    JLabel tempBullet = new JLabel(shipBullet);
    tempBullet.setBounds(ship.getX()+20,ship.getY()-20,10,20);
    playerBullets.add(tempBullet);
    cont.add((JLabel)(playerBullets.get(playerBullets.size()-1)));
    cont.setComponentZOrder(
        (JLabel)(playerBullets.get(
        playerBullets.size()-1)),0);
    score-=2;
  }
}
public void keyReleased(KeyEvent event) { }
```

```
  public void keyTyped(KeyEvent event) { }
  public static void main (String[ ] args)
  {
    new SpaceDestroyers();
  }
}
```

Figures 32-1 through 32-4 show the fierce battle for Earth.

You need as much information as possible to out-maneuver your enemies. Even though the health packs look cool, they mean nothing because the player's statistics are not displayed! In the next section, you will add this information to make the game more exciting.

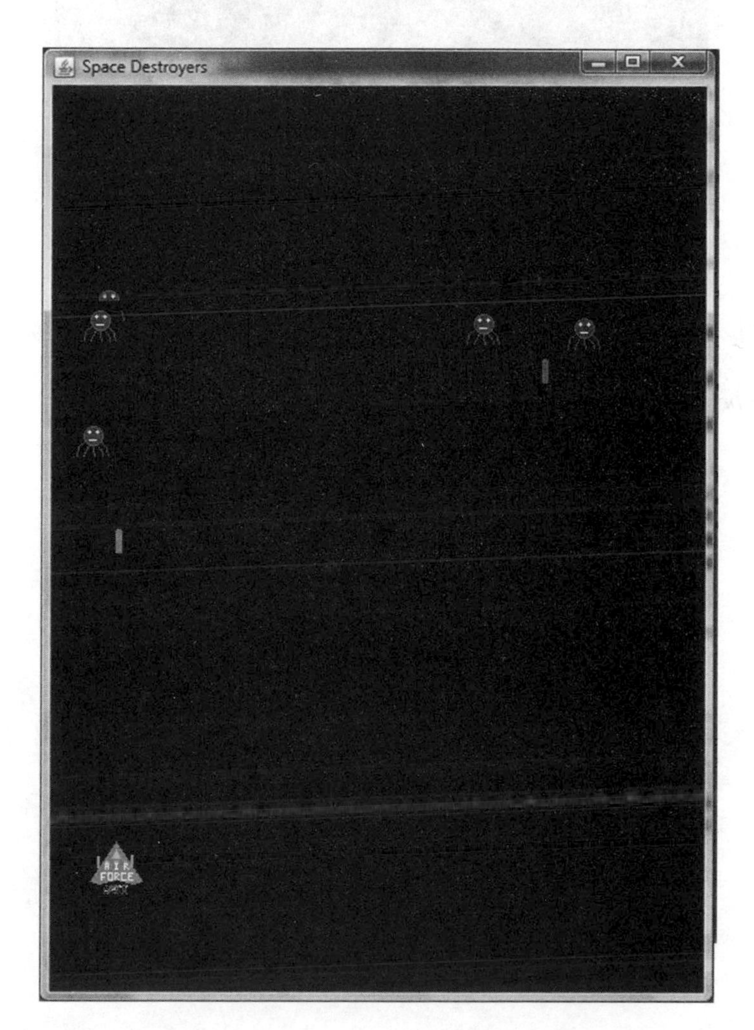

Figure 32-1 *Aliens retaliate.*

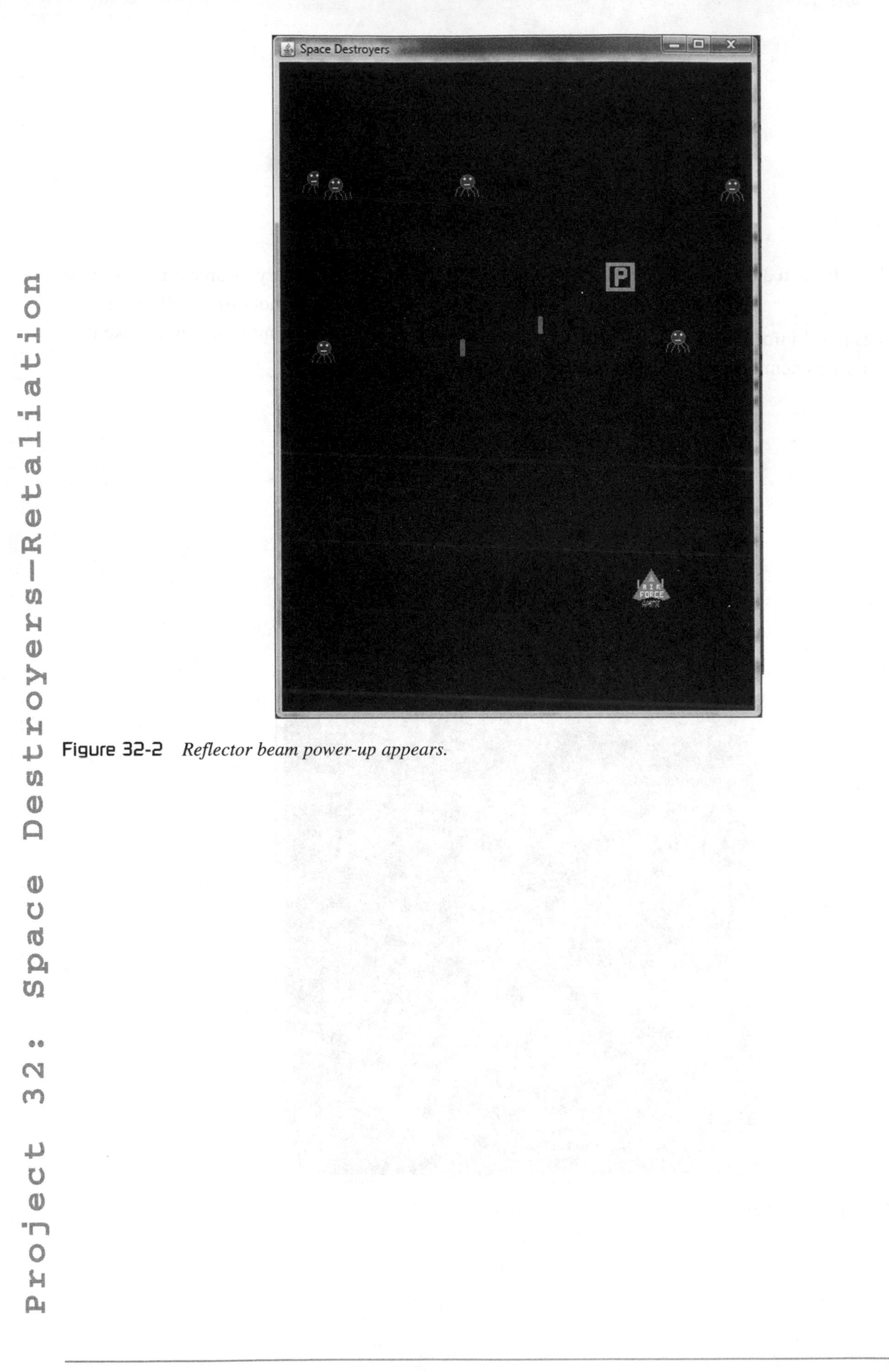

Figure 32-2 *Reflector beam power-up appears.*

Figure 32-3 *Reflector beam released.*

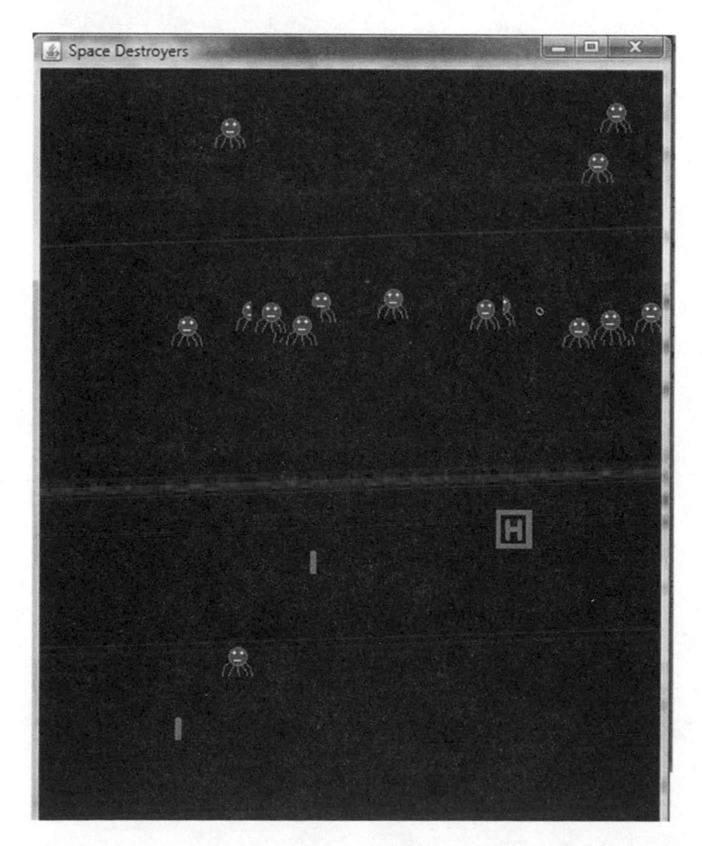

Figure 32-4 *Health pack appears.*

Project 33: Space Destroyers—Life and Death

Project

Display the score and health data. And for even more visual excitement, build scrolling title and end screens. Lock, load, and fire!

Making the game

First, create a JLabel that will display the current level, score, and health of the spaceship. To enhance visibility of the display, increase the size of the font and change its color to white.

Remember to update the JLabel after every iteration of the infinite loop!

Next, create an image that says "Space Destroyers' in a bold font. Attach that image to a JLabel. To begin the game, scroll the title image across the screen. You can do this either in a method called from the constructor or directly in the constructor.

Next, the results. To display the score, create a JLabel. Use setBounds to make this JLabel scroll across the screen. You can do this either in a separate thread or a separate method.

```
import javax.swing.*;
import javax.swing.event.*;
import java.awt.*;
import java.awt.event.*;
import java.util.*;

public class SpaceDestroyers extends JFrame implements KeyListener, MouseMotionListener
{
  //this holds the components
  Container cont;

  //this is the current level:
  int currentLevel = 1;

  //this is the number of enemies:
  int numOfEnemies = 1;

  //this is the bullet's image:
  ImageIcon shipBullet = new ImageIcon("shipBullet.PNG");
  ImageIcon enemyBullet = new ImageIcon("enemyBullet.PNG");

  //this holds the player's bullets
  ArrayList playerBullets = new ArrayList();

  //this holds the enemies
  ArrayList enemies = new ArrayList();

  //this holds the bullets of the enemies
  ArrayList enemyBullets = new ArrayList();

  //these are the ship's images
  ImageIcon shipImg = new ImageIcon("ship.PNG");
  ImageIcon shipHit = new ImageIcon("shipHit.PNG");

  //these are the images of the enemies
  ImageIcon enemyImg = new ImageIcon("enemy.PNG");
  ImageIcon enemyHit = new ImageIcon("enemyHit.PNG");

  //this is the powerup
```

```
JLabel powerup = new JLabel(new ImageIcon("powerup.PNG"));

//this is the player's ship
JLabel ship = new JLabel(shipImg);

//the attack from a powerup
JLabel powerAttack = new JLabel(new ImageIcon("powerAttack.PNG"));

//this boolean keeps track of whether the power attack was used
boolean useAttack = false;

//the health
JLabel healthPack = new JLabel(new ImageIcon("health.PNG"));

//the score:
int score = 0;

//health
int health = 500;

//the final score:
int finalScore = 0;

public SpaceDestroyers()
{
  super("Space Destroyers");
  setVisible(true);
  setDefaultCloseOperation(JFrame.EXIT_ON_CLOSE);
  setSize(500,700);

  cont = getContentPane();
  cont.setLayout(null);

  //set the background color
  cont.setBackground(Color.BLACK);
  cont.add(ship);
  ship.setBounds(225, 550,50,50);

  addKeyListener(this);
  addMouseMotionListener(this);

  populateEnemies();

  Play play= new Play();
  play.start();

  setContentPane(cont);
}
public void title()
{
  try
  {
    JLabel title = new JLabel(new ImageIcon("title.PNG"));
    cont.add(title);
    title.setBounds(-60,-300,600,200);
    do
    {
      title.setBounds(title.getX(),title.getY()+1,600,200);
      Thread.sleep(3);
    }
    while(title.getY()<700);
  }
  catch(Exception e){ }
}
public void populateEnemies()
{
  for(int i = 0; i <= numOfEnemies; i++)
  {
```

```
      JLabel tempEnemy = new JLabel(enemyImg);
      int randLocation = (int)(Math.random()*500);
      enemies.add(tempEnemy);
      cont.add((JLabel)(enemies.get(i)));
      tempEnemy.setBounds(randLocation,10,30,30);
      cont.setComponentZOrder(((JLabel)(enemies.get(i))),0);
    }
  }
  public class Play extends Thread
  {
    public void run()
    {
      while(true)
      {
        try
        {
          for(int i = 0; i <enemies.size(); i++)
          {
            JLabel tempEnemy = (JLabel)(enemies.get(i));
            int distance = (int)(Math.random()*2);
            tempEnemy.setBounds(tempEnemy.getX(),
            tempEnemy.getY()+distance,30,30);
            //check if the power attack hit the aliens
            if(useAttack)
          {
              if(powerAttack.getBounds().intersects(tempEnemy.getBounds()))
              {
                cont.remove(tempEnemy);
                i- -;
                numOfEnemies- -;
                score+=15;
              }
          }
          if(tempEnemy.getBounds().intersects(ship.getBounds()))
          {

                health—;
                cont.remove(tempEnemy);
              }
              if(tempEnemy.getY()>550)
              tempEnemy.setBounds(tempEnemy.getX(), 10, 30, 30);
              int fire = (int)(Math.random()*2500);
              if(fire<=1)
              {
                JLabel tempBullet = new JLabel(enemyBullet);
                tempBullet.setBounds(tempEnemy.getX()+5,
                tempEnemy.getY()+30,10,20);
                enemyBullets.add(tempBullet);
                cont.add((JLabel)(enemyBullets.get(enemyBullets.size()-1)));
                cont.setComponentZOrder((JLabel)(enemyBullets.get (enemyBullets.size()-
                1)),0);
              }
          }

          //chack if the player's bullets hit the aliens
          boolean breakAll = false;
          for(int i = 0; i < playerBullets.size(); i++)
          {
              JLabel temp = ((JLabel)(playerBullets.get(i)));
              temp.setBounds(temp.getX(),temp.getY()-8,10,20);

              if(temp.getY()<0)
```

```
    {
      cont.remove(temp);
      playerBullets.remove(i);
      i--;
    }
    for(int j = 0; j < enemies.size(); j++)
    {
      JLabel tempEnemy = (JLabel)  (enemies.get(j));
      if(temp.getBounds().intersects  (tempEnemy.getBounds()))
      {
        score+=1000;
        tempEnemy.setIcon(enemyHit);
        Thread.sleep(100);
        enemies.remove(j);
        cont.remove(tempEnemy);
        numOfEnemies- -;
        if(numOfEnemies<=0)
        {

          currentLevel++;

          if(currentLevel%3==0)
          {
            cont.add(powerup);
            int randLoc = (int)(Math.random()*450);
            powerup.setBounds(randLoc,0,30,30);
          }
          if(currentLevel%5==0)
          {
            cont.add(healthPack);
            int randLoc = (int)(Math.random()*450);
            healthPack.setBounds(randLoc,0,30,30);
          }
          numOfEnemies = currentLevel * currentLevel;
          populateEnemies();
          breakAll = true;
          break;
        }
      }
    }
    if(breakAll)
    break;
  }

  //move the power attack
  if(useAttack)
  {
    powerAttack.setBounds(0,powerAttack.getY()-1,500,10);
    if(powerAttack.getY()<0)
    {
      cont.remove(powerAttack);
      useAttack = false;
      currentLevel++;
      numOfEnemies = currentLevel * currentLevel;
      populateEnemies();
    }
  }

  //if it is every third round, allow the
  //powerup to be moved
  if(currentLevel%3==0)
  {
```

```
        powerup.setBounds(powerup.getX(),powerup.getY()+1,30,30);
        if(powerup.getBounds().intersects(ship.getBounds()))
    {

        useAttack = true;
        cont.add(powerAttack);
        powerAttack.setBounds(0,ship.getY(),500,10);
        cont.remove(powerup);
        powerup.setBounds(-200,-200,30,30);
    }
  }

  if(currentLevel%5==0)
  {
    healthPack.setBounds(healthPack.getX(),healthPack.getY()+1,30,30);
    if(healthPack.getBounds().intersects(ship.getBounds()))
  {
      health+=50;
      score+=100;
      cont.remove(healthPack);
      healthPack.setBounds(-100,-100,30,30);
    }
  }

  //check if the aliens' bullets hit the player
  breakAll = false;
  for(int i = 0; i <enemyBullets.size(); i++)
  {
    JLabel temp = ((JLabel)(enemyBullets.get(i)));
    temp.setBounds(temp.getX(),temp.getY()+2,10,20);
    if(temp.getY()>550)
    {
      cont.remove(temp);
      enemyBullets.remove(i);
      i- -;
    }
  if(ship.getBounds().intersects(temp.getBounds()))
  {
      ship.setIcon(shipHit);
      Thread.sleep(100);
      ship.setIcon(shipImg);
      score-=100;
      health-=50;
      cont.remove(temp);
      enemyBullets.remove(i);

      numOfEnemies—;
      if(numOfEnemies<=0)
      {
      currentLevel++;
      numOfEnemies = currentLevel * currentLevel;
      populateEnemies();
      breakAll = true;
      break;
    }
  }
  if(breakAll)
  break;
 }

 cont.repaint();
 Thread.sleep(10);
}
```

```
      catch(Exception e){ }
    }
  }
}
public void mouseMoved(MouseEvent event)
{
  ship.setBounds(event.getX()-25,event.getY()-40,50,50);
}
public void mouseDragged(MouseEvent event){ }
public void keyPressed(KeyEvent event)
{
  if(event.getKeyChar()==' ')
  {
      JLabel tempBullet = new JLabel(shipBullet);
      tempBullet.setBounds(ship.getX()+20,ship.getY()-20,10,20);
      playerBullets.add(tempBullet);
      cont.add((JLabel)(playerBullets.get(playerBullets.size()-1)));
      cont.setComponentZOrder(
      (JLabel)(playerBullets.get(
      playerBullets.size()-1)),0);
      score-=2;
    }
  }
  public void keyReleased(KeyEvent event) { }
  public void keyTyped(KeyEvent event) { }
  public static void main (String[ ] args)
  {
  new SpaceDestroyers();
  }
}
```

Figures 33-1 through 33-6 illustrate the game play of Space Destroyers.

Customizing the game

Change the spaceship's initial health—make the game super hard or easy enough for n00bs.

Add more power-ups. Make ultra-laser beams that are wider and can destroy multiple aliens at once.

Add multiple spaceships: let up to four other players compete simultaneously!

Assign different powers to the aliens – some fire lasers, some fire lightning rings.

Add dual-wielding to the player's ship. Fire up to four lasers at once!

Add a special, limited, bomb-like weapon for the spaceship that shoots shrapnel in all directions.

Add extra health to the enemies: give them enough body armor to survive the first attack by the spaceship.

Give the spaceship the power to become invisible intermittently.

Figure 33-1 *Customized title.*

Figure 33-2 *Score display.*

Figure 33-3 *Results, part 1.*

Figure 33-4 *Results, part 2.*

Figure 33-5 *Results, part 3.*

Figure 33-6 *Results, part 4.*

Section Five

Strategy Games

Project 34: Bomb Diffuser—Bomb Squad Noob

Bomb Diffuser

As the newest member of the Bomb Squad, your job is to go out into the field and risk your life defusing live explosive devices. Eye-hand coordination, quick thinking ... and a bit of luck is required.

Project

Learn the inner workings of a bomb by creating an inert one.

Making the game

First, create the JFrame. Set it to 500 by 550 pixels and set the background color to gray. Next, you will need to draw the image of the bomb and detonator. To do this, open Microsoft Paint, specify the precise size of your image by clicking "Image" and then click "Attributes," as illustrated in Figure 34-1.

Choose a light gray background color for the bomb and detonator that blends in with the background of the JFrame. This way, if the game is resized, the gap in between the image and the background will not be noticed.

There are two key parts of this image: the bomb and the detonator. Start by drawing the bomb at the top portion of the screen. Use the Rectangle and Pen tools, to make things easier. Don't forget to add the detonation cord and the text "TNT!"

The next step is to add the detonator at the bottom portion of the screen. First, create a gray rectangle. Give it the appearance of being 3D by adding perspective lines. Create spaces for a clue, the time left, and the disarm code. Later, you will add components that go into these spaces.

Once you are done drawing the image of the bomb, go ahead and add it to the JFrame. Set it to position 0,0.

```
import java.awt.*;
import java.awt.event.*;
import javax.swing.*;
import javax.swing.event.*;

public class BombDiffuser extends JFrame
{
  //the bomb label
  JLabel bomb = new JLabel(new ImageIcon("bomb.PNG"));

  //the container of the components
  Container cont;
```

```
    public BombDiffuser()
    {
      super("Bomb Diffuser");
      setSize(500,550);
      setVisible(true);
      setDefaultCloseOperation(JFrame.EXIT_ON_CLOSE);

      cont = getContentPane();
      cont.setLayout(null);
      cont.setBackground(Color.gray);

      //add the background image
      cont.add(bomb);
      bomb.setBounds(0,0,500,500);
      setContentPane(cont);
    }

    public static void main (String[ ] args)
    {
      new BombDiffuser();
    }
}
```

Figure 34-4 illustrates the completed bomb.

You've graduated ... it's time to stop working on inert bombs and start diffusing live ones!

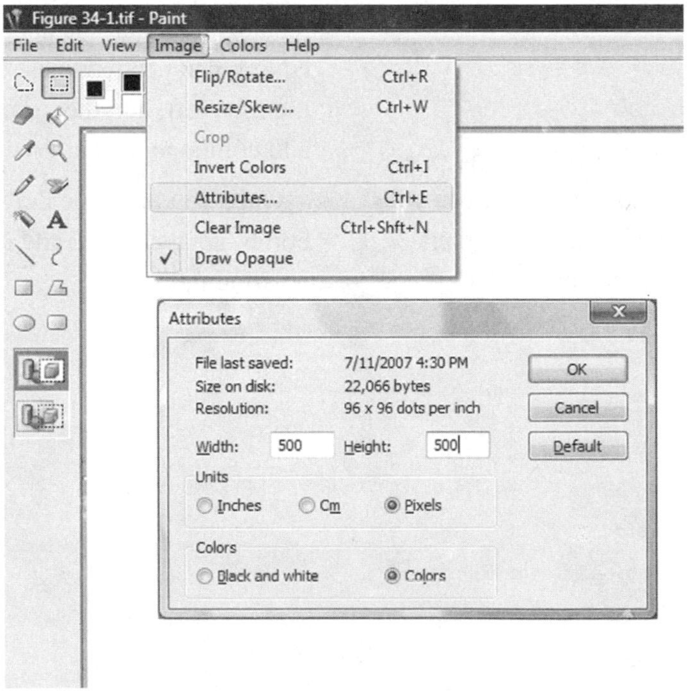

Figure 34-1 *Image size.*

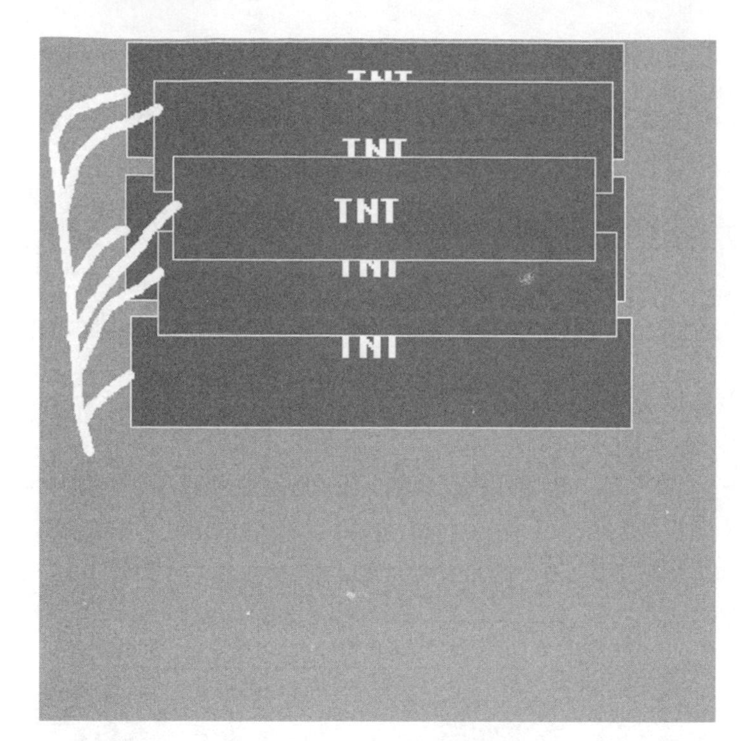

Figure 34-2 *Bomb design.*

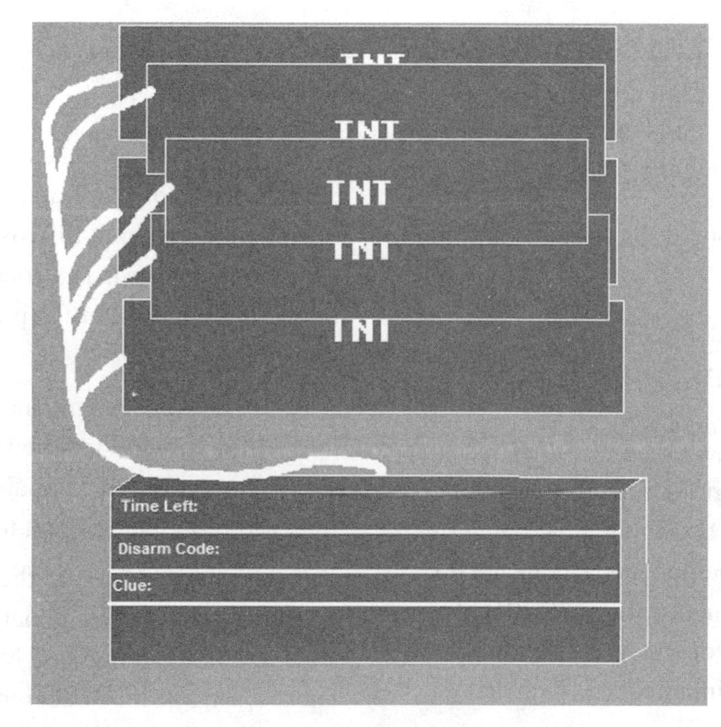

Figure 34-3 *Completed bomb before placement on the JFrame.*

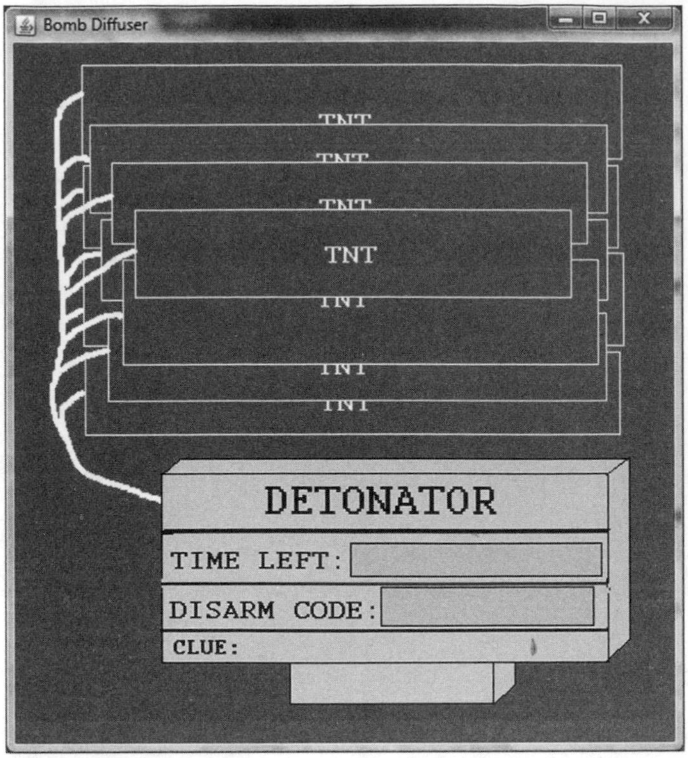

Figure 34-4 *Bomb displayed in the JFrame.*

Project 35: Bomb Diffuser—Expert Diffuser

Project

Here's where you put your skills to use to prevent detonation.

Making the game

The first part of this project consists of adding four components to the background bomb image: time, clue, disarm button, and text input area. Begin by initializing these four components. Create the text input area with a component called a TextField. Create it the same way you create a JButton or JLabel. You can get the inputted text with the following command:

```
field.getText();
```

Next, use the setBounds method to position the components. You can find the exact position by opening the image in Microsoft Paint and hovering over the target area. The coordinates are displayed on the bottom right of the program, as illustrated in Figure 35-1.

After adding the components, create a variable that will represent the time left. Use a Thread to subtract one from that variable every second. Don't forget to reset the new text to the JLabel!

Now it's time to set up the process of diffusing the bomb. First, create a method that initializes the code to a random number. Next, if the JButton is clicked, change the JLabel that represents the clue to display whether the player's guess is too high or too low.

```
import java.awt.*;
import java.awt.event.*;
import javax.swing.*;
import javax.swing.event.*;
public class BombDiffuser extends JFrame implements ActionListener
{
  //the time before detonation
  int timeLeft = 12;
  //the JLabel that displays the time left
  JLabel time = new JLabel(timeLeft+"");
  //the bomb label
  JLabel bomb = new JLabel(new ImageIcon("bomb.PNG"));
  //the clue:
  JLabel clue = new JLabel();
  //the code:
  int code = 0;
  //where the disarm code is entered:
  TextField attempt = new TextField(20);
  //the button when you think you have the correct combo
  JButton disarm = new JButton("DISARM");
  //the container of the components
  Container cont;
  //the thread
  Countdown count;
  public BombDiffuser()
  {
    super("Bomb Diffuser");
    setSize(500,550);
    setVisible(true);
    setDefaultCloseOperation(JFrame.EXIT_ON_CLOSE);
    cont = getContentPane();
    cont.setLayout(null);
    cont.setBackground(Color.gray);
    //add the background image
    cont.add(bomb);
    bomb.setBounds(0,0,500,500);
    //set the font of the JLabel
    time.setFont(new Font("Courier", Font.BOLD, 20));
    //add the time left JLabel
    cont.add(time);
    //set the position
    time.setBounds(250,371,150,18);
    //put the label on top of the background
    cont.setComponentZOrder(time,0);
    //add the disarm button
    cont.add(disarm);
    disarm.setBounds(200,456,150,30);
    disarm.addActionListener(this);
    cont.setComponentZOrder(disarm,0);
    clue.setText("A number between 0 and "+20);
    cont.add(clue);
    clue.setBounds(167,435,250,20);
    //put the label on top of the background
    cont.setComponentZOrder(clue,0);
```

```
    //add the text field
    cont.add(attempt);
    attempt.setBounds(271,404,150,25);

    //set all the components
    cont.validate();
    setContentPane(cont);

    setCode();

    count = new Countdown();
    count.start();
  }

  public void setCode()
  {
    code = (int)(Math.random()*20);
  }

  //the counting down thread:
  public class Countdown extends Thread
  {
    public void run()
    {
      while(true)
      {
        try
        {
          timeLeft--;
          time.setText(timeLeft+"");
          Thread.sleep(1000);
        }
        catch(Exception e){ }
      }
    }
  }
  public void actionPerformed(ActionEvent event)
  {
    if(Integer.parseInt(attempt.getText())>code)
    {
      clue.setText("CAUTION: Attempted Code is TOO HIGH");
    }
    if(Integer.parseInt(attempt.getText())<code)
    {
      clue.setText("CAUTION: Attempted Code is TOO LOW");
    }
    if(attempt.getText().equals(""+code))
    {
      //nothing happens ... yet!
    }
  }

  public static void main (String[ ] args)
  {
    new BombDiffuser();
  }
}
```

Figures 35-2 through 35-5 illustrate the attempt to diffuse a bomb.

Move on to the next project to learn how to add the exploding images.

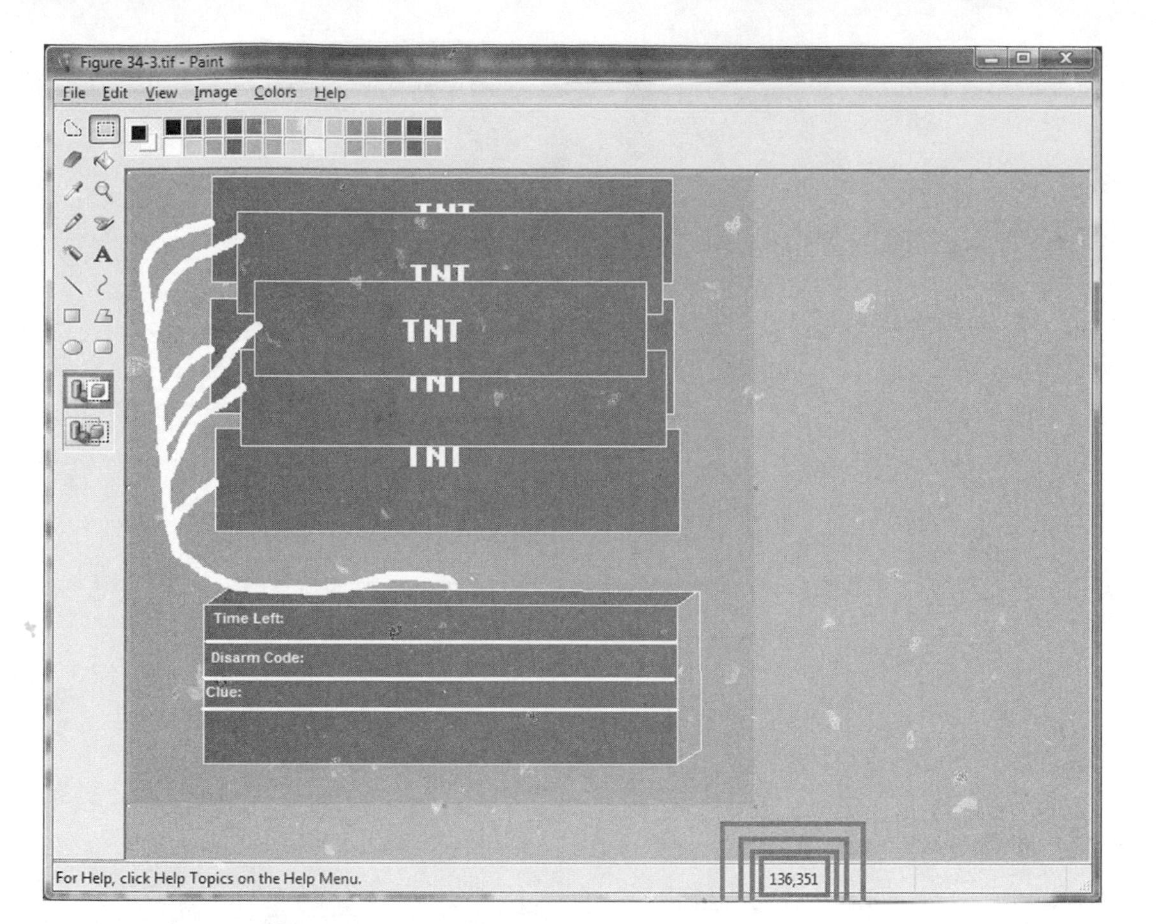

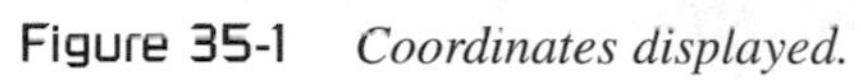
Figure 35-1 *Coordinates displayed.*

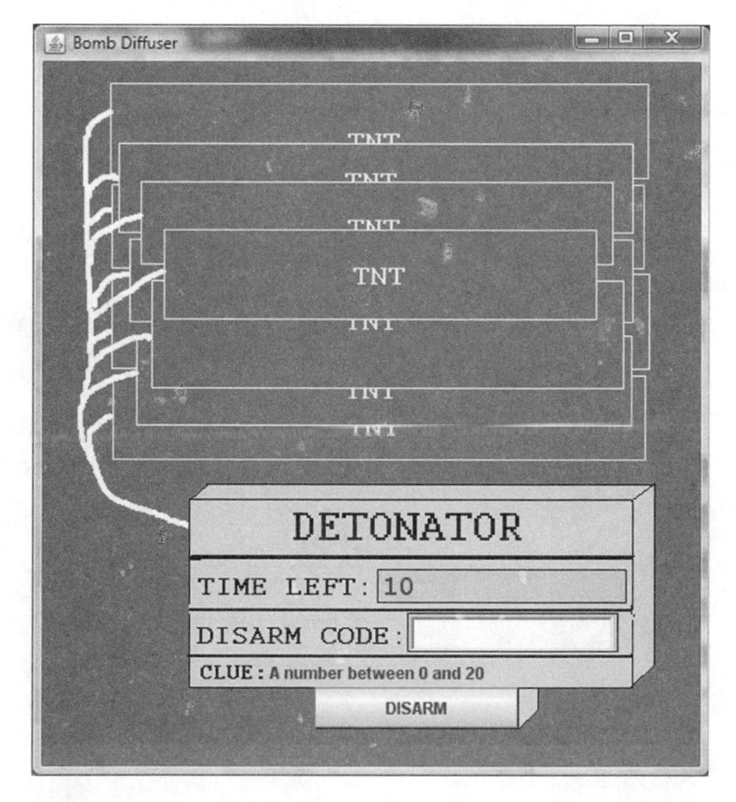

Figure 35-2 *Detonator counts down.*

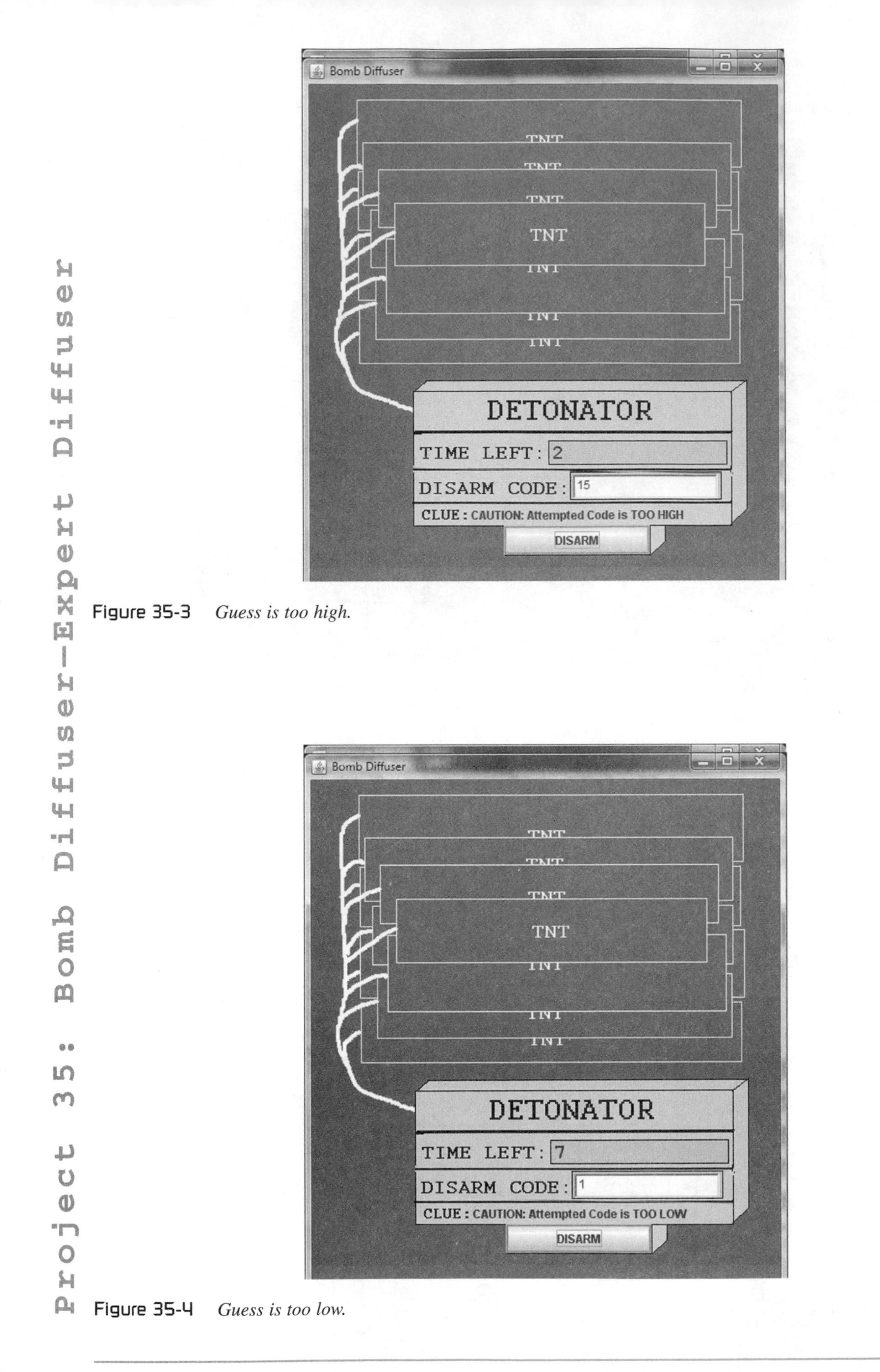

Figure 35-3 *Guess is too high.*

Figure 35-4 *Guess is too low.*

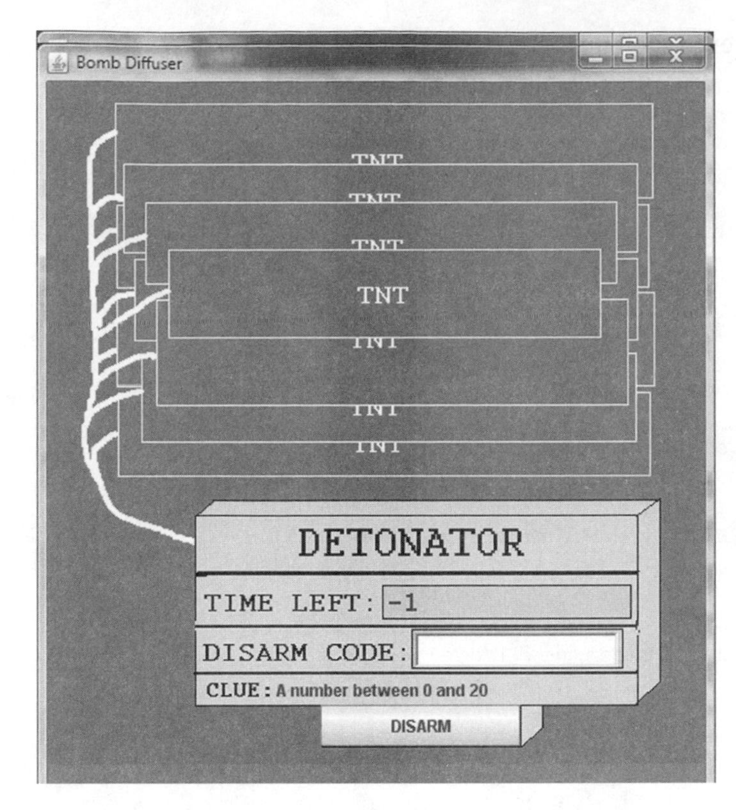

Figure 35-5 *Bomb should explode now ... but it doesn't.*

Project 36: Bomb Diffuser—Kaboom!!!

Project

Make the bomb explode. Watch out! Take cover! Run!

Making the game

To add flashing explosions, you need to create two images: the first represents a fiery explosion (see Figure 36-1); the second image is identical to the first, except the colors are inverted and the position is flipped (see Figure 36-2).

Create JLabels for each image. Then, in the countdown thread, check to see if the time left is less than or equal to zero. If so, call a method that will display the JLabels.

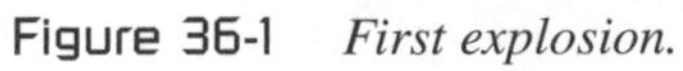
Figure 36-1 *First explosion.*

The method removes every component from the container. All that remains are the two alternating images. To alternate the images, call the setBounds method.

```
import java.awt.*;
import java.awt.event.*;
import javax.swing.*;
import javax.swing.event.*;

public class BombDiffuser extends JFrame implements ActionListener
{
  //the time before detonation
  int timeLeft = 12;
  //the JLabel that displays the time left
  JLabel time = new JLabel(timeLeft+"");
  //the bomb label
  JLabel bomb = new JLabel(new ImageIcon("bomb.PNG"));
  //the clue:
  JLabel clue = new JLabel();
  //the exploded bomb:
  JLabel exploded = new JLabel(new ImageIcon("exploded.PNG"));
  JLabel exploded2 = new JLabel(new ImageIcon("exploded2.PNG"));

  //the code:
  int code = 0;

  //where the disarm code is entered:
  TextField attempt = new TextField(20);

  //the button when you think you have the correct combo
  JButton disarm = new JButton("DISARM");

  //the container of the components
  Container cont;

  //the thread
  Countdown count;

  public BombDiffuser()
  {
    super("Bomb Diffuser");
    setSize(500,550);
    setVisible(true);
    setDefaultCloseOperation(JFrame.EXIT_ON_CLOSE);

    cont = getContentPane();
    cont.setLayout(null);
    cont.setBackground(Color.gray);

    //add the background image
    cont.add(bomb);
    bomb.setBounds(0,0,500,500);

    //set the font of the JLabel
    time.setFont(new Font("Courier", Font.BOLD, 20));
    //add the time left JLabel
    cont.add(time);
    //set the position
    time.setBounds(250,371,150,18);
    //put the label on top of the background
    cont.setComponentZOrder(time,0);

    //add the disarm button
    cont.add(disarm);
    disarm.setBounds(200,456,150,30);
```

```
    disarm.addActionListener(this);
    cont.setComponentZOrder(disarm,0);

    clue.setText("A number between 0 and 20");
    cont.add(clue);
    clue.setBounds(167,435,250,20);
    //put the label on top of the background
    cont.setComponentZOrder(clue,0);

    //add the text field
    cont.add(attempt);
    attempt.setBounds(271,404,150,25);

    //add the explosion offscreen
    cont.add(exploded);
    exploded.setBounds(-1000,-1000,500,550);
    cont.setComponentZOrder(exploded,0);
    cont.add(exploded2);
    exploded2.setBounds(-1000,-1000,500,550);
    cont.setComponentZOrder(exploded2,0);

    //set all the components
    cont.validate();
    setContentPane(cont);

    setCode();

    count = new Countdown();
    count.start();
  }

  public void setCode()
  {
    code = (int)(Math.random()*20);
  }

  //the counting down thread:
  public class Countdown extends Thread
  {
    public void run()
    {
      while(true)
      {
        try
        {
          if(timeLeft>0)
          {
            timeLeft--;
            time.setText(timeLeft+"");
          }
          else
          {
            break;
          }
          Thread.sleep(1000);
        }
        catch(Exception e){ }
      }
      //call the method that displays the blasts
      explode();
    }

    public void explode()
    {
      //first, remove everything else:
```

```
      cont.remove(time);
      cont.remove(bomb);
      cont.remove(clue);
      cont.remove(attempt);
      cont.remove(disarm);
      //move the explosion images to the correct location
      exploded.setBounds(0,0,500,550);
      exploded2.setBounds(0,0,500,550);
      while(true)
      {
        try
        {
          exploded.setBounds(0,0,500,550);
          exploded2.setBounds(-1000,-1000,500,550);
          Thread.sleep(100);
          exploded2.setBounds(0,0,500,550);
          exploded.setBounds(-1000,-1000,500,550);
          Thread.sleep(100);
        }
        catch(Exception e){ }
      }
    }
  }
  public void actionPerformed(ActionEvent event)
  {
    if(Integer.parseInt(attempt.getText())>code)
    {
      clue.setText("CAUTION: Attempted Code is TOO HIGH");
    }
    if(Integer.parseInt(attempt.getText())<code)
    {
      clue.setText("CAUTION: Attempted Code is TOO LOW");
    }
    if(attempt.getText().equals(""+code))
    {
      //this code will be added later
    }
  }
  public static void main (String[ ] args)
  {
    new BombDiffuser();
  }
}
```

Figures 36-3 through 36-5 display the steps of a bomb exploding.

Ready for more action? In the next project, add levels to your game. As you become a more experienced bomb diffuser, the codes get more challenging.

Figure 36-2 *Second explosion.*

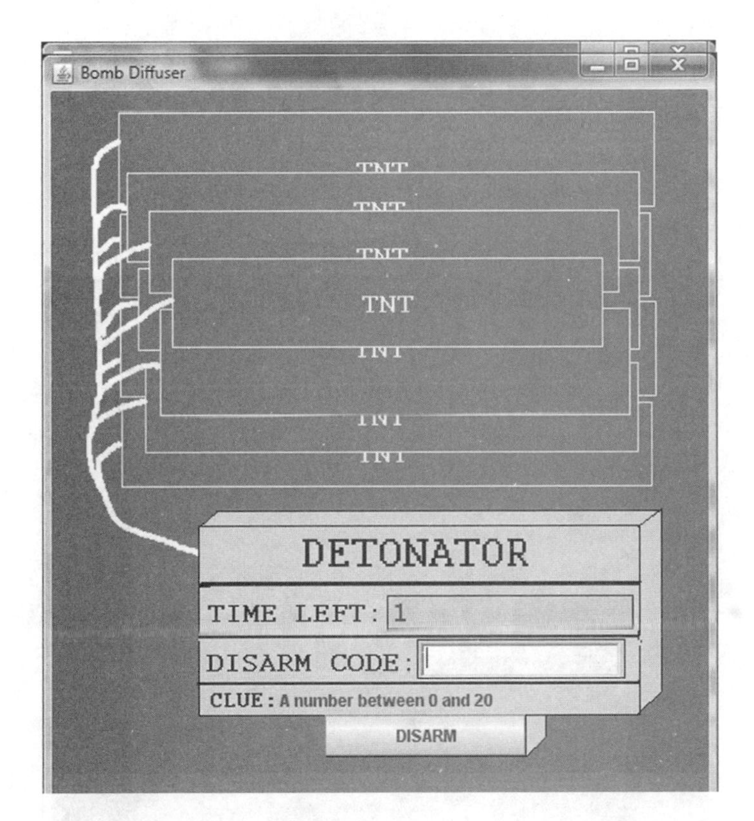

Figure 36-3 *Detonator counts down.*

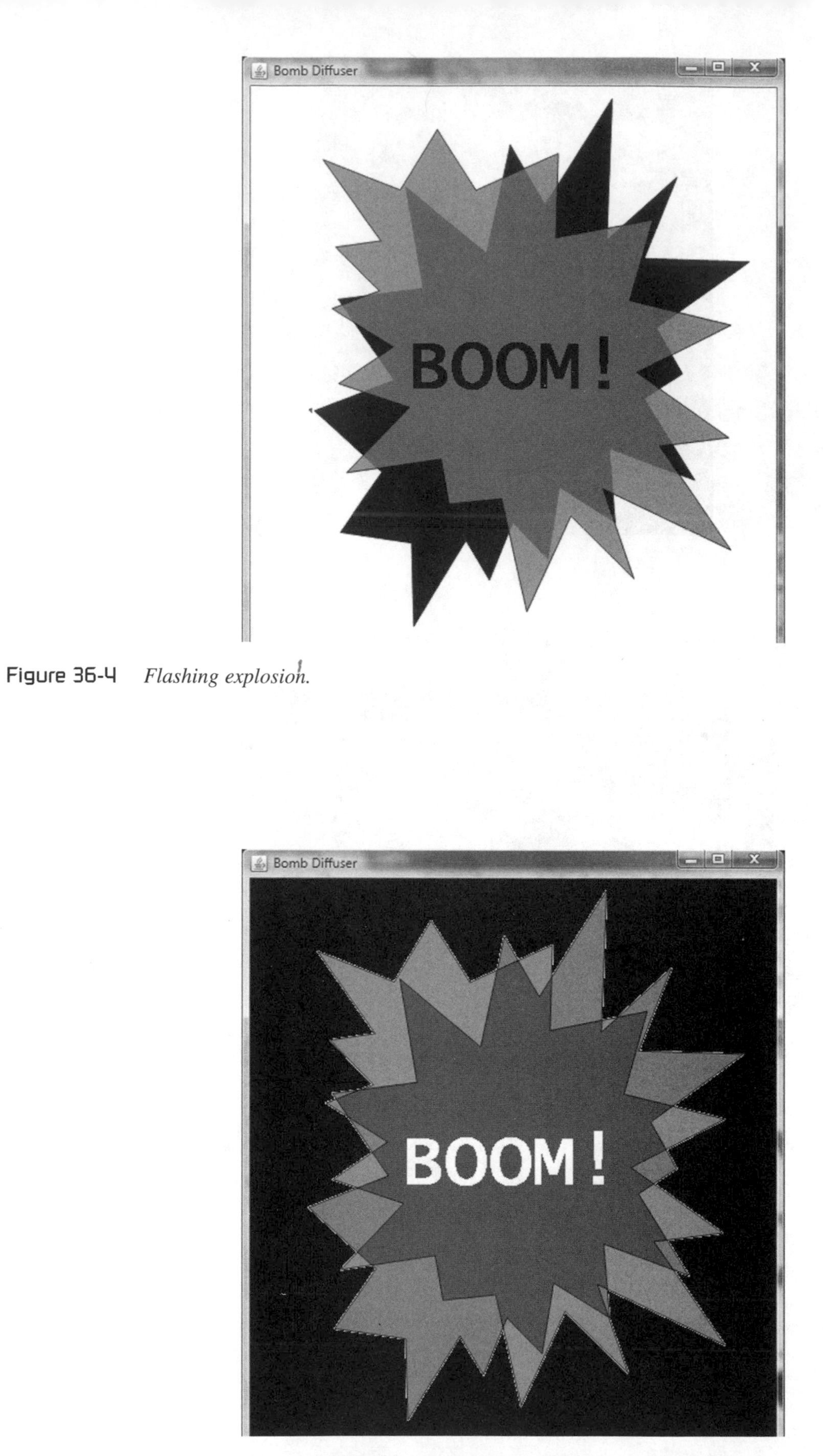

Figure 36-4 *Flashing explosion.*

Figure 36-5 *Alternate image of exploding bomb.*

Project 37: Bomb Diffuser—Rising Through the Ranks

Project

Add levels to the game. Although you will breeze past the first few challenges, don't get cocky ... the codes get more and more complex!

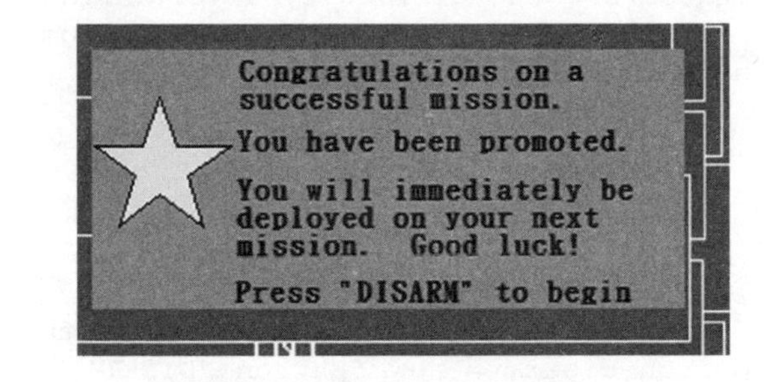

Figure 37-1 *Congratulate the player.*

Making the game

First, create a way to signal the user that he/she has diffused the bomb. To do this, create an image that congratulates the player such as the one in Figure 37-1.

Display this image in the if-statement that compares the user's input and the random code. Next, reset the code by using a higher number. You can do this by basing the code on a maximum number. Increase the maximum number every round. Because the levels get more complex, you need to give the player more time to guess the disarm code. To do this, create a multiplier that increases the starting time every round. Try playing the game. Once you beat a level, the detonator still explodes the bomb! Why? You have not yet paused the game.

Here's how you do it: create a boolean that stops the counter when its value is true. Then, you need to create a button for the player to indicate he/she is ready to start the next round. Hint: don't create a new button; use the same "disarm" button.

There is only one more thing you need to add: a score. Create a JLabel that will display the score, and increment the score by the remaining time left when a bomb is diffused. Don't forget to reset the text in the JLabel.

```
import java.awt.*;
import java.awt.event.*;
import javax.swing.*;
import javax.swing.event.*;

public class BombDiffuser extends JFrame implements ActionListener
{
  //the player's bomb diffusal rating/score
  int score = 0;
  //the time before detonation
  int timeLeft = 12;
  //the JLabel that displays the time left
  JLabel time = new JLabel(timeLeft+"");
  //the bomb label
  JLabel bomb = new JLabel(new ImageIcon("bomb.PNG"));
  //the clue:
  JLabel clue = new JLabel();
  //this display's the player's score (aka bomb diffusal rating)
  JLabel rating = new JLabel("Your Bomb Diffusal Success Rating: "+score);
  //the exploded bomb:
  JLabel exploded = new JLabel(new ImageIcon("exploded.PNG"));
  JLabel exploded2 = new JLabel(new ImageIcon("exploded2.PNG"));
```

```
//boolean that waits for the button to be pressed again
boolean waiting = false;
//this will be displayed after the player disarms the bomb
JLabel promote = new JLabel(new ImageIcon("promote.PNG"));
//these are the difficulty level variables
//the multiplier for the number of digits:
int digitsMultiplier = 1;
int maxNum = 10;
//the code:
int code = 0;
//where the disarm code is entered:
TextField attempt = new TextField(20);
//the button when you think you have the correct combo
JButton disarm = new JButton("DISARM");
//the container of the components
Container cont;
//the thread
Countdown count;
public BombDiffuser()
{
  super("Bomb Diffuser");
  setSize(500,550);
  setVisible(true);
  setDefaultCloseOperation(JFrame.EXIT_ON_CLOSE);
  cont = getContentPane();
  cont.setLayout(null);
  cont.setBackground(Color.gray);
  //add the background image
  cont.add(bomb);
  bomb.setBounds(0,0,500,500);
  //set the font of the JLabel
  time.setFont(new Font("Courier", Font.BOLD, 20));
  //add the time left JLabel
  cont.add(time);
  //set the position
  time.setBounds(250,371,150,18);
  //put the label on top of the background
  cont.setComponentZOrder(time,0);
  //set the completion label offscreen
  cont.add(promote);
  promote.setBounds(-300,-300,276,122);
  cont.setComponentZOrder(promote,0);
  //add the score to the screen
  cont.add(rating);
  rating.setBounds(5,480,500,30);
  rating.setForeground(Color.white);
  rating.setFont(new Font("Courier",Font.BOLD,19));
  cont.setComponentZOrder(rating,0);
  //add the disarm button
  cont.add(disarm);
  disarm.setBounds(200,456,150,30);
  disarm.addActionListener(this);
  cont.setComponentZOrder(disarm,0);
  clue.setText("A number between 0 and "+maxNum);
  cont.add(clue);
```

```
    clue.setBounds(167,435,250,20);
    //put the label on top of the background
    cont.setComponentZOrder(clue,0);

    //add the text field
    cont.add(attempt);
    attempt.setBounds(271,404,150,25);

    //add the explosion offscreen
    cont.add(exploded);
    exploded.setBounds(-1000,-1000,500,550);
    cont.setComponentZOrder(exploded,0);
    cont.add(exploded2);
    exploded2.setBounds(-1000,-1000,500,550);
    cont.setComponentZOrder(exploded2,0);

    //set all the components
    cont.validate();
    setContentPane(cont);

    setCode();

    count = new Countdown();
    count.start();
  }
  public void setCode()
  {
    code = (int)(Math.random()*maxNum);
  }
  //the counting down thread:
  public class Countdown extends Thread
  {
    public void run()
    {
      while(true)
      {
        try
        {
          if(!waiting)
          {
            if(timeLeft>0)
            {
              timeLeft—;
              time.setText(timeLeft+"");
            }
            else
            {
              break;
            }
          }
          Thread.sleep(1000);
        }
        catch(Exception e){}
      }
      //call the method that displays the blasts
      explode();
    }

    public void explode()
    {
      //first, remove everything else:
      cont.remove(time);
      cont.remove(bomb);
      cont.remove(clue);
```

```
      cont.remove(rating);
      cont.remove(promote);
      cont.remove(attempt);
      cont.remove(disarm);

      //move the explosion images to the correct location
      exploded.setBounds(0,0,500,550);
      exploded2.setBounds(0,0,500,550);

      while(true)
      {
        try
        {
          exploded.setBounds(0,0,500,550);
          exploded2.setBounds(-1000,-1000,500,550);
          Thread.sleep(100);

          exploded2.setBounds(0,0,500,550);
          exploded.setBounds(-1000,-1000,500,550);
          Thread.sleep(100);
        }
        catch(Exception e){ }
      }
    }
  }
  public void actionPerformed(ActionEvent event)
  {
    if(waiting)
    {
      promote.setBounds(-300,-300,276,122);
      setCode();
      clue.setText("A number between 0 and "+maxNum);
      timeLeft = 15*digitsMultiplier;
      waiting = false;

    }
    else
    {
      if(Integer.parseInt(attempt.getText())>code)
      {
        clue.setText("CAUTION: Attempted Code is TOO HIGH");
      }
      if(Integer.parseInt(attempt.getText())<code)
      {
        clue.setText("CAUTION: Attempted Code is TOO LOW");
      }
      if(attempt.getText().equals(""+code))
      {

        score+=timeLeft;
        rating.setText("Your Bomb Diffusal Success Rating: "+score);
        promote.setBounds(150,100,276,122);
        waiting = true;
        digitsMultiplier++;
        maxNum*=5;
      }
    }
  }

  public static void main (String[ ] args)
  {
    new BombDiffuser();
  }
}
```

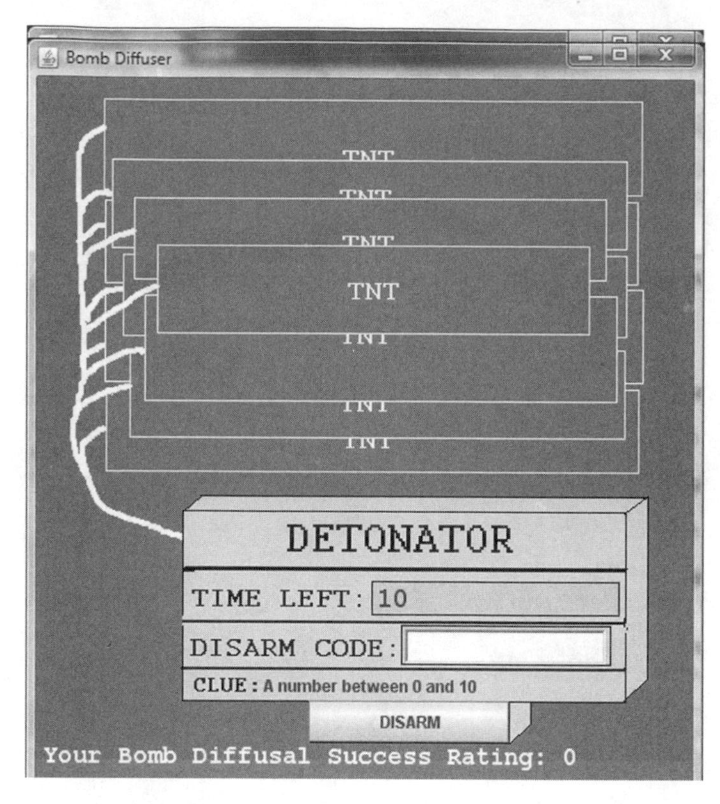

Figure 37-2 *Detonator counts down.*

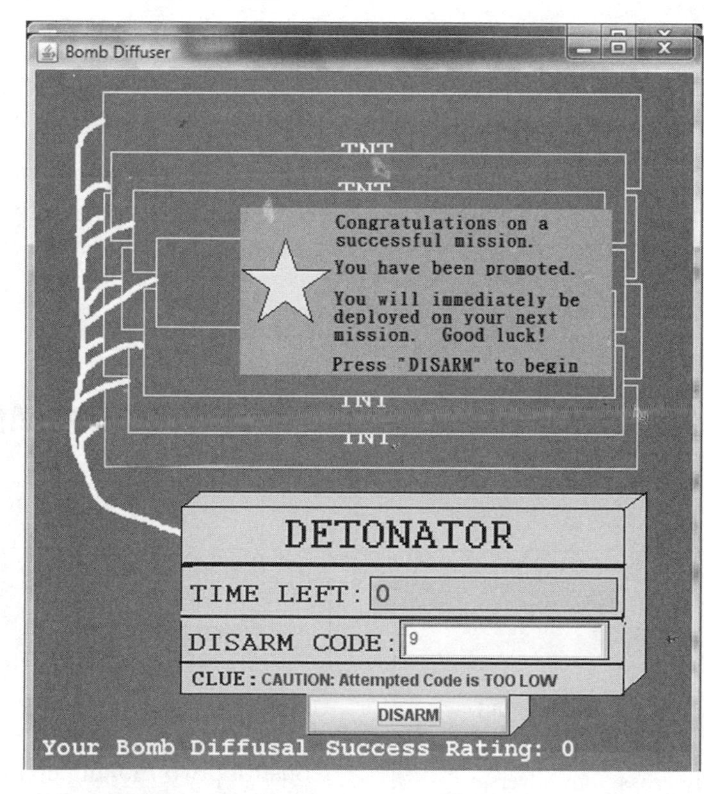

Figure 37-3 *Bomb successfully diffused.*

Figure 37-4 **#$%!!!... out of time!*

Figures 37-2 through 37-4 illustrate the stress of bomb diffusion.

Customizing the game

Place two bombs on the screen at the same time.

Alter the difficulty of the game by changing the time given for diffusion.

Add letters and/or symbols to the code.

Make the user work to break the code: offer simple mathematical equations.

To increase the stress level, have the timer not only display seconds, but milliseconds as well.

Give the player a limited number of guesses. If he/she can't guess correctly, BOOM! The bomb detonates.

Project 38: Trapper—Men on the Move

Trapper

In this multiplayer strategy game, battle it out with others by trapping them in the trail you leave behind. Don't accidentally cross your adversary's path or back up into your own trail—you'll be trapped!

Project

Construct two moving characters that are controlled by each opponent.

Making the game

Start by creating the JFrame; setting it to 500 by 500 pixels. Now, create 10×10 pixel images for each player. Make a JLabel for each image. Use the setBounds method to position the JLabels on opposite sides of the field.

Construct two threads: one to control the first player's direction; the other to control the direction of the second player. Because the direction of each player is controlled by the keyboard, each thread must implement keyListener. Add the keyListener methods to both threads. In the methods, check to see if the "A," "S," "W," or "D" keys are pressed. These keys control the direction of player one. The "J," "K," "I," and "L" keys control the second player's direction. Now, to keep track of each player's direction, employ the same technique you did in Radical Racing: use different int values to represent different directions. Once these variables are created, change the variable in the keyPressed method.

Once the players' orientations are established, they need to move around the arena. In each thread's infinite loop, move the players by using the setBounds method.

```
import javax.swing.*;
import javax.swing.event.*;
import java.awt.*;
import java.awt.event.*;

public class Trapper extends JFrame
{
  //the two JLabels that represent each player
  JLabel p1 = new JLabel(new ImageIcon("p1.PNG"));
  JLabel p2 = new JLabel(new ImageIcon("p2.PNG"));

  boolean keepPlaying = true;
  //these variables keep track of the direction
  int UP = 1, RIGHT = 2, DOWN = 3, LEFT = 4;
  int p1Direction = RIGHT;
  int p2Direction = LEFT;

  Container cont;

  public Trapper()
  {
    super("Trapper");
    setSize(500,500);
    setDefaultCloseOperation(JFrame.EXIT_ON_CLOSE);
    setVisible(true);

    cont = getContentPane();

    cont.setLayout(null);

    cont.add(p1);
    cont.setComponentZOrder(p1,0);
    p1.setBounds(20,245,10,10);
    cont.add(p2);
    cont.setComponentZOrder(p2,0);
    p2.setBounds(455,245,10,10);

    P1Move p1Thread = new P1Move();
    p1Thread.start();

    P2Move p2Thread = new P2Move();
    p2Thread.start();

    setContentPane(cont);
```

```
}

public class P1Move extends Thread implements KeyListener
{
  public void run()
  {
    addKeyListener(this);
    while(keepPlaying)
    {
      try
      {
        if(p1Direction==UP)
          p1.setBounds(p1.getX(),p1.getY()-5,10,10);
        if(p1Direction==DOWN)
          p1.setBounds(p1.getX(),p1.getY()+5,10,10);
        if(p1Direction==RIGHT)
          p1.setBounds(p1.getX()+5,p1.getY(),10,10);
        if(p1Direction==LEFT)
          p1.setBounds(p1.getX()-5,p1.getY(),10,10);

        cont.validate();
        Thread.sleep(75);
      }
      catch(Exception e){ }
    }
  }

  public void keyPressed(KeyEvent e)
  {
    if(e.getKeyChar()=='a')
      p1Direction = LEFT;
    if(e.getKeyChar()=='s')
      p1Direction = DOWN;
    if(e.getKeyChar()=='d')
      p1Direction = RIGHT;
    if(e.getKeyChar()=='w')
      p1Direction = UP;
  }

  public void keyTyped(KeyEvent e){ }
  public void keyReleased(KeyEvent e){ }
}
public class P2Move extends Thread implements KeyListener
{
  public void run()
  {
    addKeyListener(this);
    while(keepPlaying)
    {
      try
      {
        if(p2Direction==UP)
          p2.setBounds(p2.getX(),p2.getY()-5,10,10);
        if(p2Direction==DOWN)
          p2.setBounds(p2.getX(),p2.getY()+5,10,10);
        if(p2Direction==RIGHT)
          p2.setBounds(p2.getX()+5,p2.getY(),10,10);
        if(p2Direction==LEFT)
          p2.setBounds(p2.getX()-5,p2.getY(),10,10);

        cont.validate();
        Thread.sleep(75);
      }
```

```
          catch(Exception e){ }
        }
      }

      public void keyPressed(KeyEvent e)
      {
        if(e.getKeyChar()=='j')
          p2Direction = LEFT;
        if(e.getKeyChar()=='k')
          p2Direction = DOWN;
        if(e.getKeyChar()=='l')
          p2Direction = RIGHT;
        if(e.getKeyChar()=='i')
          p2Direction = UP;
      }

      public void keyTyped(KeyEvent e){ }
      public void keyReleased(KeyEvent e){ }
    }

    public static void main (String[ ] args)
    {
      new Trapper();
    }
}
```

Figures 38-1 and 38-2 show the initial movement of the characters.

Keep going to add trails that the characters leave behind as they move. It's time to start trapping!

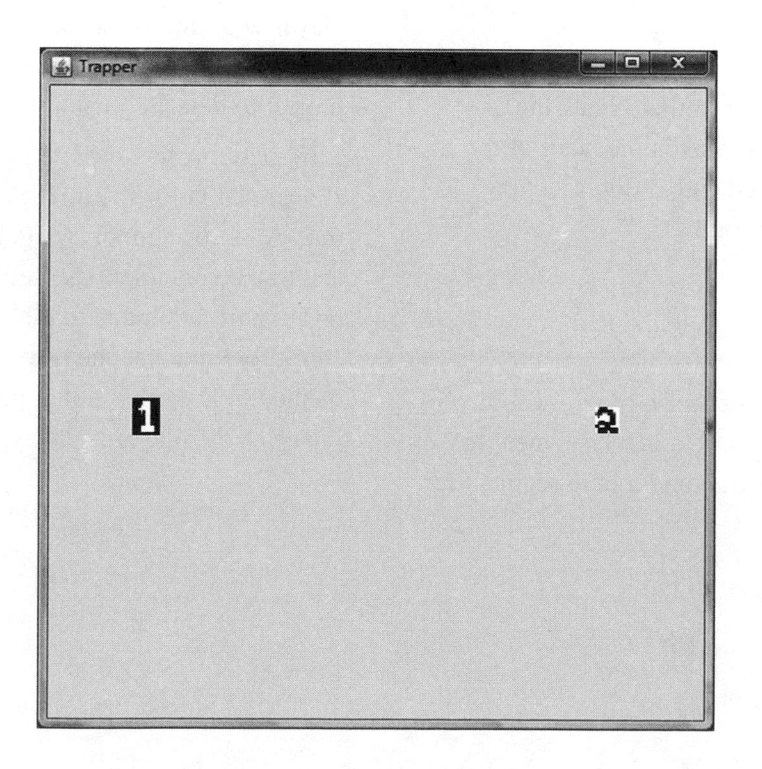

Figure 38-1 *Arena.*

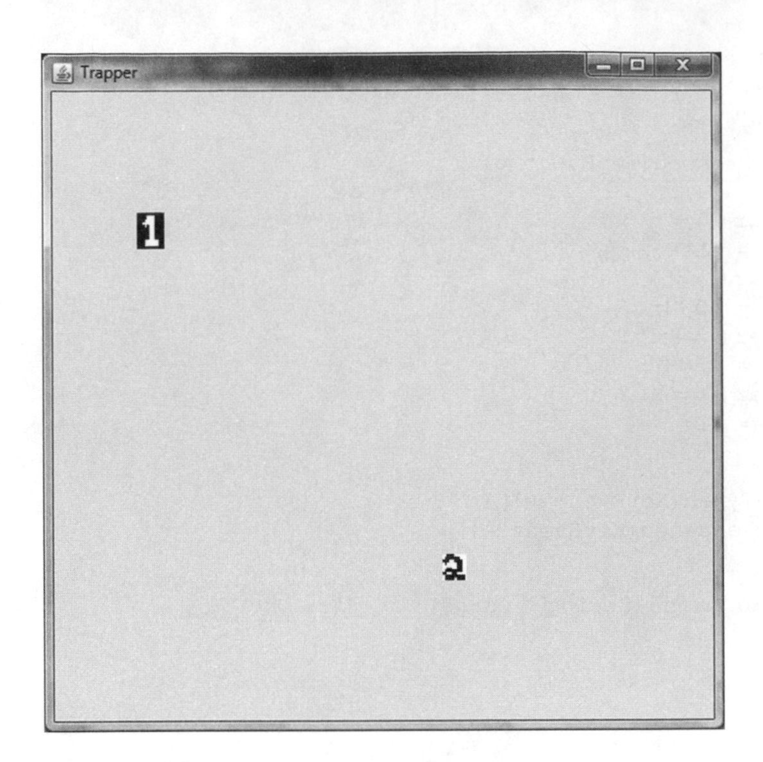

Figure 38-2 *Characters move around.*

Project 39: Trapper—Setting the Trap

Project

You're now ready to add trails to track the characters' movements. This is the heart of the game. Don't get trapped and ... don't let your opponent escape your trap!

Making the game

Each player's trail consists of a progressive line of 10 by 10 pixel images. Use a black segment to represent player one's trail and a blue segment for player two. In order to store all of the images, use an ArrayList. You will use this feature in the next project to detect collisions into the trail.

In each thread's infinite loop, create a temporary JLabel that holds the image of the segment of the trail. Add this temporary JLabel to the ArrayList and to the container. Use setBounds to set the position of the image to the location of the player. Don't forget to use the setComponentZOrder method to keep the trail at position "1" so the lead image of the player is always displayed.

```
import javax.swing.*;
import javax.swing.event.*;
import java.awt.*;
import java.awt.event.*;
import java.util.*;

public class Trapper extends JFrame
{
  //the two JLabels that represent each player
```

```
JLabel p1 = new JLabel(new ImageIcon("p1.PNG"));
JLabel p2 = new JLabel(new ImageIcon("p2.PNG"));

//these variables keep track of the direction
int UP = 1, RIGHT = 2, DOWN = 3, LEFT = 4;
int p1Direction = RIGHT;
int p2Direction = LEFT;

//this ArrayLists holds the trails of each player
ArrayList trailList = new ArrayList();

Container cont;

public Trapper()
{
  super("Trapper");
  setSize(500,500);
  setDefaultCloseOperation(JFrame.EXIT_ON_CLOSE);
  setVisible(true);

  cont = getContentPane();

  cont.setLayout(null);

  cont.add(p1);
  cont.setComponentZOrder(p1,0);
  p1.setBounds(20,245,20,20);
  cont.add(p2);
  cont.setComponentZOrder(p2,0);
  p2.setBounds(455,245,20,20);

  P1Move p1Thread = new P1Move();
  p1Thread.start();

  P2Move p2Thread = new P2Move();
  p2Thread.start();

  setContentPane(cont);
}

public class P1Move extends Thread implements KeyListener
{
  public void run()
  {
    addKeyListener(this);
    while(true)
    {
      try
      {
        //add the trail
        JLabel temp = new JLabel(new ImageIcon("p1trail.PNG"));;
        trailList.add(temp);
        cont.add(temp);
        cont.setComponentZOrder(temp,1);
        temp.setBounds(p1.getX(),p1.getY(),10,10);

        if(p1Direction==UP)
          p1.setBounds(p1.getX(),p1.getY()-5,20,20);
        if(p1Direction==DOWN)
          p1.setBounds(p1.getX(),p1.getY()+5,20,20);
        if(p1Direction==RIGHT)
          p1.setBounds(p1.getX()+5,p1.getY(),20,20);
        if(p1Direction==LEFT)
          p1.setBounds(p1.getX()-5,p1.getY(),20,20);

        cont.validate();
        Thread.sleep(75);
      }
```

```
      catch(Exception e){ }
    }
  }

  public void keyPressed(KeyEvent e)
  {
    if(e.getKeyChar()==’a’)
      p1Direction = LEFT;
    if(e.getKeyChar()==’s’)
      p1Direction = DOWN;
    if(e.getKeyChar()==’d’)
      p1Direction = RIGHT;
    if(e.getKeyChar()==’w’)
      p1Direction = UP;
  }
  public void keyTyped(KeyEvent e){ }
  public void keyReleased(KeyEvent e){ }
}
public class P2Move extends Thread implements KeyListener
{
  public void run()
  {
    addKeyListener(this);
    while(true)
    {
      try
      {
        //add the trail:
        JLabel temp = new JLabel(new ImageIcon("p2trail.PNG"));;
        trailList.add(temp);
        cont.add(temp);
        cont.setComponentZOrder(temp,1);
        temp.setBounds(p2.getX(),p2.getY(),10,10);

        if(p2Direction==UP)
          p2.setBounds(p2.getX(),p2.getY()-5,20,20);
        if(p2Direction==DOWN)
          p2.setBounds(p2.getX(),p2.getY()+5,20,20);
        if(p2Direction==RIGHT)
          p2.setBounds(p2.getX()+5,p2.getY(),20,20);
        if(p2Direction==LEFT)
          p2.setBounds(p2.getX()-5,p2.getY(),20,20);

        cont.validate();
        Thread.sleep(75);
      }

      catch(Exception e){ }
    }
  }

  public void keyPressed(KeyEvent e)
  {
    if(e.getKeyChar()==’j’)
      p2Direction = LEFT;
    if(e.getKeyChar()==’k’)
      p2Direction = DOWN;
    if(e.getKeyChar()==’l’)
      p2Direction = RIGHT;
    if(e.getKeyChar()==’i’)
      p2Direction = UP;
  }
```

```
    public void keyTyped(KeyEvent e){ }
    public void keyReleased(KeyEvent e){ }
  }

  public static void main (String[ ] args)
  {
    new Trapper();
  }
}
```

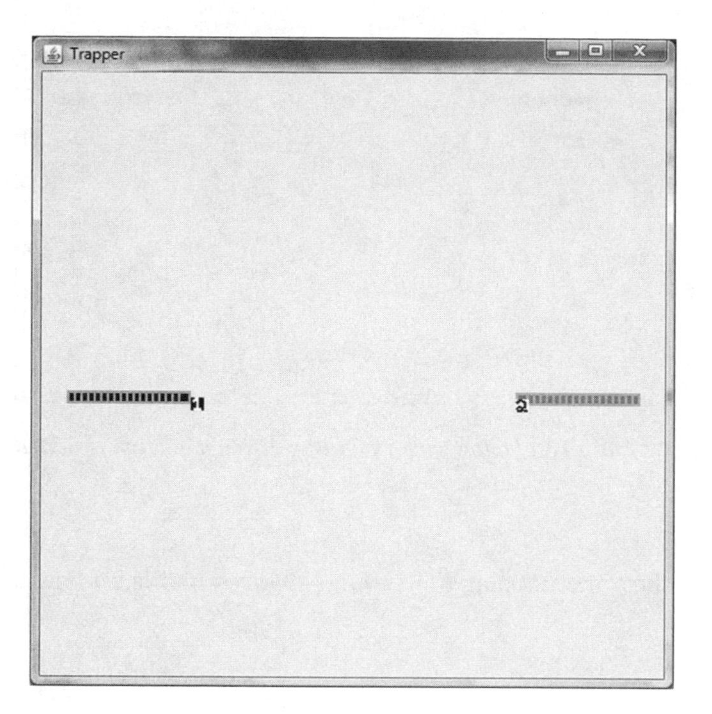

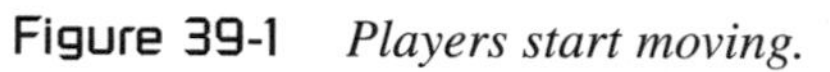

Figure 39-1 *Players start moving.*

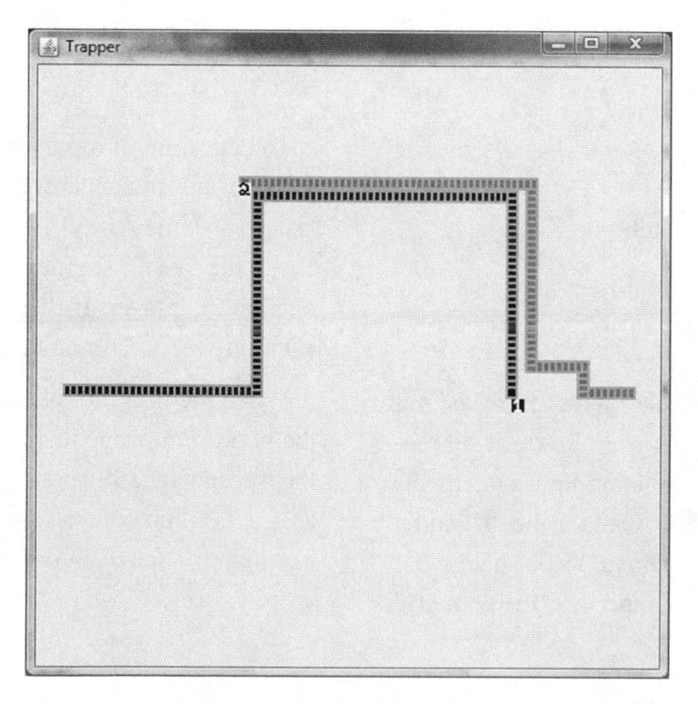

Figure 39-2 *Players leave a trail.*

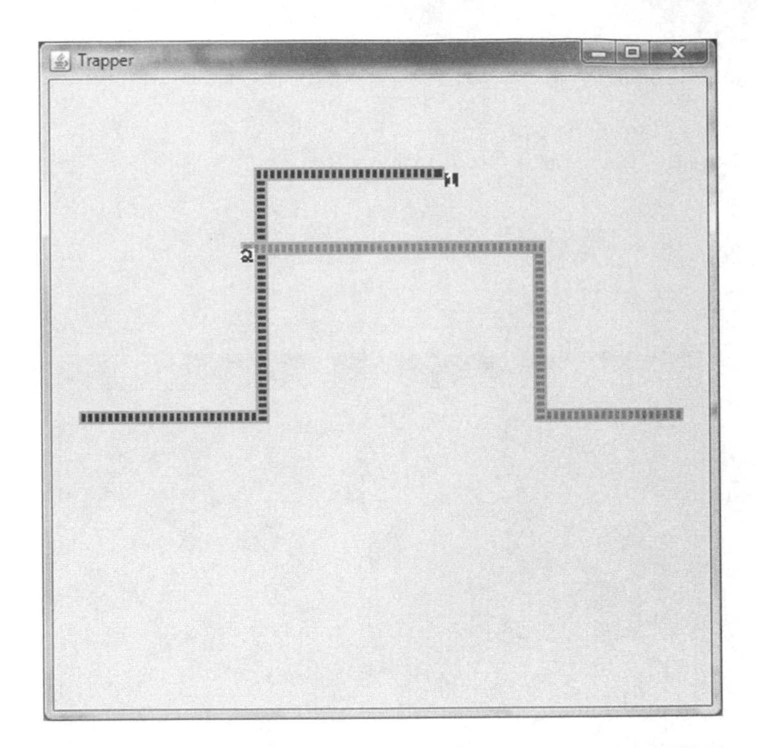

Figure 39-3 *Player 1 wins ... but it is a hollow victory until collision detection is added.*

Figures 39-1 through 39-3 show the trapping trails that follow each player.

Move on to learn how to add collision detection.

Project 40: Trapper—Trapped!

Project

Add collision detection. Watch out!

Making the game

The collision detection checks two events: leaving the grid and colliding with a trail. To determine if a player leaves the grid, use a simple if-statement in each infinite loop. Check whether the "x" and "y" values of the player is above 500 or below 0. If so, use a JOptionPane to announce the winner.

To determine if a player collides with a trail, use a `for loop` to check every JLabel —except the last 10—in the ArrayList. You do not need to check the last 10 segments because they always overlap the player. If there is a collision, use a JOptionPane to announce the winner.

When the program is executed, you will notice the pop-up message displays continuously because the threads are still running. To stop the display, use a `boolean` that controls the infinite loop. Set the `boolean` to "false" in order to terminate the threads.

```
import javax.swing.*;
import javax.swing.event.*;
import java.awt.*;
import java.awt.event.*;
import java.util.*;

public class Trapper extends JFrame
{
  //the two JLabels that represent each player
  JLabel p1 = new JLabel(new ImageIcon("p1.PNG"));
  JLabel p2 = new JLabel(new ImageIcon("p2.PNG"));

  boolean keepLooping = true;

  //these variables keep track of the direction
  int UP = 1, RIGHT = 2, DOWN = 3, LEFT = 4;
  int p1Direction = RIGHT;
  int p2Direction = LEFT;

  //this ArrayLists holds the trails of each player
  ArrayList trailList = new ArrayList();

  Container cont;
  public Trapper()
  {
    super("Trapper");
    setSize(500,500);
    setDefaultCloseOperation(JFrame.EXIT_ON_CLOSE);
    setVisible(true);

    cont = getContentPane();

    cont.setLayout(null);

    cont.add(p1);
    cont.setComponentZOrder(p1,0);
    p1.setBounds(20,245,10,10);
    cont.add(p2);
    cont.setComponentZOrder(p2,0);
    p2.setBounds(455,245,10,10);

    P1Move p1Thread = new P1Move();
    p1Thread.start();

    P2Move p2Thread = new P2Move();
    p2Thread.start();

    setContentPane(cont);
  }

  public class P1Move extends Thread implements KeyListener
  {
    public void run()
    {
      addKeyListener(this);
      while(keepLooping)
      {
        try
        {
          //check to see if p2 hits the walls
          if(p1.getX()>500 || p1.getX()<0 || p1.getY()>500 || p1.getY()<0)
          {
            p1Lose();
          }
          //check to see if p2 hits the trails
          for(int i = 0; i < trailList.size()-10; i++)
          {
```

```
        JLabel tempTrail = (JLabel) trailList.get(i);
        if(p1.getBounds().intersects(tempTrail.getBounds()))
        {
          p1Lose();
        }
      }

      //add the trail
      JLabel temp = new JLabel(new ImageIcon("p1trail.PNG"));;
      trailList.add(temp);
      cont.add(temp);
      cont.setComponentZOrder(temp,1);
      temp.setBounds(p1.getX(),p1.getY(),10,10);

      if(p1Direction==UP)
        p1.setBounds(p1.getX(),p1.getY()-5,10,10);
      if(p1Direction==DOWN)
        p1.setBounds(p1.getX(),p1.getY()+5,10,10);
      if(p1Direction==RIGHT)
        p1.setBounds(p1.getX()+5,p1.getY(),10,10);
      if(p1Direction==LEFT)
        p1.setBounds(p1.getX()-5,p1.getY(),10,10);

      cont.validate();
      Thread.sleep(75);
    }
    catch(Exception e){ }
  }
}
public void p1Lose()
{
  keepLooping = false;
  JOptionPane.showMessageDialog(null,"Player 2 Wins!!!");
}
public void keyPressed(KeyEvent e)
{
  if(e.getKeyChar()=='a')
    p1Direction = LEFT;
  if(e.getKeyChar()=='s')
    p1Direction = DOWN;
  if(e.getKeyChar()=='d')
    p1Direction = RIGHT;
  if(e.getKeyChar()=='w')
    p1Direction = UP;
}
public void keyTyped(KeyEvent e){ }
public void keyReleased(KeyEvent e){ }
}
public class P2Move extends Thread implements KeyListener
{
  public void run()
  {
    addKeyListener(this);
    while(keepLooping)
    {
      try
      {
        //check to see if p2 hits the walls
        if(p2.getX()>500 || p2.getX()<0 || p2.getY()>500 || p2.getY()<0)
        {
          p2Lose();
```

```
        }
        //check to see if p2 hits the trails
        for(int i = 0; i < trailList.size()-10; i++)
        {
          JLabel tempTrail = (JLabel) trailList.get(i);
          if(p2.getBounds().intersects(tempTrail.getBounds()))
          {
            p2Lose();
          }
        }

        //add the trail:
        JLabel temp = new JLabel(new ImageIcon("p2trail.PNG"));;
        trailList.add(temp);
        cont.add(temp);
        cont.setComponentZOrder(temp,1);
        temp.setBounds(p2.getX(),p2.getY(),10,10);

        if(p2Direction==UP)
          p2.setBounds(p2.getX(),p2.getY()-5,10,10);
        if(p2Direction==DOWN)
          p2.setBounds(p2.getX(),p2.getY()+5,10,10);
        if(p2Direction==RIGHT)
          p2.setBounds(p2.getX()+5,p2.getY(),10,10);
        if(p2Direction==LEFT)
          p2.setBounds(p2.getX()-5,p2.getY(),10,10);

        cont.validate();
        Thread.sleep(75);
      }
      catch(Exception e){ }
    }

  }
  public void p2Lose()
  {
    keepLooping = false;
    JOptionPane.showMessageDialog(null,"Player 1 Wins!!!");
  }

  public void keyPressed(KeyEvent e)
  {
    if(e.getKeyChar()=='j')
      p2Direction = LEFT;
    if(e.getKeyChar()=='k')
      p2Direction = DOWN;
    if(e.getKeyChar()=='l')
      p2Direction = RIGHT;
    if(e.getKeyChar()=='i')
      p2Direction = UP;
  }

  public void keyTyped(KeyEvent e){ }
  public void keyReleased(KeyEvent e){ }
}

public static void main (String[ ] args)
{
  new Trapper();
}
}
```

Figures 40-1 and 40-2 depict a strategy of the trails.

Read the next project to uncover ways to add a more striking background and a bolder way to alert players of wins and losses.

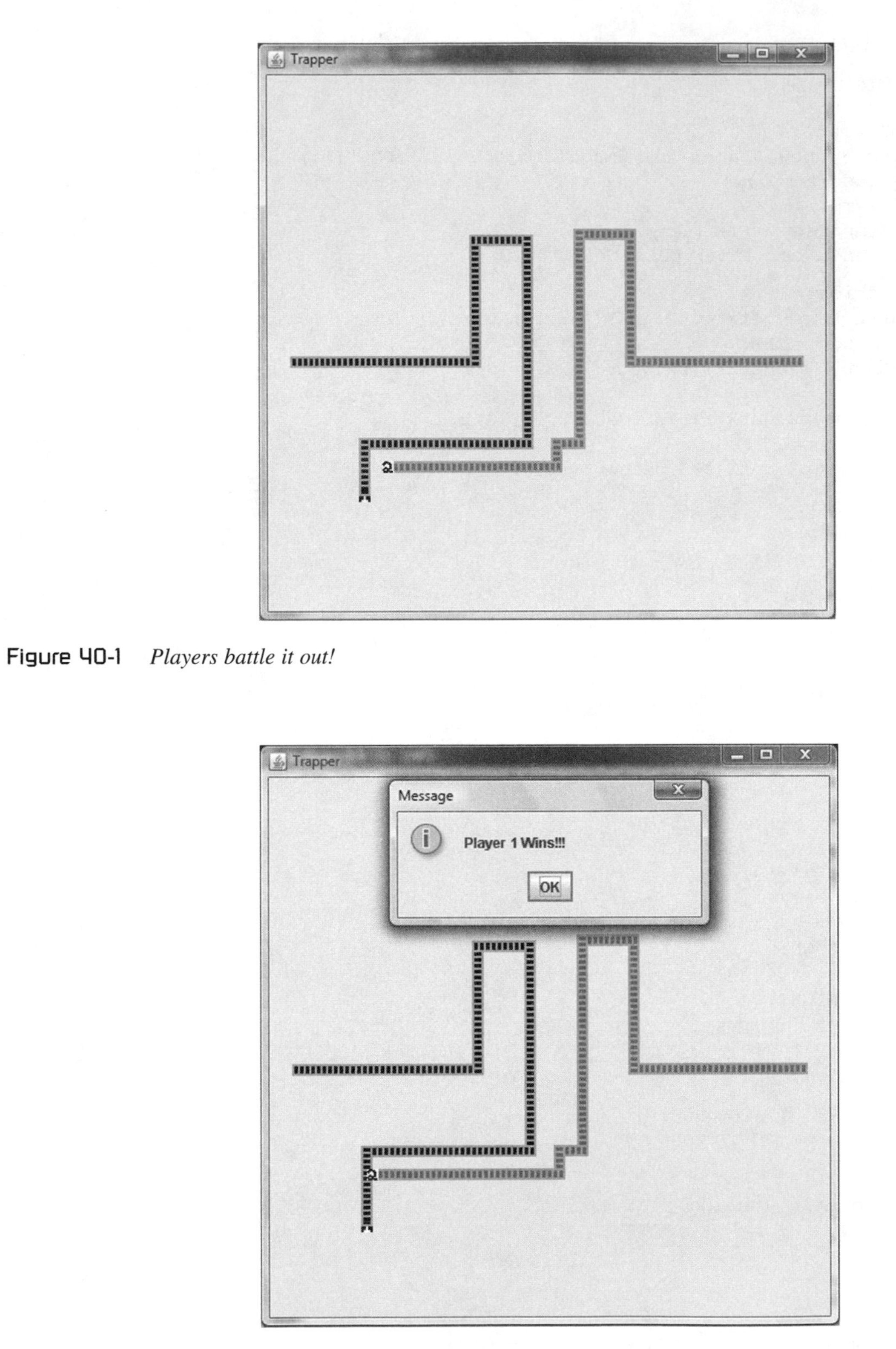

Figure 40-1 *Players battle it out!*

Figure 40-2 *Player One successfully traps Player Two.*

Project 41: Trapper—Showdown

Project

Add an eye-catching background image and attention-grabbing graphics to notify players of the winner.

Making the game

Let your imagination go wild. Draw a 500 by 500 pixel background image in Microsoft Paint. Create a JLabel for it and use the setBounds method to center it. Don't forget to use the setComponentZOrder method to send the background image behind the other icons.

Now, draw an image—any image you want—to notify the players of the winner. Display this image instead of the JOptionPane when a player wins.

Once a player wins, add a fun way of "rewinding" the game. You can "undo" the players' moves by using a `for loop` and removing the trail segments. To do this, create a `for loop` that counts down from the number of components in the container minus 4. This leaves the background image and the actual player images. To find the number of components in the container, use the following:

```
cont.getComponentCount()
```

Remember to refresh the container after removing each trail segment. In addition, create a JLabel that holds the same text as the JOptionPanes from the previous project.

```
import javax.swing.*;
import javax.swing.event.*;
import java.awt.*;
import java.awt.event.*;
import java.util.*;

public class Trapper extends JFrame
{
  //the two JLabels that represent each player
  JLabel p1 = new JLabel(new ImageIcon("p1.PNG"));
  JLabel p2 = new JLabel(new ImageIcon("p2.PNG"));
  boolean keepPlaying = true;
  //these variables keep track of the direction
  int UP = 1, RIGHT = 2, DOWN = 3, LEFT = 4;
  int p1Direction = RIGHT;
  int p2Direction = LEFT;

  //this ArrayLists holds the trails of each player
  ArrayList trailList = new ArrayList();

  Container cont;

  public Trapper()
  {
    super("Trapper");
    setSize(500,500);
    setDefaultCloseOperation(JFrame.EXIT_ON_CLOSE);
    setVisible(true);

    cont = getContentPane();

    cont.setLayout(null);

    //this is the background JLabel:
    JLabel bg = new JLabel(new ImageIcon("bg.png"));
```

```
    cont.add(bg);
    bg.setBounds(0,0,500,500);

    cont.add(p1);
    cont.setComponentZOrder(p1,0);
    p1.setBounds(20,245,10,10);
    cont.add(p2);
    cont.setComponentZOrder(p2,0);
    p2.setBounds(455,245,10,10);

    P1Move p1Thread = new P1Move();
    p1Thread.start();

    P2Move p2Thread = new P2Move();
    p2Thread.start();

    setContentPane(cont);
  }
  public class P1Move extends Thread implements KeyListener
  {
    public void run()
    {
      addKeyListener(this);
      while(keepPlaying)
      {
        try
        {
          //check to see if p2 hits the walls
          if(p1.getX()>500 || p1.getX()<0 || p1.getY()>500 || p1.getY()<0)
          {
            p1Lose();
          }
          //check to see if p2 hits the trails
          for(int i = 0; i < trailList.size()-10; i++)
          {
            JLabel tempTrail = (JLabel) trailList.get(i);
            if(p1.getBounds().intersects(tempTrail.getBounds()))
            {
              p1Lose();
            }
          }
          //add the trail
          JLabel temp = new JLabel(new ImageIcon("p1trail.PNG"));;
          trailList.add(temp);
          cont.add(temp);
          cont.setComponentZOrder(temp,1);
          temp.setBounds(p1.getX(),p1.getY(),10,10);

          if(p1Direction==UP)
            p1.setBounds(p1.getX(),p1.getY()-5,10,10);
          if(p1Direction==DOWN)
            p1.setBounds(p1.getX(),p1.getY()+5,10,10);
          if(p1Direction==RIGHT)
            p1.setBounds(p1.getX()+5,p1.getY(),10,10);
          if(p1Direction==LEFT)
            p1.setBounds(p1.getX()-5,p1.getY(),10,10);

          cont.validate();
          Thread.sleep(75);
        }
        catch(Exception e){ }
      }
    }

    public void p1Lose()
```

```
    {
      try
      {
        keepPlaying = false;
        JLabel winner = new JLabel("Player 2 (blue) Wins!!!");
        cont.add(winner);
        cont.setComponentZOrder(winner,0);
        winner.setFont(new Font("arial",Font.BOLD,30));
        winner.setBounds(75,50,400,100);
        for(int i = cont.getComponentCount()-4; i > 0; i--)
        {
          cont.remove(i);
          cont.repaint();
          Thread.sleep(25);
        }
      }
      catch(Exception e){ }
    }
    public void keyPressed(KeyEvent e)
    {
      if(e.getKeyChar()=='a')
        p1Direction = LEFT;
      if(e.getKeyChar()=='s')
        p1Direction = DOWN;
      if(e.getKeyChar()=='d')
        p1Direction = RIGHT;
      if(e.getKeyChar()=='w')
        p1Direction = UP;
    }
    public void keyTyped(KeyEvent e){ }
    public void keyReleased(KeyEvent e){ }
  }
  public class P2Move extends Thread implements KeyListener
  {
    public void run()
    {
      addKeyListener(this);
      while(keepPlaying)
      {
        try
        {
          //check to see if p2 hits the walls
          if(p2.getX()>500 || p2.getX()<0 || p2.getY()>500 || p2.getY()<0)
          {
            p2Lose();
          }
          //check to see if p2 hits the trails
          for(int i = 0; i < trailList.size()-10; i++)
          {
            JLabel tempTrail = (JLabel) trailList.get(i);
            if(p2.getBounds().intersects(tempTrail.getBounds()))
            {
              p2Lose();
            }
          }

          //add the trail:
          JLabel temp = new JLabel(new ImageIcon("p2trail.PNG"));;
          trailList.add(temp);
          cont.add(temp);
```

```
            cont.setComponentZOrder(temp,1);
            temp.setBounds(p2.getX(),p2.getY(),10,10);
            if(p2Direction==UP)
              p2.setBounds(p2.getX(),p2.getY()-5,10,10);
            if(p2Direction==DOWN)
              p2.setBounds(p2.getX(),p2.getY()+5,10,10);
            if(p2Direction==RIGHT)
              p2.setBounds(p2.getX()+5,p2.getY(),10,10);
            if(p2Direction==LEFT)
              p2.setBounds(p2.getX()-5,p2.getY(),10,10);
            cont.validate();
            Thread.sleep(75);
          }
          catch(Exception e){ }
        }
      }
      public void p2Lose()
      {
        try
        {
          keepPlaying = false;
          JLabel winner = new JLabel("Player 1 (black) Wins!!!");
          cont.add(winner);
          cont.setComponentZOrder(winner,0);
          winner.setFont(new Font("arial",Font.BOLD,30));
          winner.setBounds(75,50,400,100);
          for(int i = cont.getComponentCount()-4; i > 0; i--)
          {
            cont.remove(i);
            cont.repaint();
            Thread.sleep(25);
          }
        }
        catch(Exception e){ }
      }
      public void keyPressed(KeyEvent e)
      {
        if(e.getKeyChar()=='j')
          p2Direction = LEFT;
        if(e.getKeyChar()=='k')
          p2Direction = DOWN;
        if(e.getKeyChar()=='l')
          p2Direction = RIGHT;
        if(e.getKeyChar()=='i')
          p2Direction = UP;
      }
      public void keyTyped(KeyEvent e){ }
      public void keyReleased(KeyEvent e){ }
    }
    public static void main (String[ ] args)
    {
      new Trapper();
    }
}
```

Figure 41-1 and 41-2 illustrate the ultimate game play of Trapper.

Customizing the game

Dramatically increase the speed of one player while slowing down the other. Strategy note: each speed has its own advantages and disadvantages.

Modify the size of the players: supersized or mini!

Create levels: best out of three, five, or eleven rounds wins.

Add two additional characters so up to four players can compete at once.

No friends? Add artificial intelligence to battle the computer.

Construct the game so players are limited to right turns only ... or reverse direction only.

Randomize the keys in the middle of the game. The players have no idea which way the image will turn!

Figure 41-1 *Background image in place as players compete.*

Figure 41-2 *Player One wins; trails begin to disappear.*

Section Six

Retro Games

Project 42: Oiram—The Platform

Oiram

Help Oiram capture the stars and stomp out his enemies by leaping from platform to platform! Be warned, however . . . this is a multi-level game. As Oiram defeats enemies, more are waiting to appear in the next round.

Project

Begin by creating the platform in which Oiram battles.

New Building Blocks

Two Dimensional Arrays

Two dimensional arrays

Two dimensional arrays are like normal arrays, except they hold additional information in another dimension. Figure 42-1 explains the similarities and differences.

To make a two dimensional array, create a normal array with a second set of brackets:

```
int twoDArray [ ][ ] =
{{ 1,2,3},{ 4,5,6},{ 7,8,9}}
```

You can access elements in two dimensional arrays the same way you access elements in normal arrays:

```
int element =
twoDArray[ <rows>][ <columns>] ;
```

Use the following code to find the number of rows and columns in the two dimensional array:

```
int rows = twoDArray.length;
int columns = twoDArray[ 0] .length;
```

Making the game

There are three different areas of the game board: land, air, and ladders. Each of these areas occupies a 50 by 50 pixel zone on the board. To keep track of each of these zones, use a two dimensional array with "#" representing land, "|" representing ladder, and a "space" representing air. You must draw these images before you add the zones to the JFrames. Each image should be 50 by 50 pixels. Figures 42-2 through 43-4 represent the three different zones.

To display the board in a JFrame, iterate through the array in the constructor by using two `for loops`. On each iteration, use if-statements to verify whether the current area is air, land, or a ladder. Use the setBounds method to display the correct image.

```
import javax.swing.*;
import javax.swing.event.*;
import java.awt.*;
import java.awt.event.*;
import java.util.*;
import java.awt.geom.*;
public class Oiram extends JFrame
{
Container cont;
  //the 2 dimensional array
  String arena[ ][ ] =
  {{" "," "," "," "," "," "," "," "," "," "},
  {" "," "," "," "," "," "," "," "," "," "},
  {" "," "," "," "," "," ","|","#","#"," "},
  {" ","#","#","#"," "," ","|","#"," "," "},
  {" "," "," "," "," "," ","|","#"," "," "},
  {" ","#","#","#","#","#","#","#","#"," "},
  {" "," "," "," "," "," "," "," "," "," "},
  {"#","#","#","#"," "," ","#","#","#","#"},
  {" "," "," "," "," "," "," "," "," "," "},
  {"#","#","#","#","#","#","#","#","#","#"}};
  public Oiram()
  {
    super("Oiram");
    setSize(500,500);
    setVisible(true);
    setDefaultCloseOperation(JFrame.EXIT_ON_CLOSE);
    cont = getContentPane();
    cont.setLayout(null);
    cont.setBackground(Color.BLACK);
    //generate the board:
    for(int i = 0; i < arena.length; i++)
    {
      for(int j = 0; j < arena[0].length; j++)
      {
        JLabel lbl = null;
        if(arena[j][i].equals("#"))
        {
        lbl = new JLabel(new ImageIcon("ground.png"));
        }
        else if(arena[j][i].equals(" "))
        {
        lbl = new JLabel(new ImageIcon("air.png"));
        }
        else if(arena[j][i].equals("|"))
        {
        lbl = new JLabel(new ImageIcon("ladder.png"));
        }
        cont.add(lbl);
        lbl.setBounds(i*50,j*50,50,50);
      }
    }
    repaint();
    cont.validate();
    setContentPane(cont);
    }
  public static void main (String[ ] args)
  {
    new Oiram();
  }
  }
```

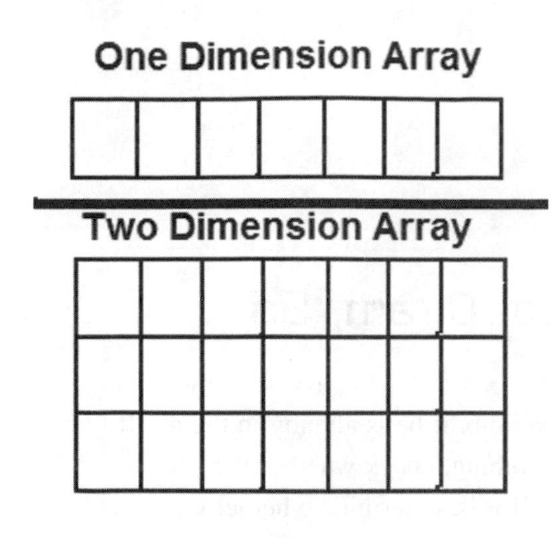

Figure 42-1 *Two types of arrays.*

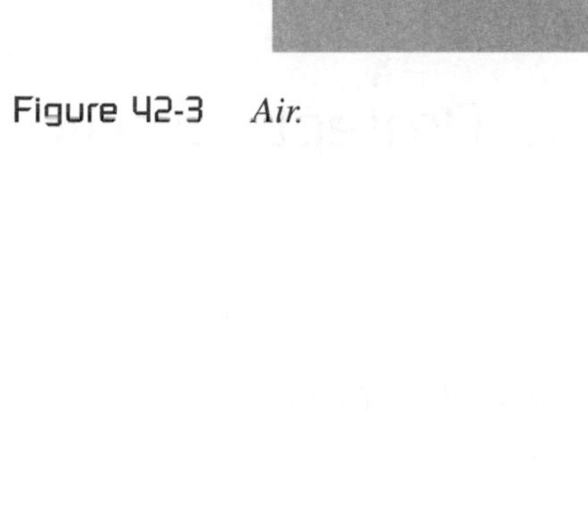

Figure 42-3 *Air.*

Figure 42-2 *Land.*

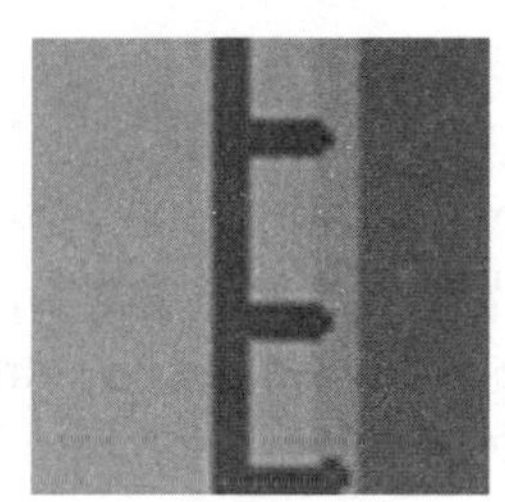

Figure 42-4 *Ladder.*

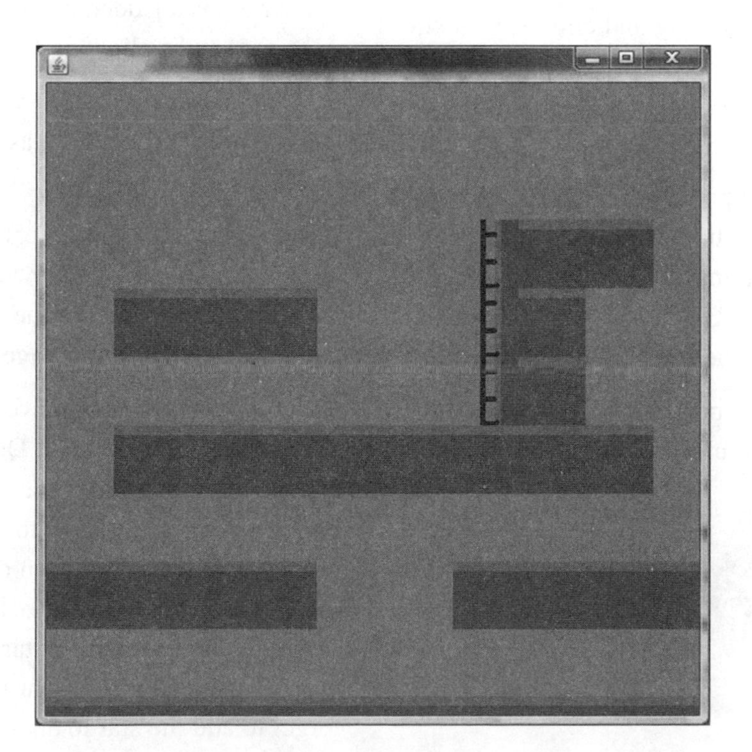

Figure 42-5 *Platform.*

Figure 42-5 shows Oiram's world: the platform.

In the next project, set Oiram free ... and give him stars to collect!

Project 43: Oiram—Go, Oiram, Go

Project

Put Oiram in his home and give him the ability to move around. Add stars for him to chase and capture.

Making the game

Create Oiram. Start by drawing a 50 by 50 pixel picture of him. Use a JLabel and the setBounds method to place Oiram in the bottom left corner. Figure 43-1 displays Oiram's starting location.

Oiram's world consists of moving on the ground, jumping in the air, and climbing ladders. Create booleans to represent jumping and climbing. Next, add a keyListener. In the keyTyped method, check whether the "A," "W," or "D" key is pressed to determine Oiram's movements.

When the "A" key is pressed, Oiram moves left. Remember to check the boolean that represents whether Oiram is climbing. If Oiram is climbing, set the boolean to false. If Oiram is not climbing, move him 50 pixels to the left.

When the "W" key is pressed, Oiram jumps (or, if he is on a ladder, climbs). Oiram should not

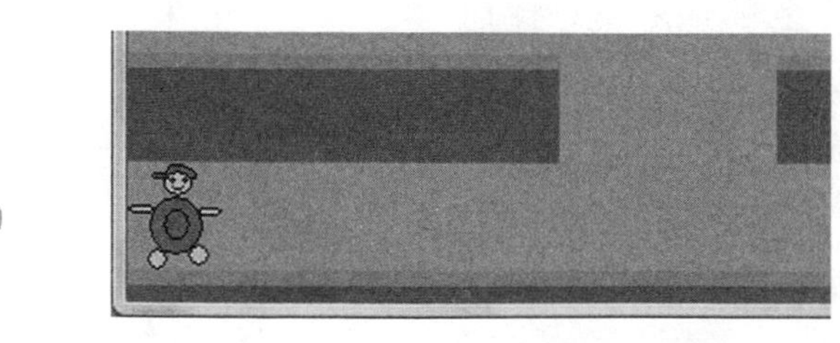

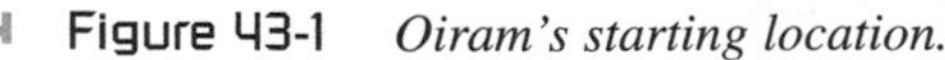

Figure 43-1 *Oiram's starting location.*

jump/climb, if he is already in the air. If Oiram is not climbing, check whether the spot above Oiram is air. If it is, determine whether Oiram is jumping and if the location below him is air. If the location below him is not air and he is not jumping, move Oiram up 50 pixels. If Oiram is climbing, move him up 50 pixels and again check whether he is in the air. If Oiram is in the air, move him 50 pixels to the right because he has reached the top of the ladder. Then, set the climbing boolean to false.

When the "D" key is pressed, Oiram moves right. First, check whether Oiram is inside the arena. If he is, check whether the space to Oiram's right is air. If so, move Oiram 50 pixels to the right. If the location to the right is a ladder, move Oiram 50 pixels to the right and set the climbing variable to true.

Now, run the game. You will notice that when Oiram climbs up a ladder, as shown in Figure 43-2, Oiram looks weird.

To fix this, create a new 50 by 50 pixel image of Oiram climbing the ladder. Only display it when the climbing variable is true. Figure 43-3 illustrates the new and improved image.

Now that Oiram can move around, you need to add stars for him to catch. Do this by creating a method that randomly generates the location of stars. Use two `for loops` to iterate through the array of the board. If the location of the board contains air, use a random number between 0 and 9 to give a one in ten chance for a star to appear. If the number is 0, generate a star in that location. Don't forget to add the star to an ArrayList. Call this method from the constructor.

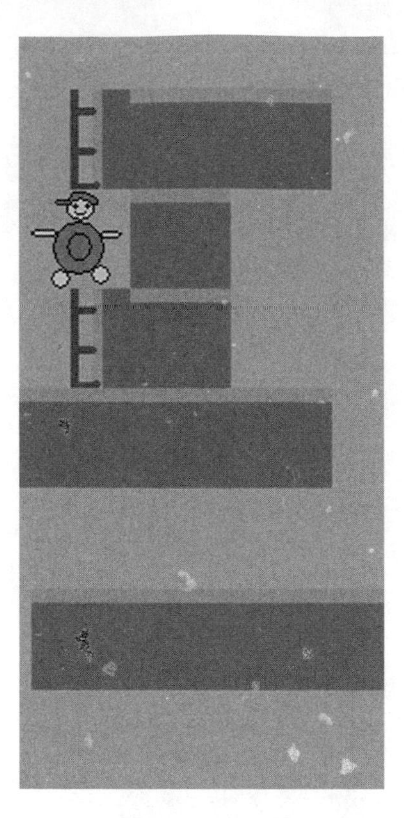

Figure 43-2 *Oiram, looking weird, climbs.*

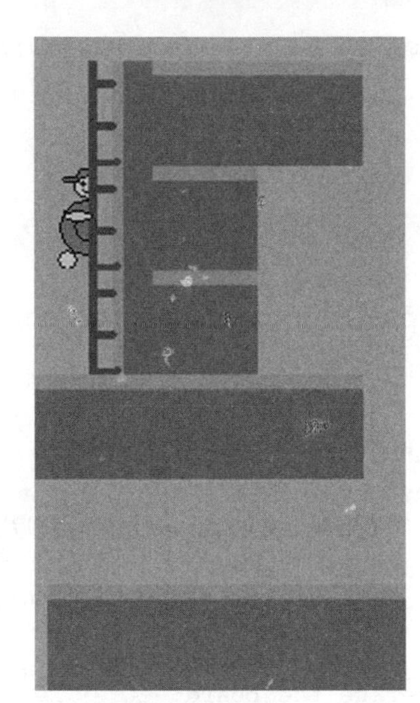

Figure 43-3 *Oiram, not looking weird, climbs.*

There are only two things left to do: make Oiram fall and let him collect the stars. To do this, create a Thread called Runner. In the infinite loop, iterate through the ArrayList of stars. If Oiram intersects a star, remove the star from both the container and ArrayList. Next, create an if-statement that checks whether Oiram is jumping. If he is jumping, move Oiram up 50 pixels and then end his jump by setting the boolean to false. If Oiram is not jumping, check whether the spot below him is air. If it is, lower Oiram 50 pixels. Don't forget to add the Thread.sleep method to your infinite loop.

```
import javax.swing.*;
import javax.swing.event.*;
import java.awt.*;
import java.awt.event.*;
import java.util.*;
import java.awt.geom.*;

public class Oiram extends JFrame implements KeyListener
{
  Container cont;
  //the two dimensional array:
  String arena[ ][ ] =
  {{" "," "," "," "," "," "," "," "," "," "},
  {" "," "," "," "," "," "," "," "," "," "},
  {" "," "," "," "," "," ","|","#","#"," "},
  {" ","#","#","#"," "," ","|","#"," "," "},
  {" "," "," "," "," "," ","|","#"," "," "},
  {" ","#","#","#","#","#","#","#","#"," "},
  {" "," "," "," "," "," "," "," "," "," "},
  {"#","#","#","#"," "," ","#","#","#","#"},
  {" "," "," "," "," "," "," "," "," "," "},
  {"#","#","#","#","#","#","#","#","#","#"}};
```

```
ArrayList stars = new ArrayList();

JLabel character = new JLabel(new ImageIcon("oiram.png"));
boolean jumping = false;
boolean climbing = false;
Runner runner;
public Oiram()
{
  super("Oiram");
  setSize(500,500);
  setVisible(true);
  setDefaultCloseOperation(JFrame.EXIT_ON_CLOSE);

  cont = getContentPane();
  cont.setLayout(null);

  addKeyListener(this);
  cont.setBackground(Color.BLACK);

  cont.add(character);
  character.setBounds(0,400,50,50);

  generateStars();

  //generate the board:
  for(int i = 0; i < arena.length; i++)
  {
    for(int j = 0; j < arena[0].length; j++)
    {
      JLabel lbl = null;
      if(arena[j][i].equals())
      {
        lbl = new JLabel(new ImageIcon("ground.png"));
      }
      else if(arena[j][i].equals(" "))
      {
        lbl = new JLabel(new ImageIcon("air.png"));
      }
      else if(arena[j][i].equals("|"))
      {
        lbl = new JLabel(new ImageIcon("ladder.png"));
      }
      cont.add(lbl);
        lbl.setBounds(i*50,j*50,50,50);
    }
  }

  repaint();
  cont.validate();

  runner = new Runner();
  runner.start();

  setContentPane(cont);
}

//add the stars
public void generateStars()
{
  for(int i = 1; i < arena.length; i++)
  {
    for(int j = 0; j < arena[0].length; j++)
    {
      //check if the current location is air
      if(arena[i][j].equals(" "))
      {
```

```
        //this random number gives a 1 in 10 chance of a star being placed here
        int placeOrNot = (int)(Math.random()*10);
        if(placeOrNot==0)
        {
          JLabel star = new JLabel(new ImageIcon("star.png"));
          cont.add(star);
          star.setBounds(j*50,i*50,50,50);
          cont.setComponentZOrder(star,0);
          cont.setComponentZOrder(character,0);
          stars.add(star);
        }
      }
    }
  }
}
public class Runner extends Thread
{
  public void run()
  {
    while(true)
    {
      try
      {
        //if Oiram touches a star, remove it
        for(int i = 0; i < stars.size(); i++)
        {
          JLabel star = (JLabel) stars.get(i);
          if(star.getBounds().intersects(character.getBounds()))
          {
            cont.remove(star);
            stars.remove(star);
          }
        }
        //let Oiram fall
        if(!jumping)
        {
          if(arena[(character.getY()/50)+1][character.getX()/50].equals(" "))
          {
            character.setBounds(character.getX(),character.getY()+50,50,50);
          }
        }
        else
        {
          //end Oiram's jump at the next iteration
          jumping = false;
          //move Oiram up:
          if(arena[(character.getY()/50)-1][character.getX()/50].equals(" "))
          {
            character.setBounds(character.getX(),character.getY()-50,50,50);
          }

        }
        Thread.sleep(250);
      }
      catch(Exception e){ }
    }
  }
}
public void keyPressed(KeyEvent e){ }
public void keyReleased(KeyEvent e){ }
public void keyTyped(KeyEvent e)
```

```
    {
      //move left
      if(e.getKeyChar()=='a')
      {
        if(climbing)
        {
          climbing = false;
          character.setIcon(new ImageIcon("oiram.png"));
        }
        if(character.getX()>=50 &&
        arena[character.getY()/50][(character.getX()/50)-1].equals(" "))
        {
          character.setBounds(character.getX()-50,character.getY(),50,50);
        }
      }
      //move right
      if(e.getKeyChar()=='d')
      {
        if(character.getX()<=400 && arena
                [character.getY()/50][(character.getX()/50)+1].equals(" "))
        {
          character.setBounds(character.getX()+50,character.getY(),50,50);
        }
        if(arena[character.getY()/50][(character.getX()/50)+1].equals("|"))
        {
          character.setBounds(character.getX()+50,character.getY(),50,50);
          climbing = true;
          character.setIcon(new ImageIcon("onladder.png"));
        }
      }
      //move up
      if(e.getKeyChar()=='w')
      {
        if(!climbing)
        {
          if(arena[(character.getY()/50)-1][character.getX()/50].equals(" "))
          {
            if(!jumping && !arena[(character.getY()/50)+1][character.getX()/50].equals(" "))
            {
              jumping = true;
              character.setBounds(character.getX(),character.getY()-50,50,50);
            }
          }
        }
        else
        {
          character.setBounds(character.getX(), character.getY()-50,50,50);
          if(arena[character.getY()/50][character.getX()/50].equals(" "))
          {
            character.setBounds(character.getX()+50,character.getY(),50,50);
            climbing = false;
            character.setIcon(new ImageIcon("oiram.png"));
          }
        }
      }
    }
    public static void main (String[] args)
    {
      new Oiram();
    }
}
```

Figures 43-4 and 43-5 show Oiram in action.

Every Evil Genius knows a game is not complete without adversaries. Move on to the next project and learn how to make Oiram's life miserable.

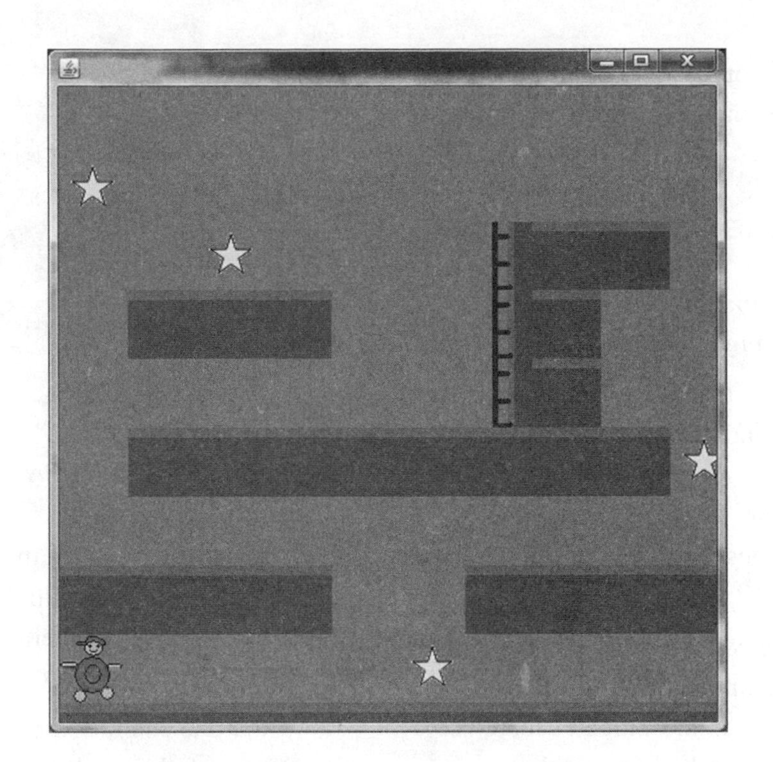

Figure 43-4 *Platform.*

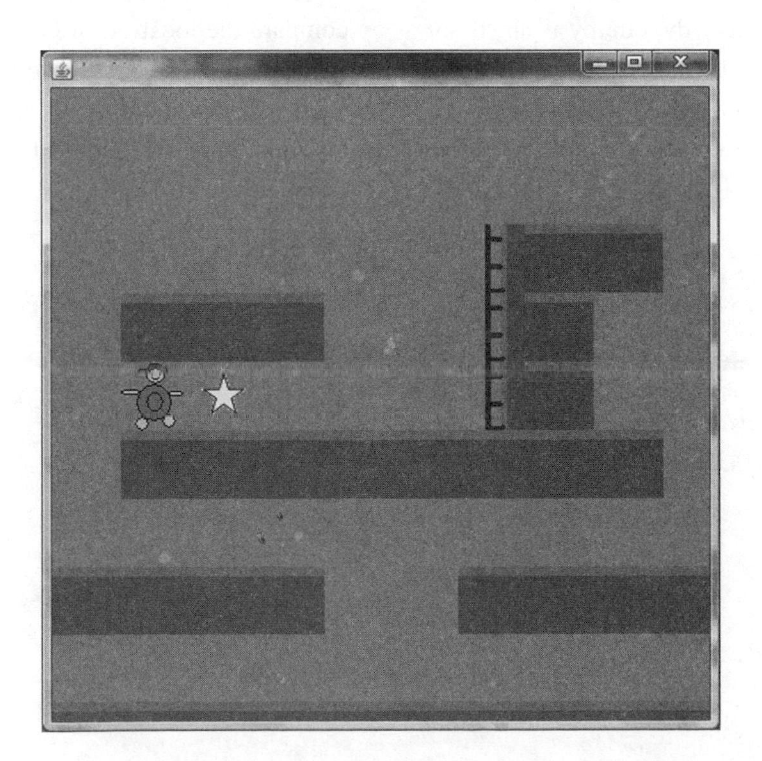

Figure 43-5 *Collecting stars.*

Project

The chase is on! Here you design Oiram's enemies. They're fast, they're relentless . . . they're deadly.

Figure 44-1 *Oiram's enemy.*

Making the game

First, draw a 50 by 50 pixel image of Oiram's enemy like the one in Figure 44-1.

Next, create a method that adds the enemies to the platform. Use a loop to generate two enemies. Place them at the very top of the board and give them a random "X" coordinate. In addition, add the enemies to an ArrayList for easy access later.

Oiram's enemies are controlled by simple logic: if there is air below them, they fall. If not, they either move left or right. In the Thread's loop, use a `for loop` to examine each enemy in the ArrayList. Use an if-statement to determine whether the location below the enemy is air. If so, lower the enemy 50 pixels. Next, generate a random number (0 or 1) by multiplying the Math.random method by two. If the number is zero, check whether the location to the right of the enemy is air. If it is, move the enemy 50 pixels to the right. If the random number is one, check whether the space to the left of the enemy is air. If so, move the enemy 50 pixels to the left.

Now that the enemies are mobile, you need to give Oiram a fighting chance. If he jumps on an enemy, the enemy dies. Use an if-statement to compare the location of Oiram and his adversary. If Oiram is 50 pixels above the enemy, remove the enemy from the container. If Oiram occupies the same location as his adversary, remove Oiram: the enemy wins.

```
import javax.swing.*;
import javax.swing.event.*;
import java.awt.*;
import java.awt.event.*;
import java.util.*;
import java.awt.geom.*;
public class Oiram extends JFrame implements KeyListener
{
  //the container:
  Container cont;
  //the 2 dimensional array
  String arena[ ][ ] =
  {{" "," "," "," "," "," "," "," "," "," "},
  {" "," "," "," "," "," "," "," "," "," "},
  {" "," "," "," "," "," ","|","#","#"," "},
  {" ","#","#","#"," "," ","|","#"," "," "},
  {" "," "," "," "," "," ","|","#"," "," "},
```

```
{" ","#","#","#","#","#","#","#","#"," "},
{" "," "," "," "," "," "," "," "," "," "},
{"#","#","#","#"," "," ","#","#","#","#"},
{" "," "," "," "," "," "," "," "," "," "},
{"#","#","#","#","#","#","#","#","#","#"}};
//this holds the stars:
ArrayList stars = new ArrayList();

//Oiram!
JLabel character = new JLabel(new ImageIcon("oiram.png"));
//whether or not Oiram is jumping/climbing
boolean jumping = false;
boolean climbing = false;
//the Thread
Runner runner;
//this holds the enemies
ArrayList enemies = new ArrayList();
public Oiram()
{
  super("Oiram");
  setSize(500,500);
  setVisible(true);
  setDefaultCloseOperation(JFrame.EXIT_ON_CLOSE);

  cont = getContentPane();
  cont.setLayout(null);

  addKeyListener(this);
  cont.setBackground(Color.BLACK);

  //add Oiram
  cont.add(character);
  character.setBounds(0,400,50,50);
  //create teh stars and enemies
  generateStars();
  generateEnemies();
  //generate the board:
  for(int i = 0; i < arena.length; i++)
  {
    for(int j = 0; j < arena[0].length; j++)
    {
      JLabel lbl = null;
      if(arena[j][i].equals("#"))
      {
      lbl = new JLabel(new ImageIcon("ground.png"));
      }
      else if(arena[j][i].equals(" "))
      {
        lbl = new JLabel(new ImageIcon("air.png"));
      }
      else if(arena[j][i].equals("|"))
      {
        lbl = new JLabel(new ImageIcon("ladder.png"));
      }
      cont.add(lbl);
      lbl.setBounds(i*50,j*50,50,50);
    }
  }

  repaint();
  cont.validate();
  //start the Thread:
  runner = new Runner();
```

```
  runner.start();

  setContentPane(cont);
  }

public void generateStars()
{
  //loop through te two dimensional array
  for(int i = 1; i < arena.length; i++)
  {
    for(int j = 0; j < arena[0].length; j++)
    {
      if(arena[i][j].equals(" "))
      {
        //give a 1 in 10 chance of placing a star
        int placeOrNot = (int)(Math.random()*10);
        if(placeOrNot==0)
        {
          //add the star
          JLabel star = new JLabel(new ImageIcon("star.png"));
          cont.add(star);
          star.setBounds(j*50,i*50,50,50);
          cont.setComponentZOrder(star,0);
          cont.setComponentZOrder(character,0);
          stars.add(star);
        }
      }
    }
  }
}
public void generateEnemies()
{
  //add the new enemies
  for(int i = 0; i < 2; i++)
  {
    JLabel enemy = new JLabel(new ImageIcon("enemy.png"));
    cont.add(enemy);
    int xLoc = (int)(Math.random()*8);
    enemy.setBounds(xLoc*50,0,50,50);
    cont.setComponentZOrder(enemy,0);
    enemies.add(enemy);
  }
}

public class Runner extends Thread
{
  public void run()
  {
    while(true)
    {
      try
      {
        //check the following on every enemy
        for(int i = 0; i < enemies.size(); i++)
        {
          JLabel enemy = (JLabel) enemies.get(i);
          //only apply the following to onscreen enemies
          if(enemy.getY()<=450 && enemy.getX()<=450)
          {
            //move the enemy down, if possible
            if(arena[(enemy.getY()/50)+1][enemy.getX()/50].equals(" "))
            {
```

```
      enemy.setBounds(enemy.getX(),enemy.getY()+50,50,50);
    }
    //move the enemy left/right
    int direction = (int)(Math.random()*2);
    if(direction==0)
    {
      if(arena[enemy.getY()/50][(enemy.getX()/50)+1].equals(" "))
      {
        enemy.setBounds(enemy.getX()+50,enemy.getY(),50,50);
      }
    }
    else
    {
      if(arena[enemy.getY()/50][(enemy.getX()/50)-1].equals(" "))
      {
        enemy.setBounds(enemy.getX()-50,enemy.getY(),50,50);
      }
    }
    //if Oiram jumps on an enemy, remove the enemy
    if(enemy.getY()-50==character.getY() && enemy.getX()==character.getX())
    {
      enemy.setBounds(1000,1000,50,50);
      cont.remove(enemy);
    }
    //if an enemy eats Oiram, display the losing image
    if(enemy.getY()==character.getY() && enemy.getX()==character.getX())
    {
      cont.remove(character);
    }
  }
}
//check the following on every star
for(int i = 0; i < stars.size(); i++)
{
  JLabel star = (JLabel) stars.get(i);
  //if Oiram captures a star, remove it
  if(star.getBounds().intersects(character.getBounds()))
  {
    cont.remove(star);
    stars.remove(star);
  }
}
//let Oiram fall
if(!jumping)
{
  if(arena[(character.getY()/50)+1][character.getX()/50].equals(" "))
  {
    character.setBounds(character.getX(),character.getY()+50,50,50);
  }
}
//let Oiram jump
else
{
  jumping = false;
  if(arena[(character.getY()/50)-1][character.getX()/50].equals(" "))
  {
    character.setBounds(character.getX(),character.getY()-50,50,50);
  }
}
```

```
        //delay
        Thread.sleep(250);
      }
      catch(Exception e){ }
    }
  }
}
public void keyPressed(KeyEvent e){ }
public void keyReleased(KeyEvent e){ }
public void keyTyped(KeyEvent e)
{
  //move Oiram left
  if(e.getKeyChar()=='a')
  {
    //check if Oiram is climbing. if so, end the climb
    if(climbing)
    {
      climbing = false;
      character.setIcon(new ImageIcon("oiram.png"));
    }
    //Move Oiram left, if possible
    if(character.getX()>=50 && arena[character.getY()/50][(character.getX()/50)-
          1].equals(" "))
    {
      character.setBounds(character.getX()-50,character.getY(),50,50);
    }
  }
  //move Oiram right
  if(e.getKeyChar()=='d')
  {
    //don't let Oiram go offscreen!'
    if(character.getX()<=400 &&
    arena[character.getY()/50][(character.getX()/50)+1].equals(" "))
    {
      character.setBounds(character.getX()+50,character.getY(),50,50);
    }
    //if Oiram hits a ladder, begin to climb
    if(arena[character.getY()/50][(character.getX()/50)+1].equals("|"))
    {
    character.setBounds(character.getX()+50,character.getY(),50,50);
    climbing = true;
    character.setIcon(new ImageIcon("onladder.png"));
    }
  }
  //jump/climb
  if(e.getKeyChar()=='w')
  {
    //if Oiram is not climbing, then jump up!
    if(!climbing)
    {
      if(arena[(character.getY()/50)-1][character.getX()/50].equals(" "))
      {
        if(!jumping && !arena[(character.getY()/50)+1][character.getX()/50].equals(" "))
        {
          jumping = true;
          character.setBounds(character.getX(),character.getY()-50,50,50);
        }
      }
    }
```

```
      //move Oiram up the ladder ...
      else
      {
      character.setBounds(character.getX(), character.getY()-50,50,50);
      //Oiram reached the top of the ladder. change his icon
      //and move hime over.
      if(arena[ character.getY()/50][ character.getX()/50] .equals(“ ”))
      {
          character.setBounds(character.getX()+50,character.getY(),50,50);
          climbing = false;
          character.setIcon(new ImageIcon(“oiram.png”));
        }
      }
    }
  }
  public static void main (String[ ] args)
  {
    new Oiram();
  }
}
```

Figures 44-2 through 44-4 illustrate enemy action.

It's not over yet for Oiram! Keep going to learn how to add more levels of increasing difficulty.

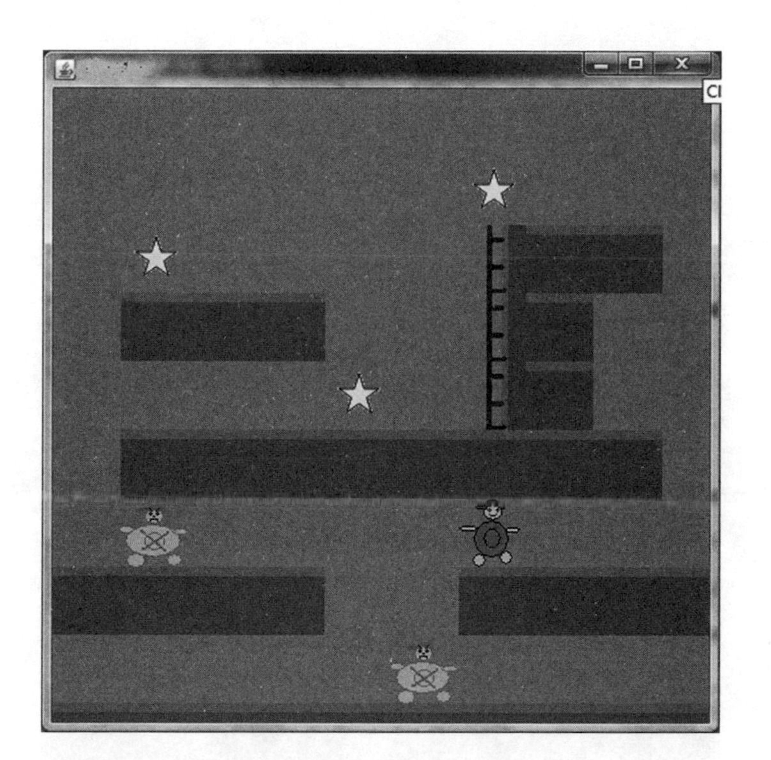

 Figure 44-2 *Enemies are alive!*

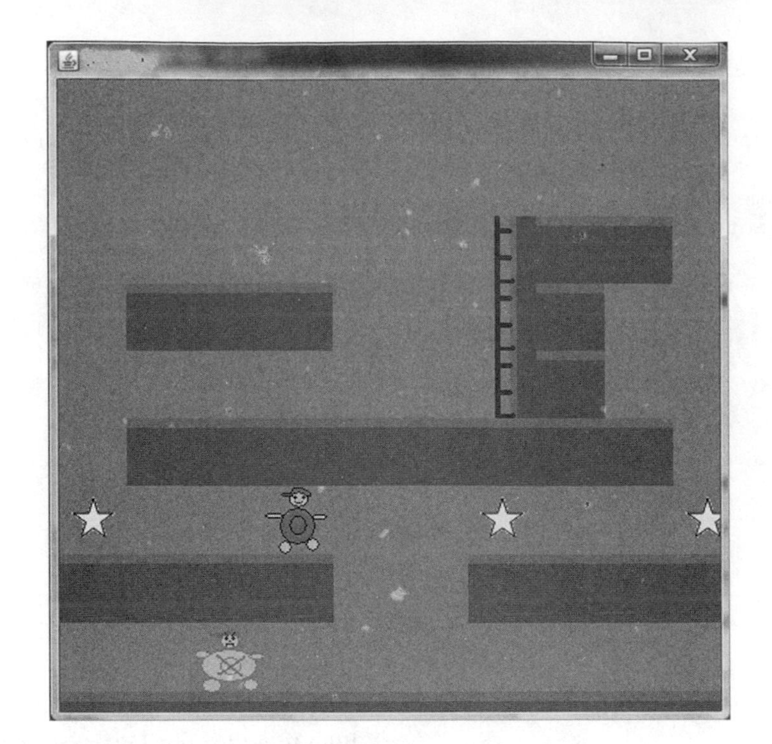

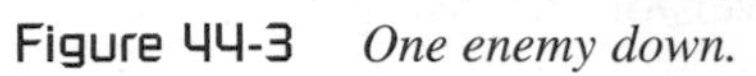

Figure 44-3 *One enemy down.*

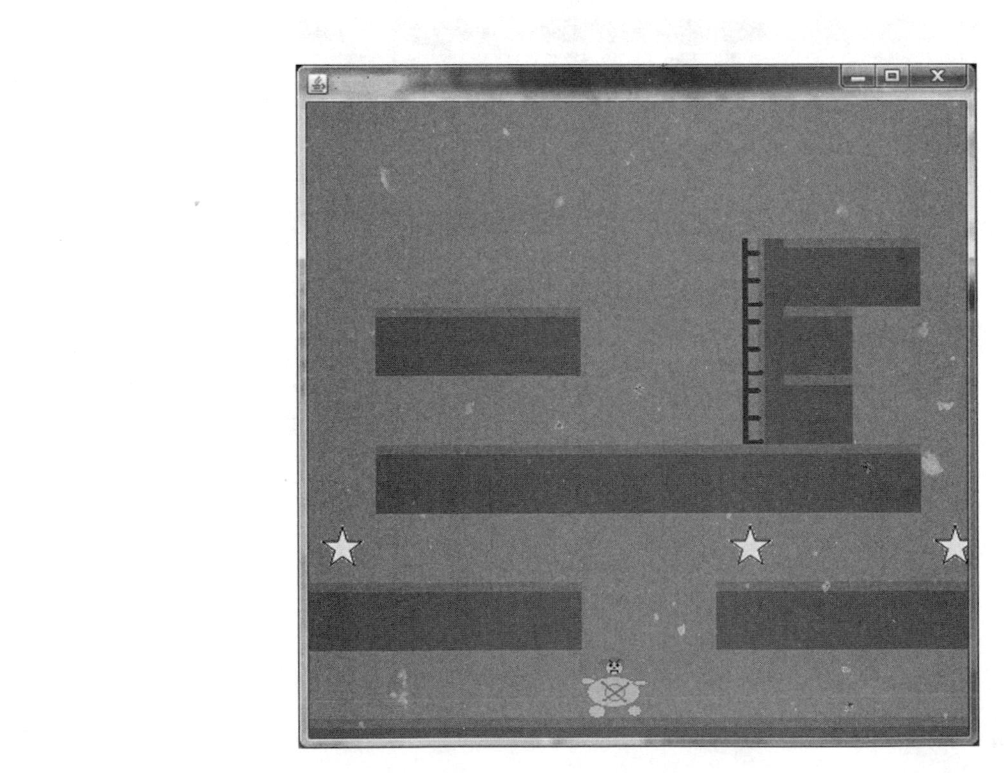

Figure 44-4 *Remaining enemy kills Oiram.*

Project 45: Oiram—Complicated World

Project

Add levels of increasing difficulty and winning/losing images to your game board.

Making the game

Oiram will fight through five levels: each level will add more enemies. To increase the number of enemies, slightly modify the method that draws them. To do this, create one global variable that keeps track of the level of play. Declare another variable inside the method that holds a random number (1 or 2). Add that random number to the level variable. Create that number of enemies.

A new level begins when Oiram defeats all of his opponents and captures all of the stars. To begin a new level, initialize two variables: one that keeps track of the number of enemies and one that keeps track of the number of stars. Each time Oiram collects a star or defeats an enemy, subtract one from the appropriate variable. In the infinite loop, check the values of both variables. Once both variables are zero, increment the level variable by one. Call the methods that generate the stars and enemies. In addition, create a JLabel that displays the current level. Remember to refresh the JLabel whenever the level changes.

To give Oiram a score, create a variable. Increment it by 100 each time Oiram captures a star. Increment it by 200 when Oiram defeats an adversary. Display the score in a JLabel.

You can also add a fun title image. Draw Oiram in a 500 by 50 pixel image. Display the JLabel that holds the image at the top of the platform.

To add winning/losing images, create two 500 by 500 pixel images. Start by displaying them offscreen. If Oiram dies or wins by getting past level five, move the appropriate image to the center of the arena. Remember to remove all other images.

```
import javax.swing.*;
import javax.swing.event.*;
import java.awt.*;
import java.awt.event.*;
import java.util.*;
import java.awt.geom.*;

public class Oiram extends JFrame implements KeyListener
{
  //the container:
  Container cont;

  //the 2 dimensional array
  String arena[ ][ ] =
  {{" "," "," "," "," "," "," "," "," "," "},
  {" "," "," "," "," "," "," "," "," "," "},
  {" "," "," "," "," "," ","|","#","#"," "},
  {" ","#","#","#"," "," ","|","#"," "," "},
  {" "," "," "," "," "," ","|","#"," "," "},
  {" ","#","#","#","#","#","#","#","#"," "},
  {" "," "," "," "," "," "," "," "," "," "},
  {"#","#","#","#"," "," ","#","#","#","#"},
  {" "," "," "," "," "," "," "," "," "," "},
  {"#","#","#","#","#","#","#","#","#","#"}};
```

```
//this holds the stars:
ArrayList stars = new ArrayList();
//Oiram!
JLabel character = new JLabel(new ImageIcon("oiram.png"));
//whether or not Oiram is jumping/climbing
boolean jumping = false;
boolean climbing = false;
//the Thread
Runner runner;
//Oiram's score'
int score = 0;
//the number of remaining stars
int starsLeft;
//this holds the enemies
ArrayList enemies = new ArrayList();
//the winning/losing images
JLabel win = new JLabel(new ImageIcon("win.png"));
JLabel lose = new JLabel(new ImageIcon("lose.png"));
//the title image
JLabel title = new JLabel(new ImageIcon("title.png"));
//the current level
int level = 1;
//the number of enemies
int enemyCount = 1;
//displays the level/score:
JLabel levelLbl = new JLabel("Level "+level+"/5");
JLabel scoreLbl = new JLabel("Score "+score);
public Oiram()
{
  super("Oiram");
  setSize(500,500);
  setVisible(true);
  setDefaultCloseOperation(JFrame.EXIT_ON_CLOSE);
  cont = getContentPane();
  cont.setLayout(null);
  addKeyListener(this);
  cont.setBackground(Color.BLACK);
  //add the win/lose images offscreen
  cont.add(win);
  win.setBounds(500,500,500,500);
  cont.add(lose);
  lose.setBounds(500,500,500,500);
  //add the level label
  cont.add(levelLbl);
  levelLbl.setFont(new Font("arial",Font.BOLD,20));
  levelLbl.setBounds(375,5,150,50);
  //add the score label
  cont.add(scoreLbl);
  scoreLbl.setFont(new Font("arial",Font.BOLD,20));
  scoreLbl.setBounds(20,5,150,50);
  //add the title image
  cont.add(title);
  title.setBounds(0,0,500,50);
  //add Oiram
  cont.add(character);
  character.setBounds(0,400,50,50);
  //create teh stars and enemies
  generateStars();
  generateEnemies();
```

```
  //generate the board:
  for(int i = 0; i < arena.length; i++)
  {
    for(int j = 0; j < arena[0].length; j++)
    {
      JLabel lbl = null;
      if(arena[j][i].equals("#"))
      {
        lbl = new JLabel(new ImageIcon("ground.png"));
      }
      else if(arena[j][i].equals(" "))
      {
        lbl = new JLabel(new ImageIcon("air.png"));
      }
      else if(arena[j][i].equals("|"))
      {
        lbl = new JLabel(new ImageIcon("ladder.png"));
      }
      cont.add(lbl);
      lbl.setBounds(i*50,j*50,50,50);
    }
  }

  repaint();
  cont.validate();
  //start the Thread:
  runner = new Runner();
  runner.start();

  setContentPane(cont);
}

public void generateStars()
{
  //loop through te two dimensional array
  for(int i = 1; i < arena.length; i++)
  {
    for(int j = 0; j < arena[0].length; j++)
    {
      if(arena[i][j].equals(" "))
      {
        //give a 1 in 10 chance of placing a star
        int placeOrNot = (int)(Math.random()*10);
        if(placeOrNot==0)
        {
          //add the star
          JLabel star = new JLabel(new ImageIcon("star.png"));
          cont.add(star);
          star.setBounds(j*50,i*50,50,50);
          cont.setComponentZOrder(star,0);
          cont.setComponentZOrder(character,0);
          stars.add(star);
          starsLeft++;
        }
      }
    }
  }
}
public void generateEnemies()
{
  //add a random number of enemies
  int increaseBy = (int)(Math.random()*2)+1;
```

```
  enemyCount = level+increaseBy;
  //add the new enemies
  for(int i = 0; i < enemyCount; i++)
  {
    JLabel enemy = new JLabel(new ImageIcon("enemy.png"));
    cont.add(enemy);
    int xLoc = (int)(Math.random()*8);
    enemy.setBounds(xLoc*50,0,50,50);
    cont.setComponentZOrder(enemy,0);
    enemies.add(enemy);
  }
}
public class Runner extends Thread
{
  public void run()
  {
    while(true)
    {
      try
      {
        //the current score
        scoreLbl.setText("Score "+score);
        //check the following on every enemy
        for(int i = 0; i < enemies.size(); i++)
        {
          JLabel enemy = (JLabel) enemies.get(i);
          //only apply the following to onscreen enemies
          if(enemy.getY()<=450 && enemy.getX()<=450)
          {
            //move the enemy down, if possible
            if(arena[(enemy.getY()/50)+1] [enemy.getX()/50].equals(" "))
            {
              enemy.setBounds(enemy.getX(), enemy.getY()+50,50,50);
            }
            //move the enemy left/right
            int direction = (int)(Math.random()*2);
            if(direction==0)
            {
              if(arena[enemy.getY()/50][(enemy.getX()/50)+1].equals(" "))
              {
                enemy.setBounds(enemy.getX()+50,enemy.getY(),50,50);
              }
            }
            else
            {
              if(arena[enemy.getY()/50][(enemy.getX()/50)-1].equals(" "))
              {
                enemy.setBounds(enemy.getX()-50,enemy.getY(),50,50);
              }
            }
            //if Oiram jumps on an enemy, remove the enemy
            if(enemy.getY()-50==character.getY() && enemy.getX()==character.getX())
            {
              enemy.setBounds(1000,1000,50,50);
              cont.remove(enemy);
              enemyCount--;
              score+=200;
            }
            //if an enemy eats Oiram, display the losing image
            if(enemy.getY()==character.getY() &&enemy.getX()==character.getX())
```

```
        {
          lose.setBounds(0,0,500,500);
          cont.setComponentZOrder(lose,0);
          for(int j=2; j < cont.getComponentCount(); j++)
          {
            cont.remove(j);
          }
          cont.validate();
        }
      }
    }

    //get to level 5 to win!
    if(level>=5)
    {
      win.setBounds(0,0,500,500);
      cont.setComponentZOrder(win,0);
      for(int i = 2; i < cont.getComponentCount(); i++)
      {
        cont.remove(i);
      }
      cont.validate();
    }
    //increase levels
    if(enemyCount<=0 && starsLeft<=0)
    {
      level++;
      generateStars();
      generateEnemies();
      levelLbl.setText("Level "+level+"/5");
    }
    //check the following on every star
    for(int i = 0; i < stars.size(); i++)
    {
      JLabel star = (JLabel) stars.get(i);
      //if Oiram captures a star, remove it
      if(star.getBounds().intersects(character.getBounds()))
      {
        score+=100;
        cont.remove(star);
        stars.remove(star);
        starsLeft--;
      }
    }
    //let Oiram fall
    if(!jumping)
    {
      if(arena[(character.getY()/50)+1][character.getX()/50].equals(" "))
      {
        character.setBounds(character.getX(),character.getY()+50,50,50);
      }
    }
    //let Oiram jump
    else
    {
      jumping = false;
      if(arena[(character.getY()/50)-1][character.getX()/50].equals(" "))
      {
        character.setBounds(character.getX(),character.getY()-50,50,50);
      }
    }
```

```
        //delay
        Thread.sleep(250);
      }
      catch(Exception e){ }
    }
  }
}
public void keyPressed(KeyEvent e){ }
public void keyReleased(KeyEvent e){ }
public void keyTyped(KeyEvent e)
{
  //move Oiram left
  if(e.getKeyChar()=='a')
  {
    //check if Oiram is climbing. if so, end the climb
    if(climbing)
    {
      climbing = false;
      character.setIcon(new ImageIcon("oiram.png"));
    }
    //Move Oiram left, if possible
    if(character.getX()>=50 &&
      arena[character.getY()/50][(character.getX()/50)-1] .equals(" "))
    {
      character.setBounds(character.getX()-50,character.getY(),50,50);
    }
  }
  //move Oiram right
  if(e.getKeyChar()=='d')
  {
    //don't let Oiram go offscreen!'
    if(character.getX()<=400 &&
    arena[character.getY()/50][(character.getX()/50)+1].equals(" "))
    {
      character.setBounds(character.getX()+50,character.getY(),50,50);
    }
    //if Oiram hits a ladder, begin to climb
    if(arena[character.getY()/50][(character.getX()/50)+1].equals("|"))
    {
      character.setBounds(character.getX()+50,character.getY(),50,50);
      climbing = true;
      character.setIcon(new ImageIcon("onladder.png"));
    }
  }
  //jump/climb
  if(e.getKeyChar()=='w')
  {
    //if Oiram is not climbing, then jump up!
    if(!climbing)
    {
      if(arena[(character.getY()/50)-1][character.getX()/50].equals(" "))
      {
        if(!jumping && !arena[(character.getY()/50)+1][character.getX()/50].equals(" "))
        {
          jumping = true;
          character.setBounds(character.getX(),character.getY()-50,50,50);
        }
      }
    }
    //move Oiram up the ladder...
```

```
        else
        {
        character.setBounds(character.getX(),character.getY()-50,50,50);
        //Oiram reached the top of the ladder. change his icon
        //and move hime over.
        if(arena[ character.getY()/50][ character.getX()/50].equals(" "))
        {
            character.setBounds(character.getX()+50,character.getY(),50,50);
            climbing = false;
            character.setIcon(new ImageIcon("oiram.png"));
          }
        }
      }
    }
    public static void main (String[ ] args)
    {
      new Oiram();
    }
}
```

Figures 45-1 through 45-3 depict the life and death of Oiram.

Customizing the game

Change the layout of the arena. Add more platforms, more ladders, etc.

Change Oiram's icon into a hot air balloon that allows him to drop bombs onto his enemies.

Add sound effects whenever Oiram collects a star or kills an enemy. Program applause for wins and boos for losses.

Add super-stars that give Oiram temporary flying abilities.

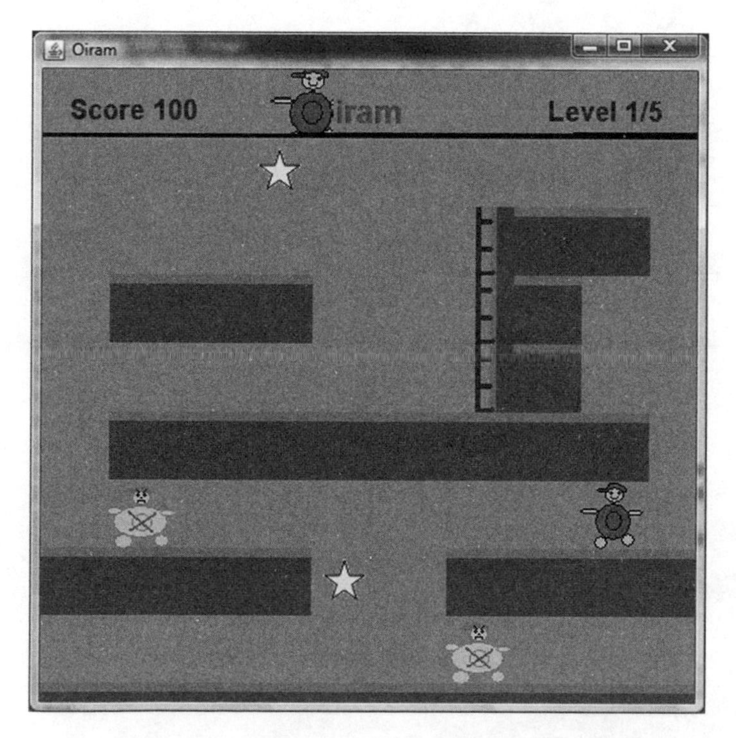

Figure 45-1 *Level and score display.*

Increase the size of the board: make each zone 10 by 10 pixels.

Clone Oiram so that there are two—or even three—of them in the arena.

Place an invisible enemy within the playing field. When close by, program the screen to "shake" to alert the player of Oiram's enemy's whereabouts.

Figure 45-2 *Oiram bites the dust.*

Figure 45-3 *Oiram prevails.*

Project 46: Java Man—Java Man's Universe

Java Man

Java and C++, the two largest programming languages, battle it out! Maneuver Java Man to collect all the dots on the field before C++ drinks him dry.

Project

Begin by creating Java Man's universe—the game board.

Making the game

To create the course, use a two dimensional array of Strings. Use "#" to represent the walls and a "space" to denote empty areas.

For example, an empty box is shown below:

```
####
#  #
#  #
####
```

```
import javax.swing.*;
import javax.swing.event.*;
import java.awt.*;
import java.awt.event.*;
import java.util.*;
import java.awt.geom.*;

public class JavaMan extends JFrame
{
  Container cont;
  //the 2 dimensional array
  String arena[ ][ ] =
  {{"#","#","#","#","#","#","#","#","#","#"},
  {"#"," "," "," ","#"," "," "," "," ","#"},
  {"#"," ","#"," ","#"," "," ","#"," ","#"},
  {"#"," ","#"," ","#"," ","#","#"," ","#"},
  {"#"," ","#"," "," "," "," ","#"," ","#"},
  {"#"," ","#","#"," "," "," ","#"," ","#"},
  {"#"," ","#"," "," ","#"," ","#"," ","#"},
  {"#"," ","#"," "," ","#"," ","#"," ","#"},
  {"#"," "," "," "," ","#"," "," "," ","#"},
  {"#","#","#","#","#","#","#","#","#","#"}};
```

A JavaMan board can be created the same way:

```
##########
#   #    #
# ## #   #
# ## #   #
# #  ##  #
# ##  #  #
# #  ##  #
# #  ##  #
#    #   #
##########
```

In the constructor, use a loop to examine each element in the array. The position of the elements in the two dimensional array should correspond with their placement on the JFrame. Draw a 50 by 50 pixel image of a solid yellow block and a 50 by 50 pixel image of a black square with a white circle inside. Use the yellow block for the "#" character and the black block for the empty space.

The use of arrays makes it easy to modify the layout of the board.

```
public JavaMan()
{
  super("JavaMan");
  setSize(500,500);
  setVisible(true);
  setDefaultCloseOperation(JFrame.EXIT_ON_CLOSE);

  cont = getContentPane();
  cont.setLayout(null);

  cont.setBackground(Color.BLACK);

  //create the board
  for(int i = 0; i < arena.length; i++)
  {
    for(int j = 0; j < arena[0].length; j++)
    {
      JLabel lbl = null;
      if(arena[i][j].equals("#"))
    {
      lbl = new JLabel(new ImageIcon("border.png"));
    }
    else
    {
      lbl = new JLabel(new ImageIcon("track_full.png"));
    }
    cont.add(lbl);
    lbl.setBounds(i*50,j*50,50,50);
    }
  }
  repaint();
  cont.validate();
  setContentPane(cont);
}
public static void main (String[ ] args)
{
  new JavaMan();
}
}
```

Figure 46-1 illustrates Java Man's universe.

In the next project, bring JavaMan to life! Prepare him for the fight of his life by making him mobile.

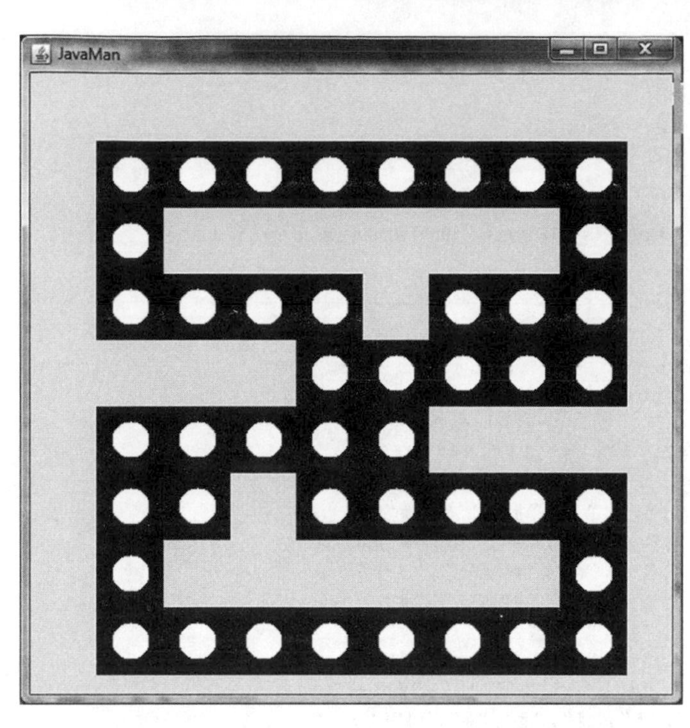

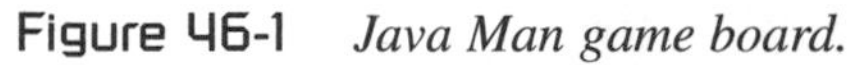
Figure 46-1 *Java Man game board.*

Project 47: Java Man—Java Man Lives!

Project

Place Java Man within the field of play. Then give Java Man the power of movement by adding keyboard controls to change his direction.

Making the game

Now that you have the course, it's time to add Java Man. In Microsoft Paint, create a 50 by 50 pixel image of a coffee mug. In the constructor, add this image before you add the board. Set it to the top left corner, position (50,50), as shown in Figure 47-1.

Make Java Man free to move. First, add a KeyListener to the class. Don't forget to add the three mandatory methods. Just like in the game Radical Racing, use variables to keep track of Java Man's orientation (1 represents up, 2 down, 3 right, and 4 left). Now, in the KeyPressed method, change JavaMan's direction.

Next, create a thread to move Java Man forward by either adding or subtracting 50 from the x or y coordinates. To prevent Java Man from passing through the walls, check his position against the value of the corresponding element in the two dimensional array. If the element is "#", move Java Man back one space.

Figure 47-1 *Java Man—starting position.*

```
import javax.swing.*;
import javax.swing.event.*;
import java.awt.*;
import java.awt.event.*;
import java.util.*;
import java.awt.geom.*;

public class JavaMan extends JFrame implements KeyListener
{
Container cont;

int UP = 0, DOWN = 1, RIGHT = 2, LEFT = 3;
int direction = RIGHT;
int positionX = 1, positionY = 1;
  String arena[ ][ ] =
  {{"#","#","#","#","#","#","#","#","#","#"},
  {"#"," "," "," ","#"," "," "," "," ","#"},
  {"#"," ","#"," ","#"," "," ","#"," ","#"},
  {"#"," ","#"," ","#"," ","#","#"," ","#"},
  {"#"," ","#"," "," "," "," ","#"," ","#"},
  {"#"," ","#","#"," "," "," ","#"," ","#"},
  {"#"," ","#"," "," ","#"," ","#"," ","#"},
  {"#"," ","#"," "," ","#"," ","#"," ","#"},
  {"#"," "," "," "," ","#"," "," "," ","#"},
  {"#","#","#","#","#","#","#","#","#","#"}};
  JLabel javaMan = new JLabel(new ImageIcon("man.PNG"));

  public JavaMan()
  {
    super("JavaMan");
    setSize(500,500);
    setVisible(true);
    setDefaultCloseOperation(JFrame.EXIT_ON_CLOSE);

    cont = getContentPane();
    cont.setLayout(null);

    addKeyListener(this);

    cont.setBackground(Color.BLACK);

    //add JavaMan
    cont.add(javaMan);
    javaMan.setBounds(50,50,50,50);

    for(int i = 0; i < arena.length; i++)
    {
      for(int j = 0; j < arena[0].length; j++)
      {
        JLabel lbl = null;
        if(arena[i][j].equals("#"))
        {
          lbl = new JLabel(new ImageIcon("border.png"));
        }
        else
        {
          lbl = new JLabel(new ImageIcon("track_full.png"));
        }
        cont.add(lbl);
        lbl.setBounds(i*50,j*50,50,50);
        System.out.println("X: "+i*50+" -- Y: "+j*50);
      }
    }

    repaint();
    cont.validate();
```

```
    Runner run = new Runner();
    run.start();

    setContentPane(cont);
  }
  public class Runner extends Thread
  {
    public void run()
    {
      while(true)
      {
        try
        {
          if(direction == RIGHT)
          {
            javaMan.setBounds(javaMan.getX()+50,javaMan.getY(),50,50);
            positionX++;
            if(arena[positionX][positionY].equals("#"))
            {
              javaMan.setBounds(javaMan.getX()-50,javaMan.getY(),50,50);
              positionX--;
            }
            cont.setComponentZOrder(javaMan,1);
          }
          if(direction == LEFT)
          {
            javaMan.setBounds(javaMan.getX()-50,javaMan.getY(),50,50);
            positionX--;
            if(arena[positionX][positionY].equals("#"))
            {
              javaMan.setBounds(javaMan.getX()+50,javaMan.getY(),50,50);
              positionX++;
            }
            cont.setComponentZOrder(javaMan,1);
          }
          if(direction == UP)
          {
          javaMan.setBounds(javaMan.getX(),javaMan.getY()-50,50,50);
            positionY--;
            if(arena[positionX][positionY].equals("#"))
            {
              javaMan.setBounds(javaMan.getX(),javaMan.getY()+50,50,50);
              positionY++;
            }
            cont.setComponentZOrder(javaMan,1);
          }
          if(direction == DOWN)
          {
            javaMan.setBounds(javaMan.getX(),javaMan.getY()+50,50,50);
            positionY++;
            if(arena[positionX][positionY].equals("#"))
            {
              javaMan.setBounds(javaMan.getX(),javaMan.getY()-50,50,50);
              positionY--;
            }
            cont.setComponentZOrder(javaMan,1);
          }
          cont.validate();
          Thread.sleep(500);
        }
        catch(Exception e){ }
```

```
      }
    }
  }
  public void keyTyped(KeyEvent e)
  {
    if(e.getKeyChar()=='w' )
      direction = UP;
    if(e.getKeyChar()=='a' )
      direction = LEFT;
    if(e.getKeyChar()=='s' )
      direction = DOWN;
    if(e.getKeyChar()=='d' )
      direction = RIGHT;
  }
  public void keyPressed(KeyEvent e){ }
  public void keyReleased(KeyEvent e){ }
  public static void main (String[ ] args)
  {
    new JavaMan();
  }
}
```

Figure 47-2 and 47-3 illustrates Java Man's mobility.

Java Man is done with training ... it's time to add the C++ adversaries. Let's get going ...

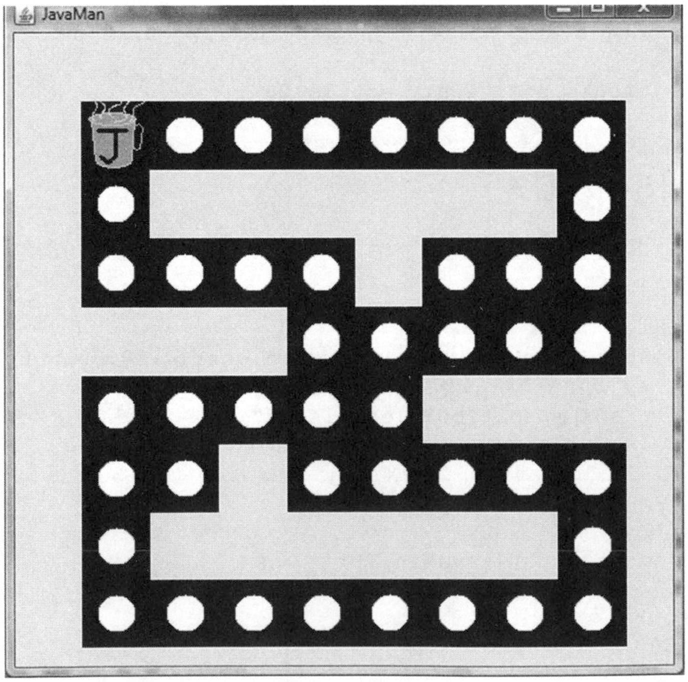

Figure 47-2 *Java Man comes to life.*

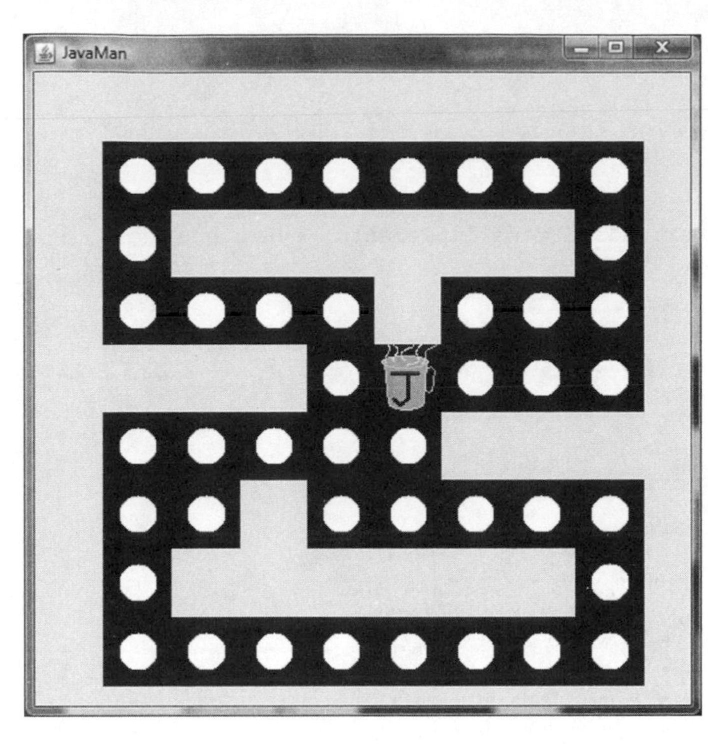

Figure 47-3 *Java Man moves!*

Project 48: Java Man—C++ Attacks

Project

Create and unleash the enemies. Although the C++ enemies blindly move around, they are swift and relentless!

Making the game

Start by drawing a 50 by 50 pixel image of the C++ culprit. In order to keep track of his clones, use an array 3 elements large. Each element will hold an enemy that occupies one corner of the board. Remember, the movements of the C++ clones are random. To create the movement, make another array 3 elements large in the thread. Each element of the new array stores the orientation of each C++ character by holding a random number between 1 and 4. Next, move each enemy in the direction of the random number (1 represents up, 2 down, 3 left, 4 right). To prevent the enemies from running through walls, use the same technique you used in the last project.

Next, add an if-statement to the loop. Compare the position of the C++ enemy and Java Man. If they are the same, Java Man loses. Add a JOptionPane that tells the player the sad news.

While Java Man is avoiding C++, he is also trying to devour the circles on the game board. When Java Man moves over a circle, replace that tile of the arena with a 50 by 50 pixel black image. Also, decrease the variable that keeps track of the number of remaining circles. When the count reaches zero, Java Man wins. Use an if-statement to check for the win. If Java Man wins, add a JOptionPane alerting the player of the victory.

```
import javax.swing.*;
import javax.swing.event.*;
import java.awt.*;
import java.awt.event.*;
import java.util.*;
import java.awt.geom.*;

public class JavaMan extends JFrame implements KeyListener
{
  Container cont;

  //the direction
  int UP = 0, DOWN = 1, RIGHT = 2, LEFT = 3;
  int direction = RIGHT;

  int score = 0;

  int positionX = 1, positionY = 1;

  //the 2 dimensional array
  String arena[ ][ ] =
  {{“#”,“#”,“#”,“#”,“#”,“#”,“#”,“#”,“#”,“#”},
  {“#”,“ ”,“ ”,“ ”,“#”,“ ”,“ ”,“ ”,“ ”,“#”},
  {“#”,“ ”,“#”,“ ”,“#”,“ ”,“ ”,“#”,“ ”,“#”},
  {“#”,“ ”,“#”,“ ”,“#”,“ ”,“#”,“#”,“ ”,“#”},
  {“#”,“ ”,“#”,“ ”,“ ”,“ ”,“ ”,“#”,“ ”,“#”},
  {“#”,“ ”,“#”,“#”,“ ”,“ ”,“ ”,“#”,“ ”,“#”},
  {“#”,“ ”,“#”,“ ”,“ ”,“#”,“ ”,“#”,“ ”,“#”},
  {“#”,“ ”,“#”,“ ”,“ ”,“#”,“ ”,“#”,“ ”,“#”},
  {“#”,“ ”,“ ”,“ ”,“ ”,“#”,“ ”,“ ”,“ ”,“#”},
  {“#”,“#”,“#”,“#”,“#”,“#”,“#”,“#”,“#”,“#”}};
  //the JLabel icons
  JLabel javaMan = new JLabel(new ImageIcon(“man.PNG”));
  JLabel enemies[ ] = {new JLabel(new ImageIcon(“monster.png”)),new JLabel(new ImageIcon
          (“monster.png”)),new JLabel(new ImageIcon(“monster.png”))};
  int dotsLeft = 44;

  public JavaMan()
  {
    super(“JavaMan”);
    setSize(500,500);
    setVisible(true);
    setDefaultCloseOperation(JFrame.EXIT_ON_CLOSE);

    cont = getContentPane();
    cont.setLayout(null);

    addKeyListener(this);

    cont.setBackground(Color.BLACK);

    //add JavaMan
    cont.add(javaMan);
    javaMan.setBounds(50,50,50,50);

    //add the board
    for(int i = 0; i < arena.length; i++)
    {
      for(int j = 0; j < arena[0].length; j++)
      {
        JLabel lbl = null;
        if(arena[i][j].equals(“#”))
        {
          lbl = new JLabel(new ImageIcon(“border.png”));
        }
        else
        {
```

```
      lbl = new JLabel(new ImageIcon("track_full.png"));
    }
    cont.add(lbl);
    lbl.setBounds(i*50,j*50,50,50);
    System.out.println("X: "+i*50+" - Y: "+j*50);
   }
  }

  //set the position of the enemies
  cont.add(enemies[0]);
  enemies[0].setBounds(400,50,50,50);
  cont.setComponentZOrder(enemies[0],1);
  cont.add(enemies[1]);
  enemies[1].setBounds(50,400,50,50);
  cont.setComponentZOrder(enemies[1],1);
  cont.add(enemies[2]);
  enemies[2].setBounds(400,400,50,50);
  cont.setComponentZOrder(enemies[2],1);

  repaint();
  cont.validate();

  Runner run = new Runner();
  run.start();

  setContentPane(cont);
}
public class Runner extends Thread
{
  public void run()
  {
    while(true)
    {
      try
      {
        //check if JavaMan wins
        if(dotsLeft<=0)
        {
          JOptionPane.showMessageDialog(null,"You Win!");
        }
        //enemy movement loop
        int dir[] = new int[3];
        for(int i = 0; i < dir.length; i++)
        {
          dir[i] = (int)(Math.random()*4);
          if(dir[i]==UP)
          {
            enemies[i].setBounds(enemies[i].getX(),enemies[i].getY()-50,50,50);
            if(arena[enemies[i].getX()/50]
              [enemies[i].getY()/50].equals("#"))
            {
              enemies[i].setBounds(enemies[i].getX(),enemies[i].getY()+50,50,50);
            }
          }
          if(dir[i]==DOWN)
          {
            enemies[i].setBounds(enemies[i].getX(),enemies[i].getY()+50,50,50);
            if(arena[enemies[i].getX()/50]
              [enemies[i].getY()/50].equals("#"))
          {
            enemies[i].setBounds(enemies[i].getX(),enemies[i].getY()-50,50,50);
          }
```

```
}
if(dir[i]==LEFT)
{
    enemies[i].setBounds(enemies[i].getX()-50,enemies[i].getY(),50,50);
    if(arena[enemies[i].getX()/50]
      [enemies[i].getY()/50].equals("#"))
    {
      enemies[i].setBounds(enemies[i].getX()+50,enemies[i].getY(),50,50);
    }
  }
  if(dir[i]==RIGHT)
  {
    enemies[i].setBounds(enemies[i].getX()+50,enemies[i].getY(),50,50);
    if(arena[enemies[i].getX()/50][enemies[i].getY()/50].equals("#"))
    {
    enemies[i].setBounds(enemies[i].getX()-50,enemies[i].getY(),50,50);
    }
  }
  //you lose!
  if(enemies[i].getX()/50==positionX && enemies[i].getY()/50==positionY)
  {
    JOptionPane.showMessageDialog(null,"You Lose!");
  }
  cont.setComponentZOrder(enemies[i],1);
}
//remove the circle
if(arena[positionX][positionY].equals(" "))
{
  arena[positionX][positionY] = ".";
  dotsLeft--;
  JLabel lbl = new JLabel(new ImageIcon("track_empty.png"));
  cont.add(lbl);
  lbl.setBounds(positionX*50,positionY*50,50,50);
  cont.setComponentZOrder(lbl,1);
  score++;
}

//move Java Man
if(direction == RIGHT)
{
  javaMan.setBounds(javaMan.getX()+50,javaMan.getY(),50,50);
  positionX++;
  if(arena[positionX][positionY].equals("#"))
  {
    javaMan.setBounds(javaMan.getX()-50,javaMan.getY(),50,50);
    positionX--;
  }
  cont.setComponentZOrder(javaMan,1);
}
if(direction == LEFT)
{
  javaMan.setBounds(javaMan.getX()-50,javaMan.getY(),50,50);
  positionX--;
if(arena[positionX][positionY].equals("#"))
{
  javaMan.setBounds(javaMan.getX()+50,javaMan.getY(),50,50);
  positionX++;
}
cont.setComponentZOrder(javaMan,1);
}
if(direction == UP)
```

```
        {
          javaMan.setBounds(javaMan.getX(),javaMan.getY()-50,50,50);
          positionY--;
          if(arena[positionX][positionY].equals("#"))
          {
            javaMan.setBounds(javaMan.getX(),javaMan.getY()+50,50,50);
            positionY++;
          }
          cont.setComponentZOrder(javaMan,1);
        }
        if(direction == DOWN)
        {
          javaMan.setBounds(javaMan.getX(),javaMan.getY()+50,50,50);
          positionY++;
          if(arena[positionX][positionY].equals("#"))
          {
            javaMan.setBounds(javaMan.getX(),javaMan.getY()-50,50,50);
            positionY--;
          }
          cont.setComponentZOrder(javaMan,1);
        }
        cont.validate();
        Thread.sleep(500);
      }
      catch(Exception e){ }
    }
  }
}

public void keyTyped(KeyEvent e)
{
    if(e.getKeyChar()=='w')
      direction = UP;
    if(e.getKeyChar()=='a')
      direction = LEFT;
    if(e.getKeyChar()=='s')
      direction = DOWN;
    if(e.getKeyChar()=='d')
      direction = RIGHT;
}
public void keyPressed(KeyEvent e){ }
public void keyReleased(KeyEvent e){ }
public static void main (String[ ] args)
{
  new JavaMan();
}
}
```

Figures 48-1 through 48-3 depict Java Man's quest for survival.

Keep reading to learn how to add more eye-catching win/lose images.

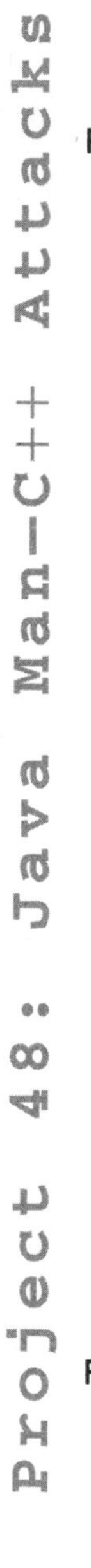

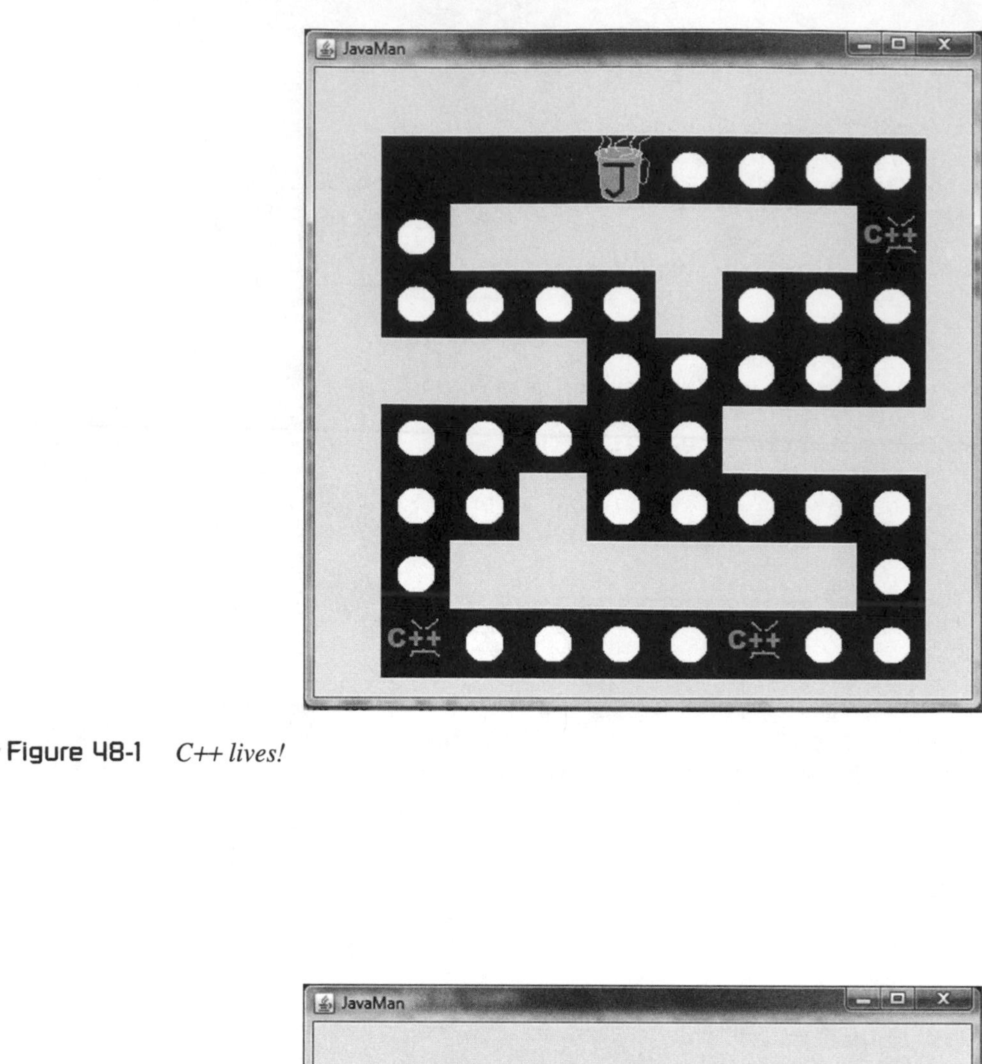

Figure 48-1 *C++ lives!*

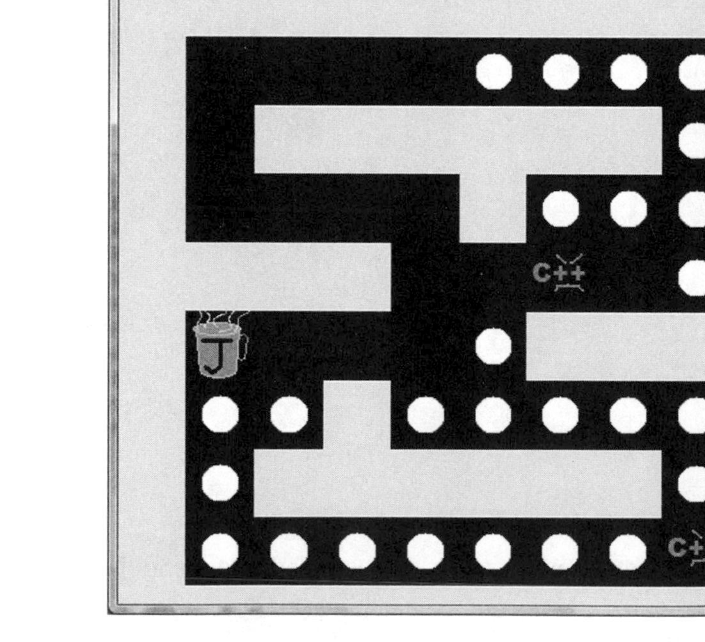

Figure 48-2 *Java Man chomps up the dots.*

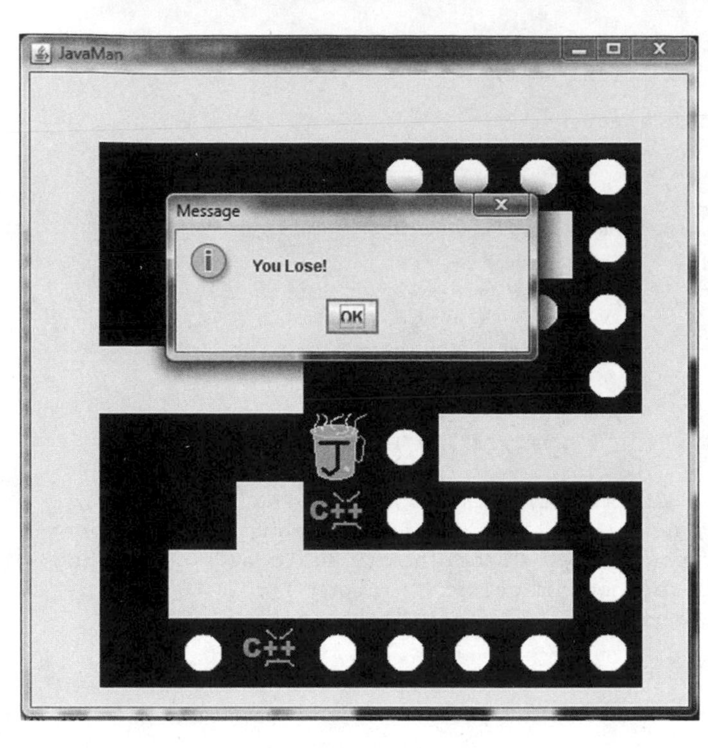

Figure 48-3 *C++ wins ... .*

Project 49: Java Man—Obituaries

Project

Turn up the graphic impact of the win/lose display.

Making the game

Replace those dull JOptionPanes with something fresh! Use Microsoft Paint to draw two 500 by 500 pixel images: one of C++ drinking spilled coffee from the fallen Java Man; the other, a defeated C++ falling into the hot cup of Java.

In the constructor, add these two images offscreen. When there is a win or loss, replace the JOptionPane code with the setBounds method. Don't forget to use a loop to remove all other components!

```
import javax.swing.*;
import javax.swing.event.*;
import java.awt.*;
import java.awt.event.*;
import java.util.*;
import java.awt.geom.*;

public class JavaMan extends JFrame implements KeyListener
{
  Container cont;
  int UP = 0, DOWN = 1, RIGHT = 2, LEFT = 3;
  //the direction of JavaMan
  int direction = RIGHT;
  int score = 0;
```

```
//JavaMan's position
int positionX = 1, positionY = 1;

//the two dimensional array
String arena[ ][ ] =
{{"#","#","#","#","#","#","#","#","#","#"},
{"#"," "," "," ","#"," "," "," "," ","#"},
{"#"," ","#"," ","#"," "," ","#"," ","#"},
{"#"," ","#"," ","#"," ","#","#"," ","#"},
{"#"," ","#"," "," "," "," ","#"," ","#"},
{"#"," ","#","#"," "," "," ","#"," ","#"},
{"#"," ","#"," "," ","#"," ","#"," ","#"},
{"#"," ","#"," "," ","#"," ","#"," ","#"},
{"#"," "," "," "," ","#"," "," "," ","#"},
{"#","#","#","#","#","#","#","#","#","#"}};
//the JLabels
JLabel javaMan = new JLabel(new ImageIcon("man.PNG"));
JLabel enemies[ ] = {new JLabel(new ImageIcon("monster.png")),new JLabel(new ImageIcon
        ("monster.png")),new JLabel(new ImageIcon("monster.png"))};
JLabel win = new JLabel(new ImageIcon("win.png"));
JLabel lose = new JLabel(new ImageIcon("lose.png"));
int dotsLeft = 44;
public JavaMan()
{
  super("JavaMan");
  setSize(500,500);
  setVisible(true);
  setDefaultCloseOperation(JFrame.EXIT_ON_CLOSE);

  cont = getContentPane();
  cont.setLayout(null);

  addKeyListener(this);

  cont.setBackground(Color.BLACK);

  cont.add(win);
  win.setBounds(500,500,500,500);
  cont.add(lose);
  lose.setBounds(500,500,500,500);

  //add JavaMan
  cont.add(javaMan);
  javaMan.setBounds(50,50,50,50);
  //create the arena
  for(int i = 0; i < arena.length; i++)
  {
    for(int j = 0; j < arena[0].length; j++)
    {
      JLabel lbl = null;
      if(arena[i][j].equals("#"))
      {
        lbl = new JLabel(new ImageIcon("border.png"));
      }
      else
      {
        lbl = new JLabel(new ImageIcon("track_full.png"));
      }
      cont.add(lbl);
      lbl.setBounds(i*50,j*50,50,50);
      System.out.println("X: "+i*50+" -- Y: "+j*50);
    }
  }
```

```
    //add the enemies
    cont.add(enemies[0]);
    enemies[0].setBounds(400,50,50,50);
    cont.setComponentZOrder(enemies[0],1);
    cont.add(enemies[1]);
    enemies[1].setBounds(50,400,50,50);
    cont.setComponentZOrder(enemies[1],1);
    cont.add(enemies[2]);
    enemies[2].setBounds(400,400,50,50);
    cont.setComponentZOrder(enemies[2],1);
    repaint();
    cont.validate();
    Runner run = new Runner();
    run.start();
    setContentPane(cont);
  }
  public class Runner extends Thread
  {
    public void run()
    {
      while(true)
      {
        try
        {
          //check if JavaMan wins
          if(dotsLeft<=0)
          {
            win.setBounds(0,0,500,500);
            cont.setComponentZOrder(win,0);
            for(int j = 2; j < cont.getComponentCount(); j++)
            {
              cont.remove(j);
            }
          }
          //control the C++
          int dir[] = new int[3];
          for(int i = 0; i < dir.length; i++)
          {
            //get the random direction
            dir[i] = (int)(Math.random()*4);
            if(dir[i]==UP)
            {
              enemies[i].setBounds(enemies[i].getX(),enemies[i].getY()-50,50,50);
              if(arena[enemies[i].getX()/50]
                [enemies[i].getY()/50].equals("#"))
            {
              enemies[i].setBounds(enemies[i].getX(),enemies[i].getY()+50,50,50);
            }
          }
          if(dir[i]==DOWN)
          {
            enemies[i].setBounds(enemies[i].getX(),enemies[i].getY()+50,50,50);
            if(arena[enemies[i].getX()/50]
              [enemies[i].getY()/50].
              equals("#"))
          {
            enemies[i].setBounds(enemies[i].getX(),enemies[i].getY()-50,50,50);
            }
            }
            if(dir[i]==LEFT)
```

```
    {
      enemies[i].setBounds(enemies[i].getX()-50,enemies[i].getY(),50,50);
      if(arena[enemies[i].getX()/50]
        [enemies[i].getY()/50].equals("#"))
      {
        enemies[i].setBounds(enemies[i].getX()+50,enemies[i].getY(),50,50);
      }
    }
    if(dir[i]==RIGHT)
    {
      enemies[i].setBounds(enemies[i].getX()+50,enemies[i].getY(),50,50);
      if(arena[enemies[i].getX()/50]
      [enemies[i].getY()/50].equals("#"))
    {
      enemies[i].setBounds(enemies[i].getX()-50,enemies[i].getY(),50,50);
      }
    }
    if(enemies[i].getX()/50==positionX && enemies[i].getY()/50==positionY)
    {
      lose.setBounds(0,0,500,500);
      cont.setComponentZOrder(lose,0);
      for(int j = 2; j < cont.getComponentCount(); j++)
      {
        cont.remove(j);
      }
    }
    cont.setComponentZOrder(enemies[i],1);
  }
  //remove the dot
  if(arena[positionX][positionY].equals(" "))
  {
    arena[positionX][positionY] = ".";
    dotsLeft--;
    JLabel lbl = new JLabel(new ImageIcon("track_empty.png"));
    cont.add(lbl);
    lbl.setBounds(positionX*50,positionY*50,50,50);cont.setComponentZOrder(lbl,1);s
      core++;
  }
    //move Java Man
    if(direction == RIGHT)
  {
    javaMan.setBounds(javaMan.getX()+50,javaMan.getY(),50,50);
    positionX++;
    if(arena[positionX][positionY].equals("#"))
    {
      javaMan.setBounds(javaMan.getX()-50,javaMan.getY(),50,50);
      positionX--;
    }
    cont.setComponentZOrder(javaMan,1);
  }
  if(direction == LEFT)
  {
    javaMan.setBounds(javaMan.getX()-50,javaMan.getY(),50,50);
    positionX--;
    if(arena[positionX][positionY].equals("#"))
    {
      javaMan.setBounds(javaMan.getX()+50,  javaMan.getY(),50,50);
      positionX++;
    }
    cont.setComponentZOrder(javaMan,1);
  }
```

```
            if(direction == UP)
            {
              javaMan.setBounds(javaMan.getX(),javaMan.getY()-50,50,50);
              positionY--;
              if(arena[positionX][positionY].equals("#"))
              {
                javaMan.setBounds(javaMan.getX(),javaMan.getY()+50,50,50);
                positionY++;
              }
              cont.setComponentZOrder(javaMan,1);
            }
            if(direction == DOWN)
            {
              javaMan.setBounds(javaMan.getX(),javaMan.getY()+50,50,50);
              positionY++;
              if(arena[positionX][positionY].equals("#"))
              {
                javaMan.setBounds(javaMan.getX(),java  Man.getY()-50,50,50);
                positionY--;
              }
              cont.setComponentZOrder(javaMan,1);
              cont.setComponentZOrder(win,1);
              cont.setComponentZOrder(lose,1);
            }
            cont.validate();
            Thread.sleep(500);
          }
          catch(Exception e){ }
        }
      }
    }
    public void keyTyped(KeyEvent e)
    {
      if(e.getKeyChar()=='w')
        direction = UP;
      if(e.getKeyChar()=='a')
        direction = LEFT;
      if(e.getKeyChar()=='s')
        direction = DOWN;
      if(e.getKeyChar()=='d')
        direction = RIGHT;
    }
    public void keyPressed(KeyEvent e){ }
    public void keyReleased(KeyEvent e){ }
    public static void main (String[ ] args)
    {
      new JavaMan();
    }
}
```

Figures 49-1 through 49-3 illustrate Java Man versus C++.

Customizing the game

Change the pathway of the board or, make it an open field.

Add more enemies.

Change the icons of the players to laser ships or geometric shapes.

Speed up or slow down the game ... or, make speed freak zones within the field.

Add special dots that give Java Man superpowers.

Instead of random movements, make the C++ chase Java Man.

Change the win/lose images.

Add fun slurping sounds when Java Man loses and splashing sounds when Java Man wins.

Add a sink hole on the board that vaporizes anything crossing its path.

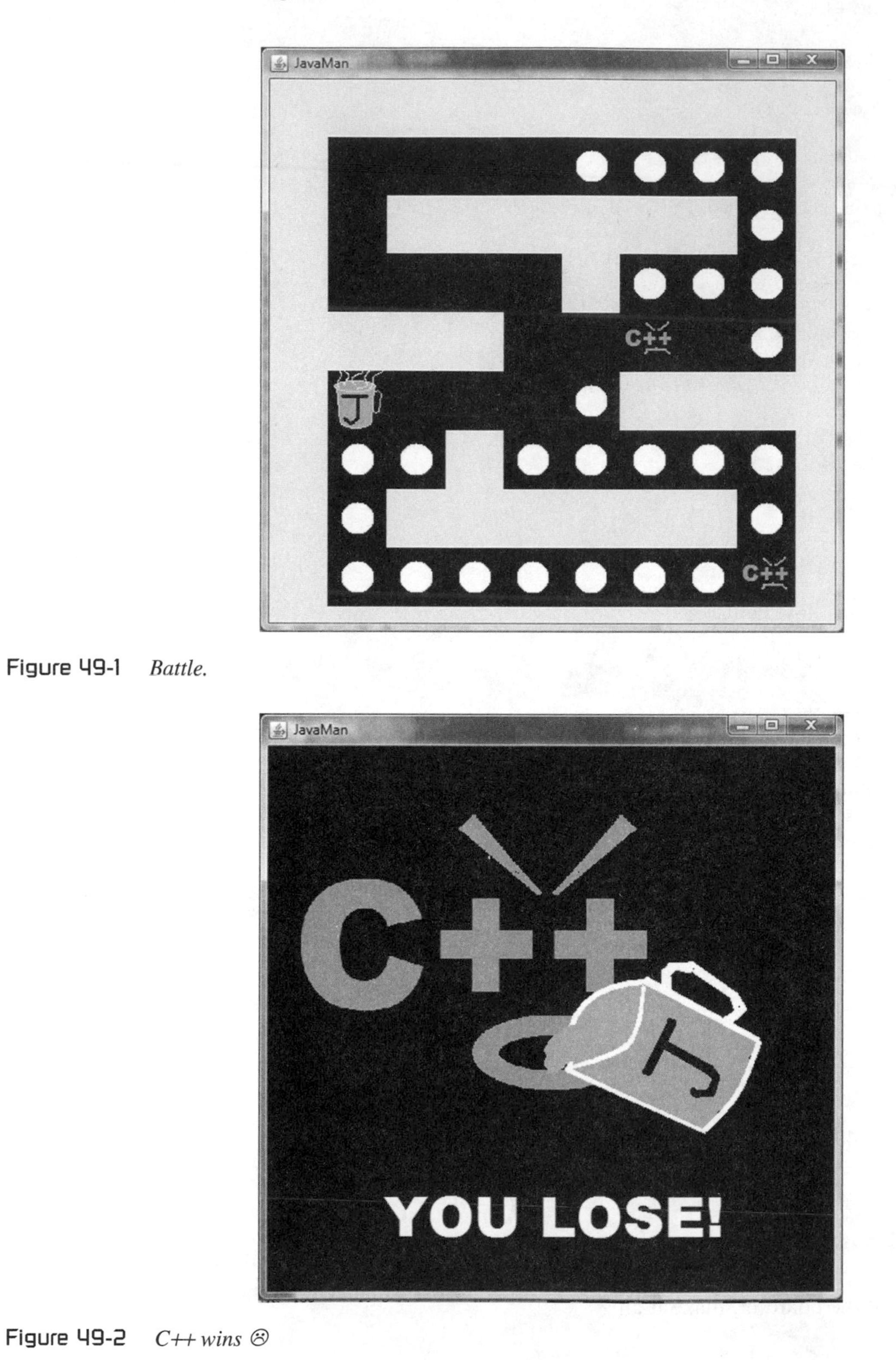

Figure 49-1 *Battle.*

Figure 49-2 *C++ wins ☹*

Figure 49-3 *Java Man is victorious!*

Section Seven

Brain Busters

Project 50: Memory—Grid Design

Memory

Although the biochemical process of memory remains a mystery, you can test yours by playing this game. A set of images flashes before your eyes. The challenge? Match the identical pictures before time runs out.

Project

Plan and develop the 16 grid game frame.

Making the game

The game, "Memory," consists of a 4 by 4 grid containing eight pairs of images. You will learn how to integrate photographs to create the images of your choosing in Project 53. For now, however, simply draw eight 100 by 100 pixel images and name them "img1.png," "img2.png," etc. In each generic image, write the image number, as shown in Figure 50-1.

Now that you have created the images, store them in a two dimensional array of JButtons that is 4 by 4 elements large. In the constructor, use two `for loops` to add the JButtons to the JFrame. Don't attach the images you created just yet ... instead, attach a new 100 by 100 pixel image, like the one shown in Figure 50-2, that displays the name of the game ("Memory").

Next, create a method that randomly assigns your images to the JButtons. To do this, use the Thread.sleep method to pause for half a second.

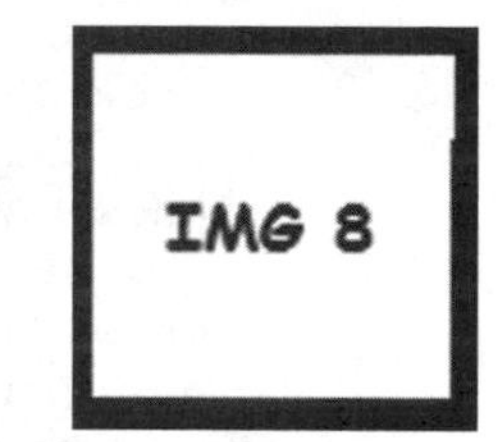

Figure 50-1 *Sample image.*

Then, initialize a normal array of ints that stores the number of times each image is used. This will prevent images from appearing more than twice. Now, use two `for loops` to go through the JButton array. At each iteration, create a random number between 1 and 8, inclusive. Using the normal array you created earlier, check whether the image corresponding to the random number has been used more than once. If so, generate a new random number. Once a satisfactory random number has been generated, increment by one the appropriate value in the single dimensional array. Don't forget to set the icon of the JButton to your new image. How do you keep track of the location of each image? Easy! Create a two dimensional global array 8 ints long. Set the appropriate element to the random number.

Now that the images are displayed, you need to hide them. Use a Thread.sleep method and pause

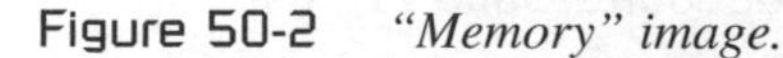

Figure 50-2 *"Memory" image.*

for 3 seconds to give players a chance to memorize the images. Then, make a new set of two `for loops`. Set the icons back to the starting image (the one that says "Memory").

```
import javax.swing.*;
import javax.swing.event.*;
import java.awt.*;
import java.awt.event.*;

public class Memory extends JFrame
{
  //the blank image
  ImageIcon blank = new ImageIcon("blank.png");

  //the buttons
  JButton buttons[ ][ ] = {{new JButton(blank),new JButton(blank),
                          new JButton(blank),new JButton(blank)},
                          {new JButton(blank),new JButton(blank),
                          new JButton(blank),new JButton(blank)},
                          {new JButton(blank),new JButton(blank),
                          new JButton(blank),new JButton(blank)},
                          {new JButton(blank),new JButton(blank),
                          new JButton(blank),new JButton(blank)}};
  //the locations of the images
  int locations[ ][ ] = {{0,0,0,0},{0,0,0,0},{0,0,0,0}, {0,0,0,0}};
  Container cont;

  public Memory()
  {
    super("Memory");
    setSize(415,500);
    setVisible(true);
    setDefaultCloseOperation(JFrame.EXIT_ON_CLOSE);

    cont = getContentPane();
    cont.setLayout(null);

    for(int i = 0; i <buttons[0].length; i++)
    {
      for(int j = 0; j <buttons.length; j++)
      {
        //add the button to the board
        cont.add(buttons[i][j]);
        buttons[i][j].setBounds(i*100,j*100,100,100);
      }
    }

    mixup();
  }
  public void mixup()
  {
    try
    {
      //pause
      Thread.sleep(500);
      //this prevent an image from being used more than twice
      int usedCount[ ] = {0,0,0,0,0,0,0,0};
      for(int i = 0; i <buttons[0].length; i++)
      {
        for(int j = 0; j <buttons.length; j++)
        {
          //create a random number
          int rand = (int)(Math.random()*8)+1;
          while(usedCount[rand-1]>1)
          {
```

```
            //find a better random number
            rand = (int)(Math.random()*8)+1;
          }
          //don't use an image more than twice!
          usedCount[rand-1]++;
          //set the image
          buttons[i][j].setIcon(new ImageIcon("img"+rand+".png"));
          //keep track of the images
          locations[i][j] = rand;
          cont.validate();
        }
      }
      //pause
      Thread.sleep(3000);
      //two for loops
      for(int i = 0; i <buttons[0].length; i++)
      {
        for(int j = 0; j <buttons.length; j++)
        {
          //change the icons back
          buttons[i][j].setIcon(blank);
          cont.validate();
        }
      }
    }
    catch(Exception e){}
  }

  public static void main (String[ ] args)
  {
    new Memory();
  }
}
```

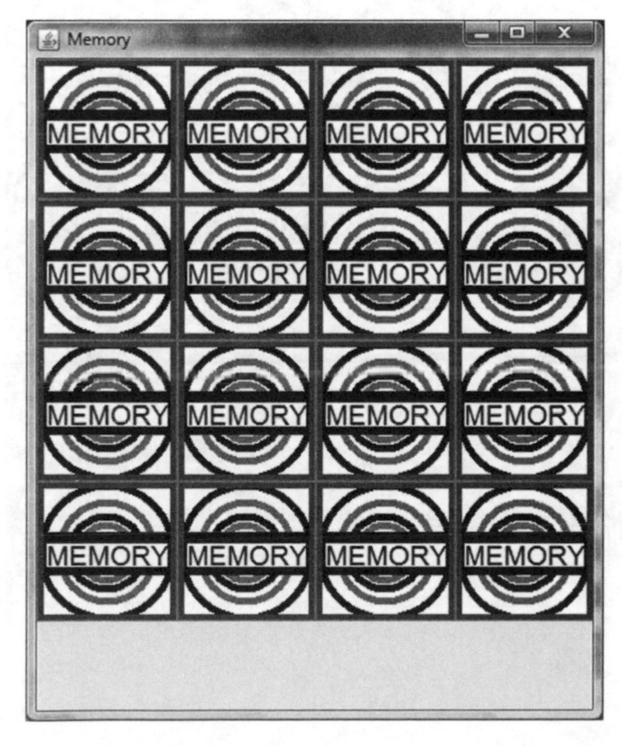

Figure 50-3 *Display of "Memory" images.*

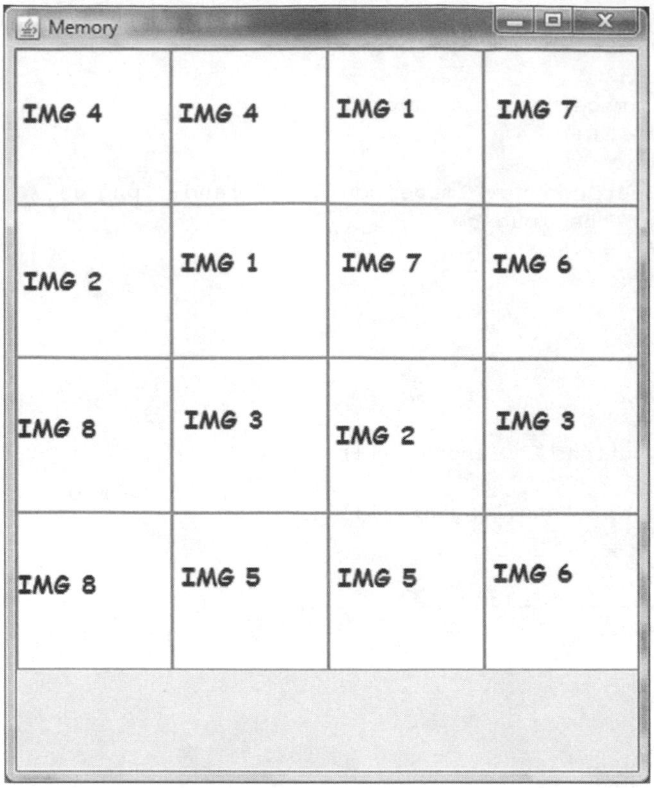

Figure 50-4 *Numbered images on the JFrame.*

Figure 50-5 *Return of "Memory" images.*

Figures 50-3 through 50-5 illustrate the initial Memory game board.

The next step: challenging the players to match the images. Read on!

Project 51: Memory—Match Time

Project

Construct the code that allows the player to test his/her memory by matching images.

Making the game

Start by implementing ActionListener. Remember to add it to each button in the loop. Keep in mind that each guess consists of two clicks: the first is the base image; the second is the matching image. Now, create two booleans: one to represent a player's first image click; another to represent the second. In the actionPerformed method, check the first boolean to confirm it is the player's first click. If so, use a `for loop` to iterate through all of the buttons. Use the e.getSource() method to check whether the button selected is the current one in the loop. If confirmed, hold the image number in a global variable. Then, set the icon of the button to the appropriate image. Don't forget to keep track of the row and column numbers of the button in global variables and to set the boolean that represents the turn number to the second position.

When it is the player's second turn, iterate through all of the buttons in the array. Use the e.getSource() method to check whether the button selected is the current one in the loop. If so, change the icon and set the button, row, and column numbers in global variables. Set the boolean that represents whether the computer needs to check for a match to "true." Also, set the other boolean that represents whether it is the player's first click to "true."

It's now time to create a Thread. In the thread's loop, use an if-statement to determine whether the boolean that represents if the board should be checked for a win is "true." If so, pause the game for half a second. Next, check if the two images selected by the player are different (i.e. incorrect guess). If so, reset the board by setting both icons to the original image. If the two images are identical (i.e. correct guess), the images will remain revealed.

```
import javax.swing.*;
import javax.swing.event.*;
import java.awt.*;
import java.awt.event.*;
public class Memory extends JFrame implements ActionListener
{
  //the blank image
  ImageIcon blank = new ImageIcon("blank.png");
  //the array of buttons
  JButton buttons[ ][ ] = {{new JButton(blank),new JButton(blank),
                           new JButton(blank),new JButton(blank)},
                           {new JButton(blank),new JButton(blank),
                           new JButton(blank),new JButton(blank)},
                           {new JButton(blank),new JButton(blank),
                           new JButton(blank),new JButton(blank)},
```

```
                    { new JButton(blank),new JButton(blank),
                      new JButton(blank),new JButton(blank)}};
//the image that is at each button
int locations[ ][ ] = {{ 0,0,0,0} ,{ 0,0,0,0} ,{ 0,0,0,0} ,{ 0,0,0,0}} ;
Container cont;
//whether the player has guessed once
boolean guessedOne = false;
boolean readyToCheck = false;
//the first guess
int firstGuess;
//the first guess pos 1 in the array
int firstGuessPos1;
//the first guess pos 2 in the array
int firstGuessPos2;
//the second guess
int secGuess;
//the second guess pos 1 in the array
int secGuessPos1;
//the second guess pos 2 in the array
int secGuessPos2;
public Memory()
{
  super("Memory");
  setSize(415,500);
  setVisible(true);
  setDefaultCloseOperation(JFrame.EXIT_ON_CLOSE);
  cont = getContentPane();
  cont.setLayout(null);
  //the Memory images
  for(int i = 0; i <buttons[ 0] .length; i++)
  {
    for(int j = 0; j <buttons.length; j++)
    {
      cont.add(buttons[ i][ j] );
      buttons[ i][ j] .setBounds(i*100,j*100,100,100);
      buttons[ i][ j] .addActionListener(this);
    }
  }
  mixup();
  Checker checker = new Checker();
  checker.start();
}
public class Checker extends Thread
{
  public void run()
  {
    while(true)
    {
      if(readyToCheck)
      {
        try
        {
        sleep(500);
        }
        catch(Exception ex){ }
        if(firstGuess!=secGuess)
        {
          //if it is wrong, reset the images
```

```
            buttons[firstGuessPos1][firstGuessPos2].setIcon(blank);
            buttons[secGuessPos1][secGuessPos2].setIcon(blank);
          }
          readyToCheck = false;
        }
      }
    }
  }
  public void mixup()
  {
    try
    {
      //pause
      Thread.sleep(500);
      //this prevent an image from being used more than twice
      int usedCount[] = {0,0,0,0,0,0,0,0};
      for(int i = 0; i <buttons[0].length; i++)
      {
        for(int j = 0; j <buttons.length; j++)
        {
          //create a random number
          int rand = (int)(Math.random()*8)+1;
          while(usedCount[rand-1]>1)
          {
            //find a better random number
            rand = (int)(Math.random()*8)+1;
          }
          //don't use an image more than twice!
          usedCount[rand-1]++;
          //set the image
          buttons[i][j].setIcon(new ImageIcon ("img"+rand+".png"));
          //keep track of the images
          locations[i][j] = rand;
          cont.validate();
        }
      }
      //pause
      Thread.sleep(3000);
      //two for loops
      for(int i = 0; i <buttons[0].length; i++)
      {
        for(int j = 0; j <buttons.length; j++)
        {
          //change the icons back
          buttons[i][j].setIcon(blank);
          cont.validate();
        }
      }
    }
    catch(Exception e){}
  }
  public void actionPerformed(ActionEvent e)
  {
    //if it is the first guess
    if(!guessedOne)
    {
      for(int i = 0; i <buttons[0].length; i++)
      {
        for(int j = 0; j <buttons.length; j++)
        {
```

```
          if(e.getSource()==buttons[i][j])
          {
            //display the image
            int picNum = locations[i][j];
            buttons[i][j].setIcon(new ImageIcon("img"+picNum+".png"));
            firstGuess = picNum;
            firstGuessPos1 = i;
            firstGuessPos2 = j;

          }
        }
      }
      guessedOne = true;
    }
    //if it is the second guess
    else
    {
      for(int i = 0; i <buttons[0].length; i++)
      {
        for(int j = 0; j <buttons.length; j++)
        {
          if(e.getSource()==buttons[i][j])
          {
            //display the image
            int picNum = locations[i][j];
            buttons[i][j].setIcon(new ImageIcon ("img"+picNum+".png"));

            secGuess = picNum;
            secGuessPos1 = i;
            secGuessPos2 = j;

            cont.validate();
          }
        }
        //let the thread check for a win
        guessedOne = false;
        readyToCheck = true;
      }

    }
  }
  public static void main (String[ ] args)
  {
    new Memory();
  }
}
```

Figures 51-1 through 51-3 displays the play sequence of Memory.

The player's satisfaction in completing the first level is short-lived as you create more levels with more intense challenges.

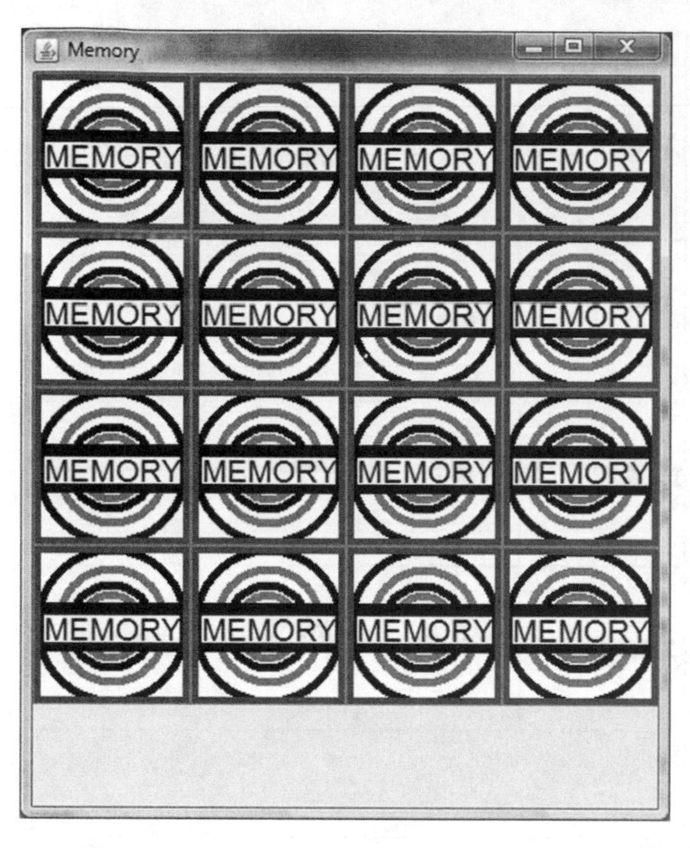

Figure 51-1 *Game begins.*

Figure 51-2 *On the verge of a win.*

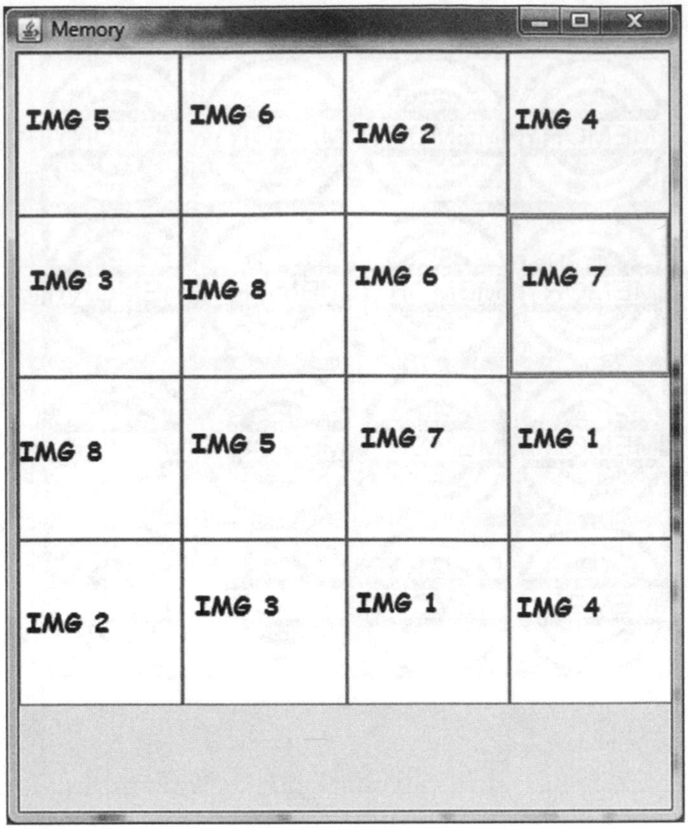

Figure 51-3 *Victory!*

Project 52: Memory—Beat the Clock

Project

Complications ... with a twist. The player must beat his/her own time from the previous round. The better you are, the harder it gets.

Making the Game

In order to create the timer, make two global variables: one that represents the allotted time given for the round, the other the time remaining in each round. Next, in the method that mixes up the icons (refer to Project 50), increment by one a new variable that keeps track of the level count. Find the new maximum time by subtracting the variable that represents the time remaining from the maximum time. Then, set the time left variable to the new maximum time.

Create a new Thread. In the infinite loop, delay for one second and then subtract one from the variable that represents the time remaining in the round. Display the time remaining in a JLabel. You can change the font to make it look like the example in Figure 52-1.

Figure 52-1 JLabel displays time remaining

Now, in the thread from Project 51, make a new if-statement that checks whether the player's guess is correct. If so, subtract one from the variable that keeps track of the number of pairs left. If that number is zero, it is time for the next round. Call the method that randomizes the images. If the time left variable is less than or equal to one, use a JOptionPane to alert the player that he/she has lost the game.

```
import javax.swing.*;
import javax.swing.event.*;
import java.awt.*;
import java.awt.event.*;

public class Memory extends JFrame implements ActionListener
{
  //the blank image
  ImageIcon blank = new ImageIcon("blank.png");

  //the array of buttons
  JButton buttons[ ][ ] = {{new JButton(blank),new JButton(blank),
                 new JButton(blank),new JButton(blank)},
                 {new JButton(blank),new JButton(blank),
                 new JButton(blank),new JButton(blank)},
                 {new JButton(blank),new JButton(blank),
                 new JButton(blank),new JButton(blank)},
                 {new JButton(blank),new JButton(blank),
                 new JButton(blank),new JButton(blank)}};
  //the image that is at each button
  int locations[ ][ ] = {{0,0,0,0},{0,0,0,0},{0,0,0,0},{0,0,0,0}};
  Container cont;
  //whether the player has guessed once
  boolean guessedOne = false;
  //the first guess
  int firstGuess;
  //the first guess pos 1 in the array
  int firstGuessPos1;
  //the first guess pos 2 in the array
  int firstGuessPos2;
  //the second guess
  int secGuess;
  //the second guess pos 1 in the array
  int secGuessPos1;
  //the second guess pos 2 in the array
  int secGuessPos2;
  //the number of pairs remaining
  int pairsLeft = 8;
  //whether the thread should check for a win
  boolean readyToCheck = false;
  //starting time
  int maxTime = 90;
  //remaining time
  int timeLeft = 0;
  //current level
  int levelCount = 0;
  JLabel time = new JLabel("Time Remaining: " + timeLeft);

  public Memory()
  {
    super("Memory");
    setSize(415,500);
    setVisible(true);
    setDefaultCloseOperation(JFrame.EXIT_ON_CLOSE);
```

```
    cont = getContentPane();
    cont.setLayout(null);
    //the Memory images
    for(int i = 0; i <buttons[0].length; i++)
    {
      for(int j = 0; j <buttons.length; j++)
      {
        cont.add(buttons[i][j]);
        buttons[i][j].setBounds(i*100,j*100,100,100);
        buttons[i][j].addActionListener(this);
      }
    }
    cont.add(time);
    time.setBounds(130,385,300,100);
    time.setFont(new Font("arial", Font.BOLD, 14));
    mixup();
    Checker checker = new Checker();
    checker.start();
    Counter count = new Counter();
    count.start();
  }
  public class Counter extends Thread
  {
    public void run()
    {
      while(true)
      {
        //count down the time
        timeLeft--;
        time.setText("Time Remaining: "+timeLeft);
        try
        {
          Thread.sleep(1000);
        }
        catch(Exception e){}
      }
    }
  }
  public class Checker extends Thread
  {
    public void run()
    {
      while(true)
      {
        if(readyToCheck)
        {
          try
          {
            sleep(500);
          }
          catch(Exception ex){}
          //if the guess is correct, remove a pair
          if(firstGuess==secGuess)
          {
            pairsLeft--;
          }
          else
          {
            //if it is wrong, reset the images
```

```
          buttons[ firstGuessPos1][ firstGuessPos2] .setIcon(blank);
          buttons[ secGuessPos1][ secGuessPos2] .setIcon(blank);
        }
        readyToCheck = false;
      }
      if(pairsLeft<=0)
      {
        //if the user wins, start the next level
        mixup();
      }
      //if the time is up, display an image
      if(timeLeft<0)
      {
        JOptionPane.showMessageDialog(null, "YOU LOSE!");break;
      }
    }
  }
}
public void mixup()
{
  levelCount++;
  maxTime = maxTime-timeLeft;
  timeLeft = maxTime;
  pairsLeft = 8;
  try
  {
    //pause
    Thread.sleep(500);
    //this prevent an image from being used more than twice
    int usedCount[ ]  = { 0,0,0,0,0,0,0,0} ;
    for(int i = 0; i <buttons[ 0] .length; i++)
    {
      for(int j = 0; j <buttons.length; j++)
      {
        //create a random number
        int rand = (int)(Math.random()*8)+1;
        while(usedCount[ rand-1] >1)
        {
          //find a better random number
          rand = (int)(Math.random()*8)+1;
        }
        //don't use an image more than twice!
        usedCount[ rand-1] ++;
        //set the image
        buttons[ i][ j] .setIcon(new ImageIcon("img"+rand+".png"));
        //keep track of the images
        locations[ i][ j]  = rand;
        cont.validate();
      }
    }
    //pause
    Thread.sleep(3000);
    //two for loops
    for(int i = 0; i <buttons[ 0] .length; i++)
    {
      for(int j = 0; j <buttons.length; j++)
      {
        //change the icons back
        buttons[ i][ j] .setIcon(blank);
        cont.validate();
```

```
            }
          }
        }
        catch(Exception e){}
      }
      public void actionPerformed(ActionEvent e)
      {
        //if it is the first guess
        if(!guessedOne)
        {
          for(int i = 0; i <buttons[ 0] .length; i++)
          {
            for(int j = 0; j <buttons.length; j++)
            {
              if(e.getSource()==buttons[ i][ j] )
              {
                //display the image
                int picNum = locations[ i][ j] ;
                buttons[ i][ j] .setIcon(new ImageIcon("img"+picNum+".png"));
                firstGuess = picNum;
                firstGuessPos1 = i;
                firstGuessPos2 = j;
              }
            }
          }
          guessedOne = true;
        }
        //if it is the second guess
        else
        {
          for(int i = 0; i <buttons[ 0] .length; i++)
          {
            for(int j = 0; j <buttons.length; j++)
            {
              if(e.getSource()==buttons[ i][ j] )
              {
                //display the image
                int picNum = locations[ i][ j] ;
                buttons[ i][ j] .setIcon(new ImageIcon("img"+picNum+".png"));

                secGuess = picNum;
                secGuessPos1 = i;
                secGuessPos2 = j;

                cont.validate();
              }
            }
            //let the thread check for a win
            guessedOne = false;
            readyToCheck = true;
          }

        }
      }
      public static void main (String[ ] args)
      {
        new Memory();
      }
    }
```

Figures 52-2 and 52-3 illustrate the rush to beat the clock.

Open up a world of possibilities by using photographs to customize your game. Next ...

Figure 52-2 *Countdown.*

YOU LOSE!

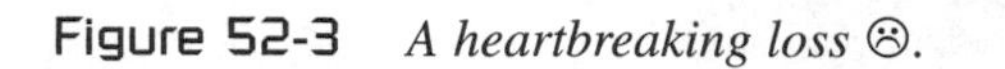

Figure 52-3 *A heartbreaking loss ☹.*

Project 53: Memory—Photo Finish

Project

Here's your chance to let your evil mind drive the players into madness. Pick a group of photos to use as the matching images. Make them as subtle as you want to increase the difficulty of the game and confuse the players. You will also insert announcement graphics to let the players know how well—or how poorly—they performed.

Making the game

Photograph eight random objects. Transfer them to your computer and open them with Microsoft Paint. Press "Ctrl+A" to select each image and drag it until "100,100" pixels is displayed on the bottom right, as shown in Figure 53-1.

Now that the image has been properly sized, select "Save As" and overwrite the generic image you created in the first project.

Now it's time to create the performance announcements that appear at the end of the game. There are three images/phrases that correspond to great, mediocre, or poor performances. Draw a fun icon and create an innovative phrase for each rank. Figures 53-2 through 53-4 illustrate a few examples.

When the player loses, use a loop to remove all components from the container. Next, create a new JLabel that displays the number of levels the player has completed. Place the JLabel in the center of the JFrame. Now, you need to know which image to display for the current level. If the player reaches level seven or above, place the

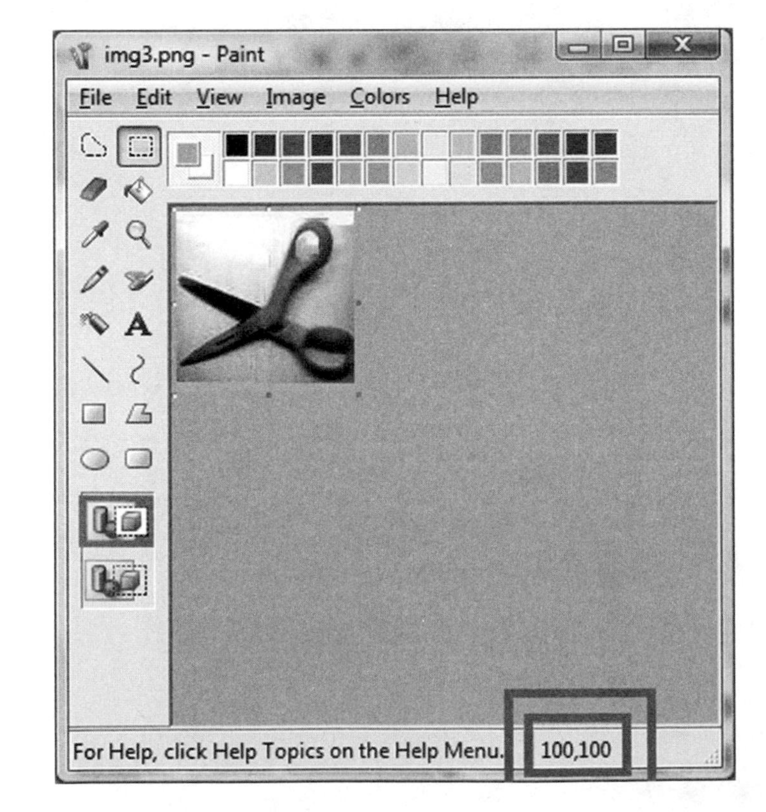

Figure 53-1 *Photograph resized to 100 by 100 pixels.*

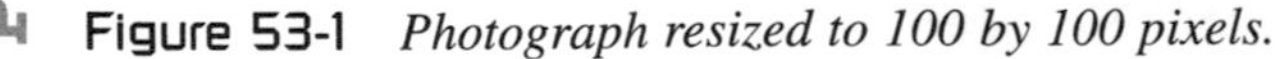

image that represents the highest rank (Figure 53-4). If the player reaches levels three through six, display the image representing the middle rank (Figure 53-3). For levels zero through five, display the image representing the lowest rank (Figure 53-2). And don't forget to use the setComponentZOrder method to place the JLabel in front of the image representing the player's standing.

Figure 53-2 *Lowest possible rank—Remember to keep trying.*

Figure 53-3 *Middle rank—Evil, but not a Genius.*

Figure 53-4 *Highest rank—Evil Genius!*

```
import javax.swing.*;
import javax.swing.event.*;
import java.awt.*;
import java.awt.event.*;

public class Memory extends JFrame implements ActionListener
{
  //the blank image
  ImageIcon blank = new ImageIcon("blank.png");

  //the array of buttons
  JButton buttons[ ][ ] = {{new JButton(blank),new JButton(blank),
                      new JButton(blank),new JButton(blank)},
                      {new JButton(blank),new JButton(blank),
                      new JButton(blank),new JButton(blank)},
                      {new JButton(blank),new JButton(blank),
                      new JButton(blank),new JButton(blank)},
                      {new JButton(blank),new JButton(blank),
                      new JButton(blank),new JButton(blank)}};
  //the image that is at each button
  int locations[ ][ ] = {{0,0,0,0},{0,0,0,0},{0,0,0,0},{0,0,0,0}};
  Container cont;
  //whether the player has guessed once
  boolean guessedOne = false;
  //the first guess
  int firstGuess;
  //the first guess pos 1 in the array
  int firstGuessPos1;
  //the first guess pos 2 in the array
  int firstGuessPos2;
  //the second guess
  int secGuess;
  //the second guess pos 1 in the array
  int secGuessPos1;
  //the second guess pos 2 in the array
  int secGuessPos2;
  //the number of pairs remaining
  int pairsLeft = 8;
  //whether the thread should check for a win
  boolean readyToCheck = false;
  //starting time
  int maxTime = 90;
  //remaining time
  int timeLeft = 0;
  //current level
  int levelCount = 0;
  JLabel time = new JLabel("Time Remaining: " + timeLeft);

  public Memory()
  {
    super("Memory");
    setSize(415,500);
    setVisible(true);
    setDefaultCloseOperation(JFrame.EXIT_ON_CLOSE);

    cont = getContentPane();
    cont.setLayout(null);

    //the Memory images
    for(int i = 0; i <buttons[0].length; i++)
    {
      for(int j = 0; j <buttons.length; j++)
      {
        cont.add(buttons[i][j]);
```

```
      buttons[i][j].setBounds(i*100,j*100,100,100);
      buttons[i][j].addActionListener(this);
    }
  }
  cont.add(time);
  time.setBounds(130,385,300,100);
  time.setFont(new Font("arial", Font.BOLD, 14));
  mixup();
  Checker checker = new Checker();
  checker.start();
  Counter count = new Counter();
  count.start();
}
public class Counter extends Thread
{
  public void run()
  {
    while(true)
    {
      //count down the time
      timeLeft--;
      time.setText("Time Remaining: "+timeLeft);
      try
      {
        Thread.sleep(1000);
      }
      catch(Exception e){}
    }
  }
}

public class Checker extends Thread
{
  public void run()
  {
    while(true)
    {
      if(readyToCheck)
      {
        try
        {
          sleep(500);
        }
        catch(Exception ex){}
        //if the guess is correct, remove a pair
        if(firstGuess==secGuess)
        {
          pairsLeft--;
        }
        else
        {
          //if it is wrong, reset the images
          buttons[firstGuessPos1][firstGuessPos2].setIcon(blank);
          buttons[secGuessPos1][secGuessPos2].setIcon(blank);
        }
        readyToCheck = false;
      }
      if(pairsLeft<=0)
      {
        //if the user wins, start the next level
```

```
            mixup();
        }
        //if the time is up, display an image
        if(timeLeft<0)
        {
          //remove the other components
          for(int i = 0; i <buttons[0].length; i++)
          {
            for(int j = 0; j <buttons.length; j++)
            {
              cont.remove(buttons[i][j]);
            }
          }
          cont.remove(time);
          JLabel levelLbl = new JLabel("- - - Levels Completed: "+ (levelCount-1)+" - - -");
          cont.add(levelLbl);
          levelLbl.setForeground(Color.white);
          levelLbl.setFont(new Font("arial narrow", Font.PLAIN, 20));
          levelLbl.setBounds(115,385,300,50);
          if(levelCount>=7)
          {
            //evil genius status
            JLabel help = new JLabel(new ImageIcon ("genius.png"));
            cont.add(help);
            help.setBounds(0,0,415,500);
          }
          else if (levelCount>=3)
          {
            //evil but not genius
            JLabel help = new JLabel(new ImageIcon("notGenius.png"));
            cont.add(help);
            help.setBounds(0,0,415,500);
          }
          else
          {
            //you need to practice
            JLabel help = new JLabel(new ImageIcon("needHelp.png"));
            cont.add(help);
            help.setBounds(0,0,415,500);
          }
          cont.setComponentZOrder(levelLbl,0);
          cont.validate();
          break;
        }
      }
    }
  }
  public void mixup()
  {
    levelCount++;
    maxTime = maxTime-timeLeft;
    timeLeft = maxTime;
    pairsLeft = 8;
    try
    {
      //pause
      Thread.sleep(500);
      //this prevent an image from being used more than twice
      int usedCount[ ] = {0,0,0,0,0,0,0,0};
```

```
    for(int i = 0; i <buttons[0].length; i++)
    {
      for(int j = 0; j <buttons.length; j++)
      {
        //create a random number
        int rand = (int)(Math.random()*8)+1;
        while(usedCount[rand-1]>1)
        {
          //find a better random number
          rand = (int)(Math.random()*8)+1;

        }
        //don't use an image more than twice!
        usedCount[rand-1]++;
        //set the image
        buttons[i][j].setIcon(new ImageIcon("img"+rand+".png"));
        //keep track of the images
        locations[i][j] = rand;
        cont.validate();
      }
    }
    //pause
    Thread.sleep(3000);
    //two for loops
    for(int i = 0; i <buttons[0].length; i++)
    {
      for(int j = 0; j <buttons.length; j++)
      {
        //change the icons back
        buttons[i][j].setIcon(blank);
        cont.validate();
      }
    }
  }
  catch(Exception e){}
}
public void actionPerformed(ActionEvent e)
{
  //if it is the first guess
  if(!guessedOne)
  {
    for(int i = 0; i <buttons[0].length; i++)
    {
      for(int j = 0; j <buttons.length; j++)
      {
        if(e.getSource()==buttons[i][j])
        {
          //display the image
          int picNum = locations[i][j];
          buttons[i][j].setIcon(new ImageIcon("img"+picNum+".png"));
          firstGuess = picNum;
          firstGuessPos1 = i;
          firstGuessPos2 = j;
        }
      }
    }
    guessedOne = true;
  }
  //if it is the second guess
  else
  {
```

```
        for(int i = 0; i <buttons[ 0] .length; i++)
        {
          for(int j = 0; j <buttons.length; j++)
          {
            if(e.getSource()==buttons[ i][ j] )
            {
              //display the image
              int picNum = locations[ i][ j] ;
              buttons[ i][ j] .setIcon(new ImageIcon("img"+picNum+".png"));
              secGuess = picNum;
              secGuessPos1 = i;
              secGuessPos2 = j;
              cont.validate();
            }
          }
          //let the thread check for a win
          guessedOne = false;
          readyToCheck = true;
        }
      }
    }
    public static void main (String[ ] args)
    {
      new Memory();
    }
  }
```

Figure 53-5 *Counter counts down ... .*

Figure 53-6 *Zero levels completed.*

Figure 53-7 *Memory failure at Level 4.*

Figure 53-8 *A true Evil Genius!*

Figures 53-5 through 53-8 illustrate the completed game of Memory.

Customizing the game

Change the images to random colors.

Change the 4×4 grid of 16 images into a super challenging 10×10 grid of 100 images.

Randomly add on or subtract time.

Shuffle the images mid-game.

Program a new set of images for each level.

Replace the images with sounds ... try matching those!

To really confuse the players, overlay sensory input. Play loud rock music throughout the test ... or the incessant sounds of a baby crying or a dog barking.

Project 54: Ian Says—Color Quad

Ian Says

Don't blink as the computer flashes a series of colors that the player must remember and repeat. Ian Says becomes more difficult at each level as the sequence grows longer and longer. How good is your recall? Make one small mistake and you start all over again.

Project

Build the game board and program the color sequencing of Level 1.

Making the game

The basic game board of Ian Says consists of four squares of different colors, each 200 by 200 pixels (see Figure 54-1).

First, create a 400 by 400 pixel JFrame. Then, program a `paint` method. In the `paint` method, draw four filled squares, each 200 by 200 pixels, as shown in Figure 54-1. Paint one square red; another, blue; another, green; the last, yellow. Next, pause the game for one second so the player will have time to prepare for the round.

Now, create a method that will randomly choose one of the four colors. The method should return a char that represents the randomly chosen color.

Using the `paint` method, call the method you just created. Change the chosen square to a brighter shade of the same color. Then, add text that displays the name of the color (see Figure 54-2).

Pause the game for a quarter of a second. Next, reset the board to the standard shade of the color by calling a new method you must create to re-initialize the board.

Figure 54-1 *Sample square.*

Figure 54-2 *"Green" displayed on the bright green square.*

```
import javax.swing.*;
import javax.swing.event.*;
import java.awt.*;
import java.awt.event.*;

public class IanSays extends JFrame
{
  public IanSays()
  {
    super("Ian Says");
    setSize(400,400);
    setDefaultCloseOperation(JFrame.EXIT_ON_CLOSE);
    setVisible(true);
  }

  public void paint(Graphics g)
  {
    super.paint(g);

    //red square
    g.setColor(new Color(200,50,50));
    g.fillRect(0,0,200,200);
    //blue square
    g.setColor(new Color(50,50,200));
    g.fillRect(200,0,200,200);
    //green square
    g.setColor(new Color(50,200,50));
    g.fillRect(0,200,200,200);
    //yellow square
    g.setColor(new Color(200,200,0));
    g.fillRect(200,200,200,200);

    try
    {
      Thread.sleep(1000);
    }
    catch(Exception e){}

    String newColor = newColor();

    //draw the appropriate new color
    if(newColor.equals("r"))
    {
      g.setColor(new Color(255,0,0));
      g.fillRect(0,0,200,200);
      g.setColor(Color.black);
      g.setFont(new Font("arial", Font.BOLD, 40));
      g.drawString("Red",50,80);
    }
    if(newColor.equals("b"))
    {
      g.setColor(new Color(0,0,250));
      g.fillRect(200,0,200,200);
      g.setColor(Color.black);
      g.setFont(new Font("arial", Font.BOLD, 40));
      g.drawString("Blue",250,80);
    }
    if(newColor.equals("g"))
    {
      g.setColor(new Color(0,255,0));
      g.fillRect(0,200,200,200);
      g.setColor(Color.black);
      g.setFont(new Font("arial", Font.BOLD, 40));
```

```
    g.drawString("Green",50,280);
  }
  if(newColor.equals("y"))
  {
    g.setColor(new Color(255,255,0));
    g.fillRect(200,200,200,200);
    g.setColor(Color.black);
    g.setFont(new Font("arial", Font.BOLD, 40));
    g.drawString("Yellow",250,280);
  }
  try
  {
    Thread.sleep(250);
  }
  catch(Exception e){}
  reset(g);
 }
 //generate a new random number
 public String newColor()
 {
   int rand = (int)(Math.random()*4);
   if(rand==0)
   {
     return "r";
   }
   if(rand==1)
   {
     return "b";
   }
   if(rand==2)
   {
     return "g";
   }
   if(rand==3)
   {
     return "y";
   }
   return " ";
 }
 public void reset(Graphics g)
 {
   //reset the squares
   g.setColor(new Color(200,50,50));
   g.fillRect(0,0,200,200);
   g.setColor(new Color(50,50,200));
   g.fillRect(200,0,200,200);
   g.setColor(new Color(50,200,50));
   g.fillRect(0,200,200,200);
   g.setColor(new Color(200,200,0));
   g.fillRect(200,200,200,200);
 }
 public static void main (String[ ]args)
 {
   new IanSays();
 }
}
```

Figures 54-3 through 54-5 illustrate the first visual cue of Ian Says.

The first part of the color sequence has been displayed. In the next project, learn how to test the player's memory skills.

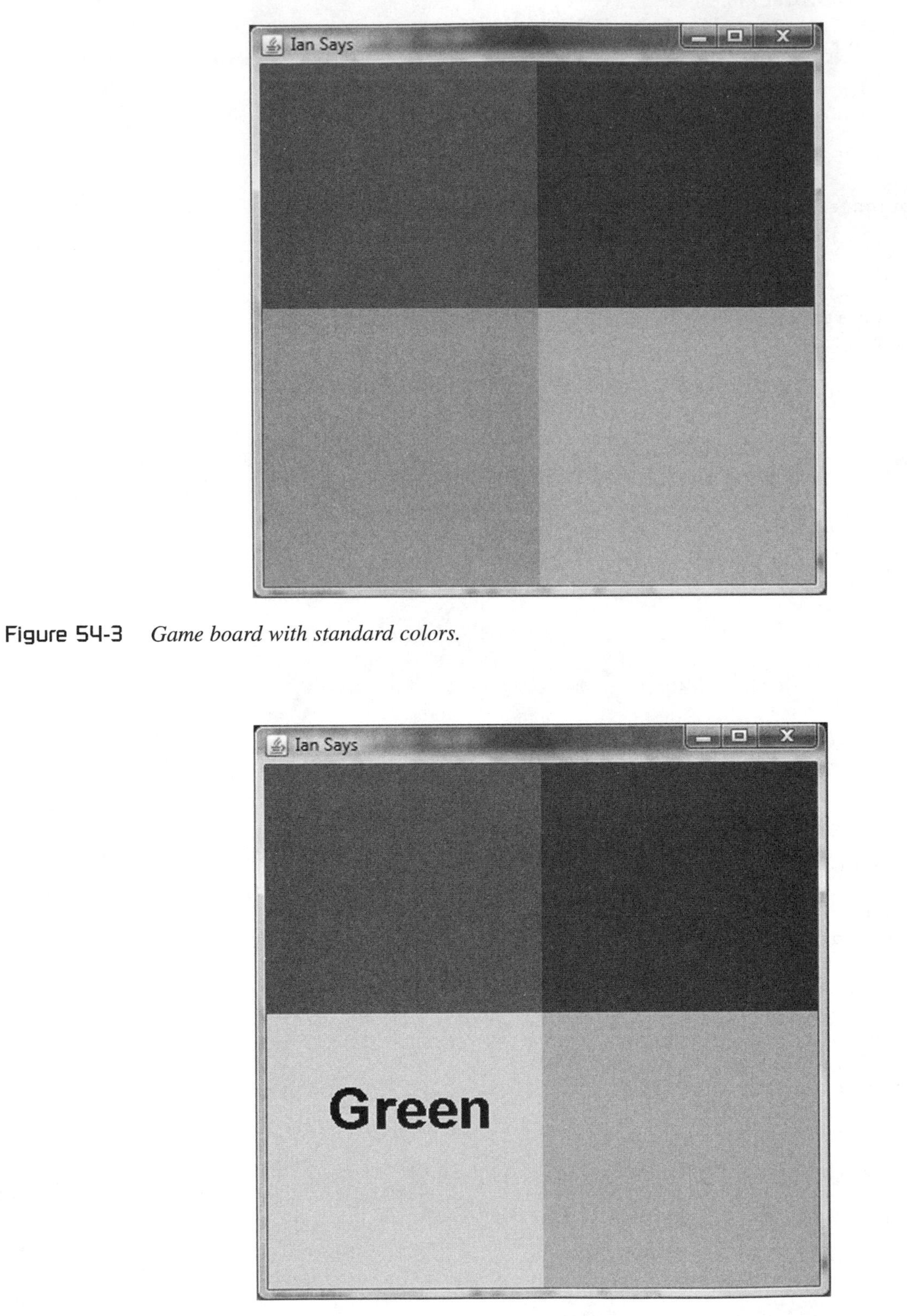

Figure 54-3 *Game board with standard colors.*

Figure 54-4 *Brightened green square.*

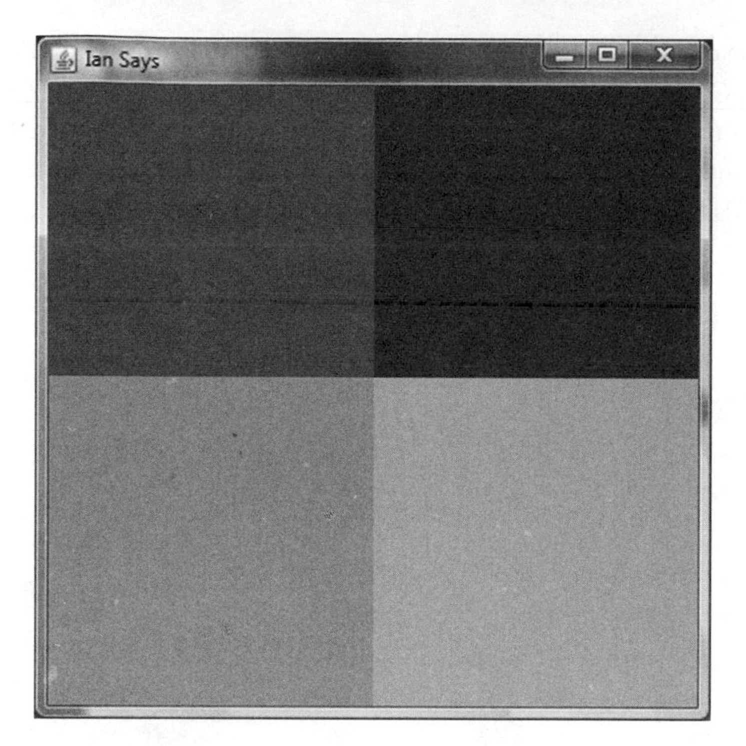

Figure 54-5 *Game board with standard colors.*

Project 55: Ian Says—Brain Drain

Project

The action begins here. Program the computer to determine whether or not the player has entered the correct sequence of colors.

Making the game

In order to let the user repeat the sequence, implement "mouseListener." Don't forget to include all five mandatory methods. In the mouseClicked method, set a boolean that represents if the computer is displaying the sequence to false. Next, increment by one a variable that represents the number of guesses the player has taken.

You must determine which square the user is clicking. If the X and Y positions of the click are less than 200, the user is clicking the top left square. If the X position is greater than 200 and the Y position is less than 200, the player is clicking the top right square. If the Y position is greater than 200 and the X position is less than 200, the player is clicking the bottom left square. If both the X an Y positions are greater than 200, the player is clicking the bottom right square. Change a variable called "lastGuess" to the square the player last clicked. In addition, add the clicked color to a String representing the player's guess. Call "repaint()."

You will need to modify the `paint` method. Check the boolean that represents if the computer is displaying a sequence. If it is not, brighten the last colored square that the player clicked by using the "lastGuess" variable.

Now, back in the `mouseClicked` method, check if the player's sequence matches that of the computer's. To do this, use the "substring" method to extract the part of the computer's sequence that the player has guessed. Compare the two Strings. If they are different, the user has guessed incorrectly. Use a JOptionPane message to tell the player he/she has lost. If the Strings are the same and the number of player guesses are the same as the number of colors in the sequence, use a JOptionPane message to tell the user he/she has correctly completed the sequence.

```
import javax.swing.*;
import javax.swing.event.*;
import java.awt.*;
import java.awt.event.*;

public class IanSays extends JFrame implements MouseListener
{
  Container cont;
  String code = "";
  String guess = "";
  int guesses = 0;
  String lastLetter;
  boolean normal = true;
  public IanSays()
  {
    super("Ian Says");
    setSize(400,400);
    setDefaultCloseOperation(JFrame.EXIT_ON_CLOSE);
    setVisible(true);

    addMouseListener(this);
  }
  public void paint(Graphics g)
  {
    super.paint(g);

    if(!normal)
    {
      reset(g);
      String newColor = lastLetter;
      //light up the correct square when clicked
      if(newColor.equals("r"))
      {
        g.setColor(new Color(255,0,0));
        g.fillRect(0,0,200,200);
      }
      if(newColor.equals("b"))
      {
        g.setColor(new Color(0,0,250));
        g.fillRect(200,0,200,200);
      }
      if(newColor.equals("g"))
      {
        g.setColor(new Color(0,255,0));
        g.fillRect(0,200,200,200);
      }
      if(newColor.equals("y"))
      {
        g.setColor(new Color(255,255,0));
        g.fillRect(200,200,200,200);
      }

      try
      {
```

```
        Thread.sleep(250);
      }
      catch(Exception ex){}
      reset(g);
    }
    else
    {
      //make the squares
      g.setColor(new Color(200,50,50));
      g.fillRect(0,0,200,200);
      g.setColor(new Color(50,50,200));
      g.fillRect(200,0,200,200);
      g.setColor(new Color(50,200,50));
      g.fillRect(0,200,200,200);
      g.setColor(new Color(200,200,0));
      g.fillRect(200,200,200,200);
      try
      {
        Thread.sleep(1000);
      }
      catch(Exception e){ }
      //iterate through the code
      for(int i = 0; i <code.length(); i++)
      {
        try
        {
          Thread.sleep(250);
        }
        catch(Exception e){ }
        //light up the next square
        char letter = code.toCharArray()[ i];
        if(letter=='r' )
        {
          g.setColor(new Color(255,0,0));
          g.fillRect(0,0,200,200);
          g.setColor(Color.black);
          g.setFont(new Font("arial", Font.BOLD, 40));
          g.drawString("Red",50,80);
        }
        if(letter=='b' )
        {
          g.setColor(new Color(0,0,250));
          g.fillRect(200,0,200,200);
          g.setColor(Color.black);
          g.setFont(new Font("arial", Font.BOLD, 40));
          g.drawString("Blue",250,80);
        }
        if(letter=='g' )
        {
          g.setColor(new Color(0,255,0));
          g.fillRect(0,200,200,200);
          g.setColor(Color.black);
          g.setFont(new Font("arial", Font.BOLD, 40));
          g.drawString("Green",50,280);
        }
        if(letter=='y' )
        {
          g.setColor(new Color(255,255,0));
          g.fillRect(200,200,200,200);
          g.setColor(Color.black);
```

```
      g.setFont(new Font("arial", Font.BOLD, 40));
      g.drawString("Yellow",250,280);
    }
    try
    {
      Thread.sleep(250);
    }
    catch(Exception e){}
    reset(g);
  }
  try
  {
    Thread.sleep(250);
  }
  catch(Exception e){ }
  String newColor = newColor();
  //display the appropriate color
  if(newColor.equals("r"))
  {
    g.setColor(new Color(255,0,0));
    g.fillRect(0,0,200,200);
    g.setColor(Color.black);
    g.setFont(new Font("arial", Font.BOLD, 40));
    g.drawString("Red",50,80);
  }
  if(newColor.equals("b"))
  {
    g.setColor(new Color(0,0,250));
    g.fillRect(200,0,200,200);
    g.setColor(Color.black);
    g.setFont(new Font("arial", Font.BOLD, 40));
    g.drawString("Blue",250,80);
  }
  if(newColor.equals("g"))
  {
    g.setColor(new Color(0,255,0));
    g.fillRect(0,200,200,200);
    g.setColor(Color.black);
    g.setFont(new Font("arial", Font.BOLD, 40));
    g.drawString("Green",50,280);
  }
  if(newColor.equals("y"))
  {
    g.setColor(new Color(255,255,0));
    g.fillRect(200,200,200,200);
    g.setColor(Color.black);
    g.setFont(new Font("arial", Font.BOLD, 40));
    g.drawString("Yellow",250,280);
  }
  try
  {
    Thread.sleep(250);
  }
  catch(Exception e){}
  reset(g);
 }
}
public String newColor()
```

```
{
  //generate the random color
  int rand = (int)(Math.random()*4);
  if(rand==0)
  {
    code+="r";
    return "r";
  }
  if(rand==1)
  {
    code+="b";
    return "b";
  }
  if(rand==2)
  {
    code+="g";
    return "g";
  }
  if(rand==3)
  {
    code+="y";
    return "y";
  }
  return " ";
}
public void reset(Graphics g)
{
  //reset the red square
  g.setColor(new Color(200,50,50));
  g.fillRect(0,0,200,200);
  //reset the blue square
  g.setColor(new Color(50,50,200));
  g.fillRect(200,0,200,200);
  //reset the green square
  g.setColor(new Color(50,200,50));
  g.fillRect(0,200,200,200);
  //reset the yellow square
  g.setColor(new Color(200,200,0));
  g.fillRect(200,200,200,200);
}

public void mouseExited(MouseEvent e){ }
public void mouseEntered(MouseEvent e){ }
public void mouseReleased(MouseEvent e){ }
public void mousePressed(MouseEvent e){ }

public void mouseClicked(MouseEvent e)
{
  normal = false;
  guesses++;
  //find out which square the player clicked
  if(e.getX()<200 && e.getY()<200)
  {
    lastLetter = "r";
    guess+="r";
  }
  if(e.getX()>200 && e.getY()<200)
  {
    lastLetter = "b";
    guess+="b";
  }
  if(e.getX()<200 && e.getY()>200)
```

```
    {
      lastLetter = “g”;
      guess+=“g”;
    }
    if(e.getX()>200 && e.getY()>200)
    {
      lastLetter = “y”;
      guess+=”y”;
    }
    repaint();
    String codeSeg = code.substring(0,guesses);
    if(!codeSeg.equals(guess))
    {
      //game over!
      JOptionPane.showMessageDialog(null,“GAME OVER!!!”);
    }
    else
    {
      //the player can proceed
      if(guesses==code.length())
      {
        JOptionPane.showMessageDialog(null,“Correct!!!”);
      }
    }
  }
  public static void main (String[ ] args)
  {
    new IanSays();
  }
}
```

Figure 55-1 *Computer's sequence.*

Figures 55-1 through 55-3 show the game's cycle.

Simple ... right? Be prepared—things are going to get very complicated very quickly.

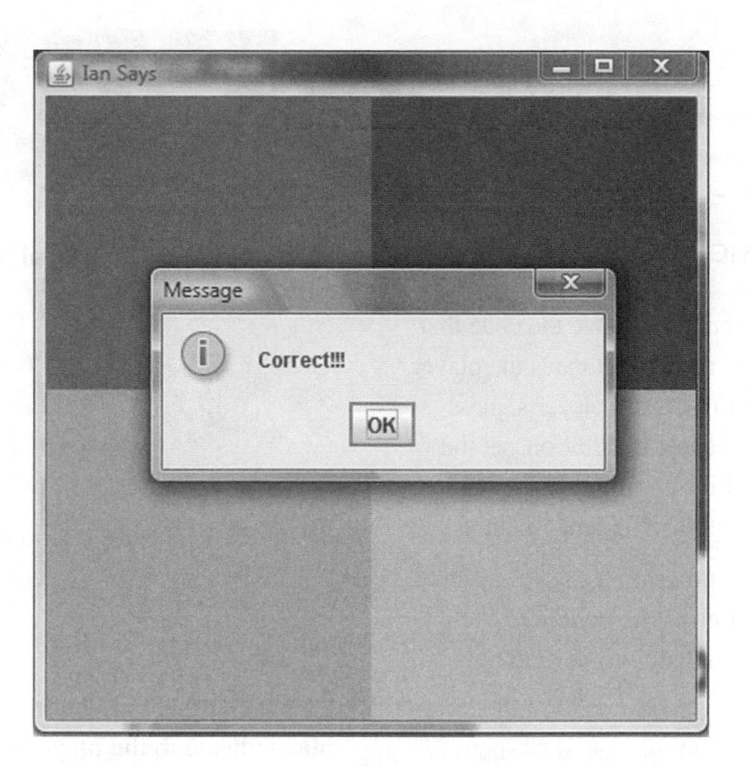

Figure 55-2 *Correctly repeated.*

Figure 55-3 *Incorrectly repeated — L-O-S-E-R!!!*

Project 56: Ian Says—More Rounds, More Frustration

Project

Build an infinite number of levels containing an ever-increasing sequence of colors to memorize.

Making the game

In the `mouseClicked` method, remove the code that creates the JOptionPane that congratulates the player. Replace it with code that resets the player's guess and the length of the sequence. In addition, set the boolean that represents if the computer is creating a sequence to true. Next, call the "repaint" method.

In the `paint` method, program a `for loop` that iterates through each letter of the code. On each iteration, use the following code to determine which color to use.

```
<code>.toCharArray()[ <element number of the code>] ;
```

The above method returns a char, which can then be checked in an if-statement to determine if it contains an "r" (red), "b" (blue), "g" (green), or "y" (yellow). Brighten the appropriate square and display the correct text to indicate its color.

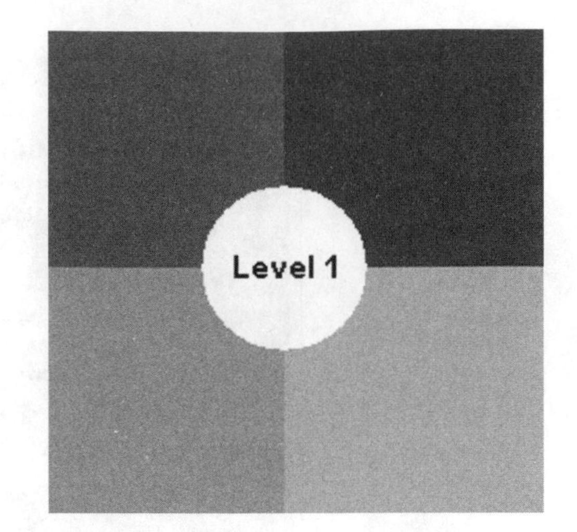

Figure 56-1 *Circle indicating level number.*

Now that there are multiple levels of play, you must indicate to the player what level he/she is on. To do this, write a method that draws a white circle surrounded by black text, like the one shown in Figure 56-1.

In the `paint` method, call the method that displays the circle and level number before the Thread.sleep methods.

```
import javax.swing.*;
import javax.swing.event.*;
import java.awt.*;
import java.awt.event.*;
public class IanSays extends JFrame implements MouseListener
{
  Container cont;
  String code = "";
  String guess = "";
  int guesses = 0;
  int levels = 0;
  String lastLetter;
  boolean normal = true;
  public IanSays()
  {
    super("Ian Says");
    setSize(400,400);
    setDefaultCloseOperation(JFrame.EXIT_ON_CLOSE);
    setVisible(true);
```

```
    Container cont = getContentPane();
    cont.setLayout(null);
    addMouseListener(this);
  }
  public void paint(Graphics g)
  {
    super.paint(g);
    if(!normal)
    {
      reset(g);
      String newColor = lastLetter;
      //light up the correct square when clicked
      if(newColor.equals("r"))
      {
        g.setColor(new Color(255,0,0));
        g.fillRect(0,0,200,200);
      }
      if(newColor.equals("b"))
      {
        g.setColor(new Color(0,0,250));
        g.fillRect(200,0,200,200);
      }
      if(newColor.equals("g"))
      {
        g.setColor(new Color(0,255,0));
        g.fillRect(0,200,200,200);
      }
      if(newColor.equals("y"))
      {
        g.setColor(new Color(255,255,0));
        g.fillRect(200,200,200,200);
      }
      drawLevel(g);
      try
      {
        Thread.sleep(250);
      }
      catch(Exception ex){}
      reset(g);
    }
    else
    {
      levels++;
      //make the squares
      g.setColor(new Color(200,50,50));
      g.fillRect(0,0,200,200);
      g.setColor(new Color(50,50,200));
      g.fillRect(200,0,200,200);
      g.setColor(new Color(50,200,50));
      g.fillRect(0,200,200,200);
      g.setColor(new Color(200,200,0));
      g.fillRect(200,200,200,200);
      drawLevel(g);
      try
      {
        Thread.sleep(1000);
      }
      catch(Exception e){ }
```

```
//iterate through the code
for(int i = 0; i <code.length(); i++)
{
  try
  {
    Thread.sleep(250);
  }
  catch(Exception e){}

  //light up the next square
  char letter = code.toCharArray()[ i];
  if(letter=='r' )
  {
    g.setColor(new Color(255,0,0));
    g.fillRect(0,0,200,200);
    g.setColor(Color.black);
    g.setFont(new Font("arial", Font.BOLD, 40));
    g.drawString("Red",50,80);
  }
  if(letter=='b' )
  {
    g.setColor(new Color(0,0,250));
    g.fillRect(200,0,200,200);
    g.setColor(Color.black);
    g.setFont(new Font("arial", Font.BOLD, 40));
    g.drawString("Blue",250,80);
  }
  if(letter=='g' )
  {
    g.setColor(new Color(0,255,0));
    g.fillRect(0,200,200,200);
    g.setColor(Color.black);
    g.setFont(new Font("arial", Font.BOLD, 40));
    g.drawString("Green",50,280);
  }
  if(letter=='y' )
  {
    g.setColor(new Color(255,255,0));
    g.fillRect(200,200,200,200);
    g.setColor(Color.black);
    g.setFont(new Font("arial", Font.BOLD, 40));
    g.drawString("Yellow",250,280);
  }

  drawLevel(g);

  try
  {
    Thread.sleep(250);
  }
  catch(Exception e){}

  reset(g);
}

try
{
  Thread.sleep(250);
}
catch(Exception e){}

String newColor = newColor();
//display the appropriate color
if(newColor.equals("r"))
```

```
    {
      g.setColor(new Color(255,0,0));
      g.fillRect(0,0,200,200);
      g.setColor(Color.black);
      g.setFont(new Font("arial", Font.BOLD, 40));
      g.drawString("Red",50,80);
    }
    if(newColor.equals("b"))
    {
      g.setColor(new Color(0,0,250));
      g.fillRect(200,0,200,200);
      g.setColor(Color.black);
      g.setFont(new Font("arial", Font.BOLD, 40));
      g.drawString("Blue",250,80);
    }
    if(newColor.equals("g"))
    {
      g.setColor(new Color(0,255,0));
      g.fillRect(0,200,200,200);
      g.setColor(Color.black);
      g.setFont(new Font("arial", Font.BOLD, 40));
      g.drawString("Green",50,280);
    }
    if(newColor.equals("y"))
    {
      g.setColor(new Color(255,255,0));
      g.fillRect(200,200,200,200);
      g.setColor(Color.black);
      g.setFont(new Font("arial", Font.BOLD, 40));
      g.drawString("Yellow",250,280);
    }
    drawLevel(g);
    try
    {
      Thread.sleep(250);
    }
    catch(Exception e){}
    reset(g);
  }
}
public void drawLevel(Graphics g)
{
  //draw the oval and text that displays the level
  g.setColor(Color.white);
  g.fillOval(160,160,80,80);
  g.setColor(Color.black);
  g.setFont(new Font("arial", Font.BOLD, 15));
  g.drawString("Level "+levels,175,205);
}
public String newColor()
{
  //generate the random color
  int rand = (int)(Math.random()*4);
  if(rand==0)
  {
    code+="r";
    return "r";
  }
  if(rand==1)
```

```
  {
    code+="b";
    return "b";
  }
  if(rand==2)
  {
    code+="g";
    return "g";
  }
  if(rand==3)
  {
    code+="y";
    return "y";
  }
  return " ";
}

public void reset(Graphics g)
{
  //reset the red square
  g.setColor(new Color(200,50,50));
  g.fillRect(0,0,200,200);
  //reset the blue square
  g.setColor(new Color(50,50,200));
  g.fillRect(200,0,200,200);
  //reset the green square
  g.setColor(new Color(50,200,50));
  g.fillRect(0,200,200,200);
  //reset the yellow square
  g.setColor(new Color(200,200,0));
  g.fillRect(200,200,200,200);

  drawLevel(g);
}

public void mouseExited(MouseEvent e){}
public void mouseEntered(MouseEvent e){}
public void mouseReleased(MouseEvent e){}
public void mousePressed(MouseEvent e){}

public void mouseClicked(MouseEvent e)
{
  normal = false;
  guesses++;
  //find out which square the player clicked
  if(e.getX()<200 && e.getY()<200)
  {
    lastLetter ="r";
    guess+="r";
  }
  if(e.getX()>200 && e.getY()<200)
  {
    lastLetter = "b";
    guess+="b";
  }
  if(e.getX()<200 && e.getY()>200)
  {
    lastLetter = "g";
    guess+="g";
  }
  if(e.getX()>200 && e.getY()>200)
  {
    lastLetter = "y";
```

```
      guess+="y";
    }
    repaint();
    String codeSeg = code.substring(0,guesses);
    if(!codeSeg.equals(guess))
    {
      JOptionPane.showMessageDialog(null,"Game Over");
    }
    else
    {
      //the player can proceed
      if(guesses==code.length())
      {
        guess = "";
        guesses = 0;
        normal = true;
        //next level!
        repaint();
      }
    }
  }
  public static void main (String[ ] args)
  {
    new IanSays();
  }
}
```

Figures 56-2 through 56-4 demonstrates sample game levels.

Add refinement to Ian Says by designing a background image with a start button that provides players time to prepare for their mental workout.

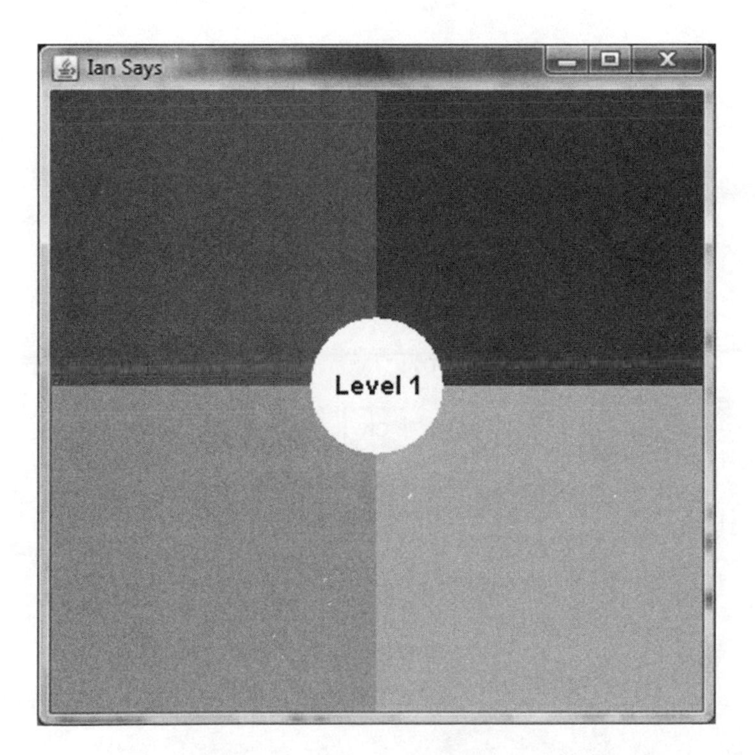

Figure 56-2 *Game begins.*

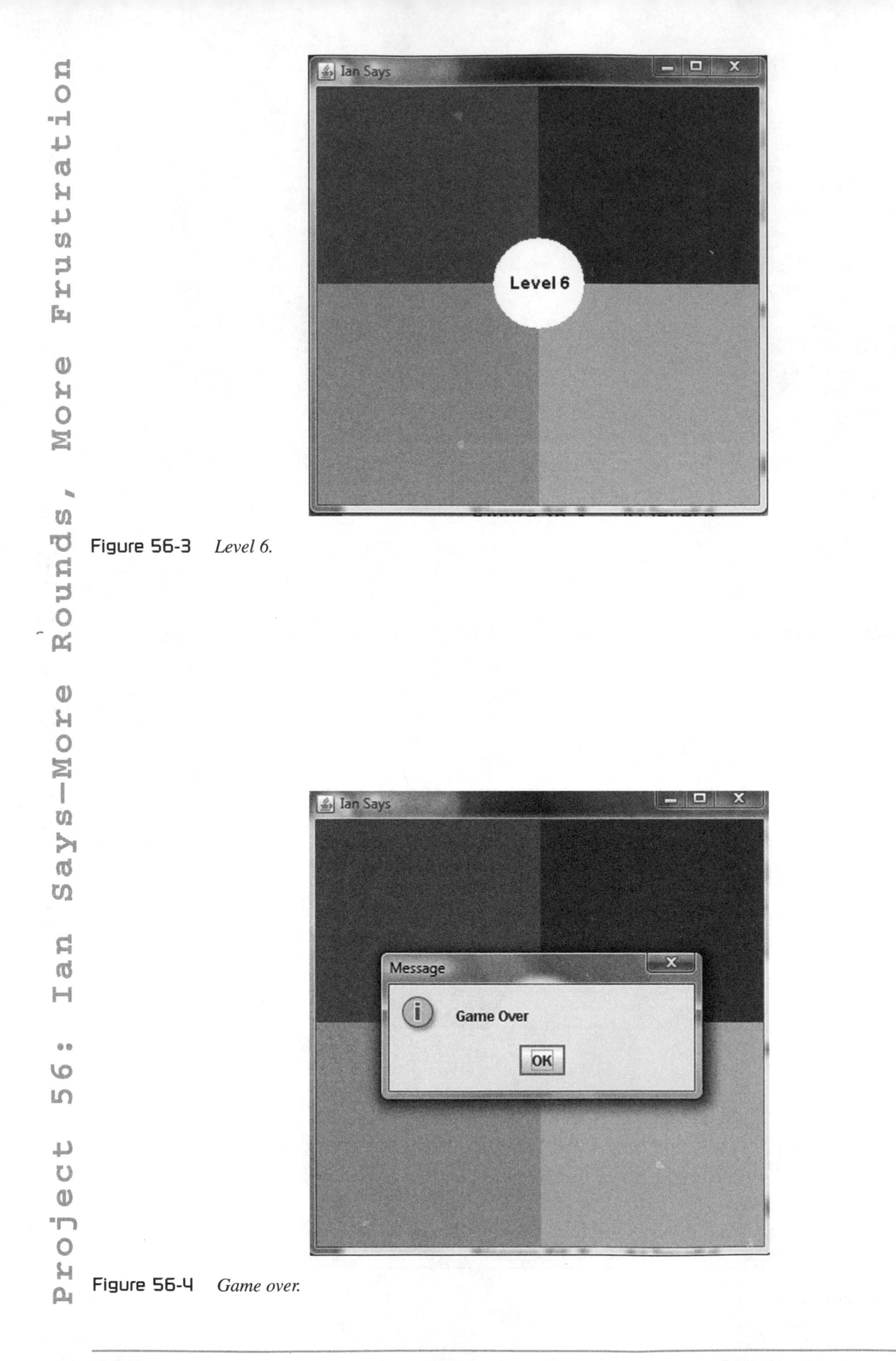

Figure 56-3 *Level 6.*

Figure 56-4 *Game over.*

Project 57: Ian Says—Play Accessories

Project

Construct a "start" image that alerts the player that the game is set to begin. Also, design images and messages to appear as the player completes levels.

Making the game

In Microsoft Paint, create a 400 by 400 pixel starting image, like the one shown in Figure 57-1.

In the constructor, add the image to the JFrame. Add a JButton on top of the image. Don't forget to add an ActionListener. Next, create a boolean that represents if the game should start and set it to false. In the JButton's actionPerformed method, set the boolean to true and move the button offscreen. In the paint method, check if the boolean is true. If so, begin the game.

At the end of the game, replace the JOptionPane with text. Display different phrases corresponding to the length of the sequence the player mastered.

Figure 57-1 *Starting image.*

For example, if the player fails to remember any colors of the sequence (☹), display the text in Figure 57-2. If the player remembers a 1 to 5 color sequence, display the text in Figure 57-3. If the user remembers a sequence greater than 5, display the text in Figure 57-4.

Figure 57-2 *No sequential memory.*

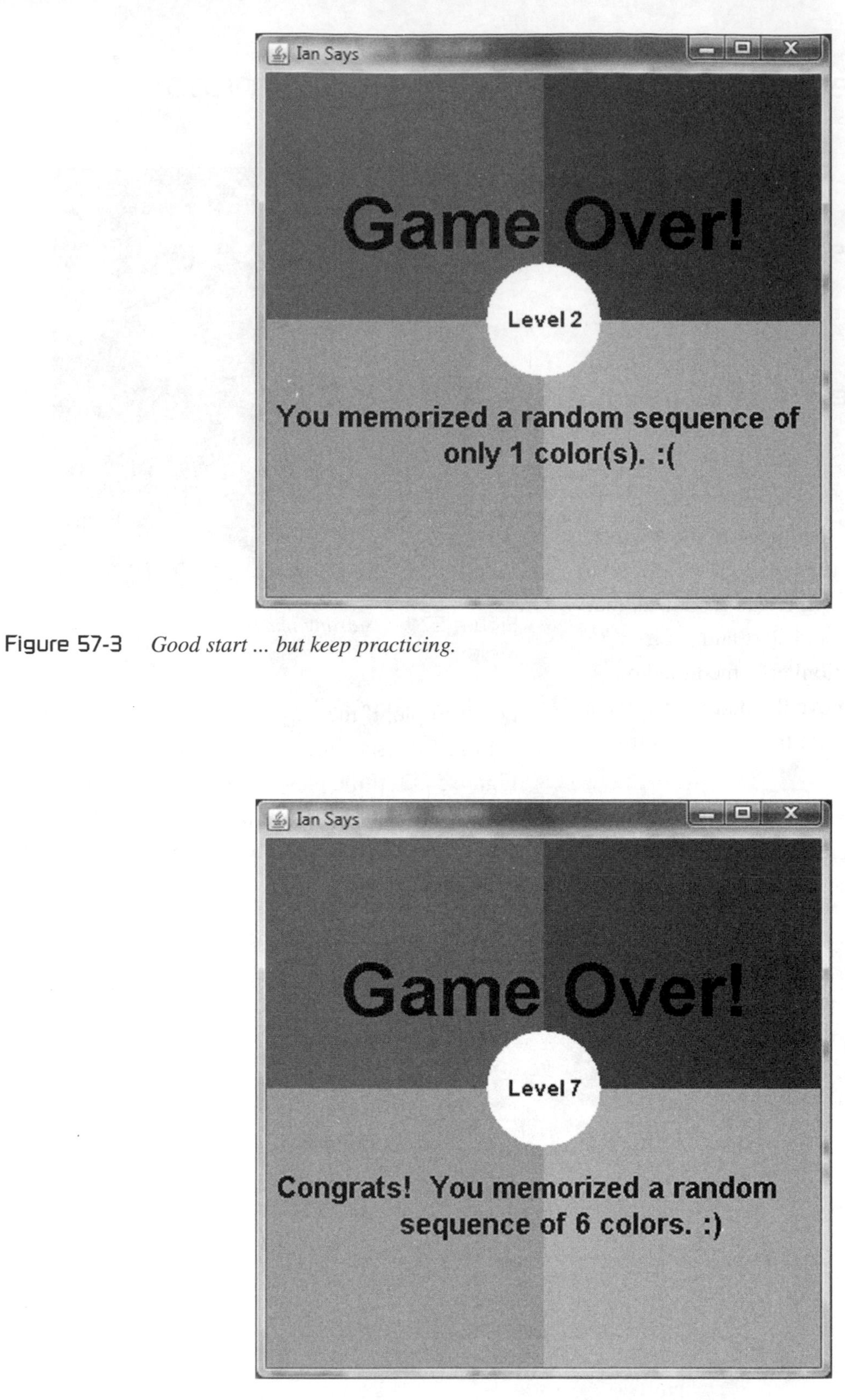

Figure 57-3 *Good start ... but keep practicing.*

Figure 57-4 *Solid memory ... the basis of a true Evil Genius.*

```
import javax.swing.*;
import javax.swing.event.*;
import java.awt.*;
import java.awt.event.*;
public class IanSays extends JFrame implements MouseListener, ActionListener
{
  Container cont;
  String code = “”;
  String guess = “”;
  int guesses = 0;
  int levels = 0;
  String lastLetter;
  boolean over = false;
  boolean normal = true;
  boolean begin = false;
  JButton start = new JButton(“START”);
  public IanSays()
  {
    super(“Ian Says”);
    setSize(400,400);
    setDefaultCloseOperation(JFrame.EXIT_ON_CLOSE);
    setVisible(true);
    Container cont = getContentPane();
    cont.setLayout(null);

    cont.add(start);
    start.addActionListener(this);
    start.setFont(new Font(“arial”,Font.BOLD,20));
    start.setBounds(145,250,100,50);
    JLabel title = new JLabel(new ImageIcon(“title.png”));
    cont.add(title);
    title.setBounds(0,0,400,400);

    addMouseListener(this);
  }

  public void paint(Graphics g)
  {
    super.paint(g);
    if(!begin)
      return;
    if(over)
    {
      //GAME OVER!
      reset(g);
      g.setColor(Color.black);
      g.setFont(new Font(“Arial”, Font.BOLD,50));
      g.drawString(“Game Over!”,60,150);
      g.setFont(new Font(“Arial”, Font.BOLD,20));
      if((levels-1)==0)
      {
        g.drawString(“You didn’t memorize anything!!!”,50,285);
      }
      else if((levels-1)<=5)
      {
        g.drawString(“You memorized a random sequence of”,15,275);
        g.drawString(“only ”+(levels-1)+” color(s).:(”,130,300);
      }
      else
      {
        g.drawString(“Congrats! You memorized a random”,15,275);
        g.drawString(“sequence of ”+(levels-1)+” colors.:)”,100,300);
```

```
    }
  }
  else
  {
    if(!normal)
    {
      reset(g);
      String newColor = lastLetter;
      //light up the correct square when clicked
      if(newColor.equals("r"))
      {
        g.setColor(new Color(255,0,0));
        g.fillRect(0,0,200,200);
      }
      if(newColor.equals("b"))
      {
        g.setColor(new Color(0,0,250));
        g.fillRect(200,0,200,200);
      }
      if(newColor.equals("g"))
      {
        g.setColor(new Color(0,255,0));
        g.fillRect(0,200,200,200);
      }
      if(newColor.equals("y"))
      {
        g.setColor(new Color(255,255,0));
        g.fillRect(200,200,200,200);
      }

      drawLevel(g);

      try
      {
        Thread.sleep(250);
      }
      catch(Exception ex){}

      reset(g);
    }
    else
    {
      levels++;
      //make the squares
      g.setColor(new Color(200,50,50));
      g.fillRect(0,0,200,200);
      g.setColor(new Color(50,50,200));
      g.fillRect(200,0,200,200);
      g.setColor(new Color(50,200,50));
      g.fillRect(0,200,200,200);
      g.setColor(new Color(200,200,0));
      g.fillRect(200,200,200,200);

      drawLevel(g);
      try
      {
        Thread.sleep(1000);
      }
      catch(Exception e){}

      //iterate through the code
      for(int i = 0; i <code.length(); i++)
      {
```

```
    try
    {
      Thread.sleep(250);
    }
    catch(Exception e){ }

  //light up the next square
  char letter = code.toCharArray()[ i];
  if(letter=='r')
  {
    g.setColor(new Color(255,0,0));
    g.fillRect(0,0,200,200);
    g.setColor(Color.black);
    g.setFont(new Font("arial", Font.BOLD, 40));
    g.drawString("Red",50,80);
  }
  if(letter=='b')
  {
    g.setColor(new Color(0,0,250));
    g.fillRect(200,0,200,200);
    g.setColor(Color.black);
    g.setFont(new Font("arial", Font.BOLD, 40));
    g.drawString("Blue",250,80);
  }
  if(letter=='g')
  {
    g.setColor(new Color(0,255,0));
    g.fillRect(0,200,200,200);
    g.setColor(Color.black);
    g.setFont(new Font("arial", Font.BOLD, 40));
    g.drawString("Green",50,280);
  }
  if(letter=='y')
  {
    g.setColor(new Color(255,255,0));
    g.fillRect(200,200,200,200);
    g.setColor(Color.black);
    g.setFont(new Font("arial", Font.BOLD, 40));
    g.drawString("Yellow",250,280);
  }

  drawLevel(g);

  try
  {
    Thread.sleep(250);
  }
  catch(Exception e){ }

  reset(g);
}

try
{
  Thread.sleep(250);
}
catch(Exception e){ }
String newColor = newColor();
//display the appropriate color
if(newColor.equals("r"))
{
  g.setColor(new Color(255,0,0));
  g.fillRect(0,0,200,200);
```

```
        g.setColor(Color.black);
        g.setFont(new Font("arial", Font.BOLD, 40));
        g.drawString("Red",50,80);
      }
      if(newColor.equals("b"))
      {
        g.setColor(new Color(0,0,250));
        g.fillRect(200,0,200,200);
        g.setColor(Color.black);
        g.setFont(new Font("arial", Font.BOLD, 40));
        g.drawString("Blue",250,80);
      }
      if(newColor.equals("g"))
      {
        g.setColor(new Color(0,255,0));
        g.fillRect(0,200,200,200);
        g.setColor(Color.black);
        g.setFont(new Font("arial", Font.BOLD, 40));
        g.drawString("Green",50,280);
      }
      if(newColor.equals("y"))
      {
        g.setColor(new Color(255,255,0));
        g.fillRect(200,200,200,200);
        g.setColor(Color.black);
        g.setFont(new Font("arial", Font.BOLD, 40));
        g.drawString("Yellow",250,280);
      }
      drawLevel(g);
      try
      {
        Thread.sleep(250);
      }
      catch(Exception e){}
      reset(g);
    }
  }
}
public void drawLevel(Graphics g)
{
  //draw the oval and text that displays the level
  g.setColor(Color.white);
  g.fillOval(160,160,80,80);
  g.setColor(Color.black);
  g.setFont(new Font("arial", Font.BOLD, 15));
  g.drawString("Level "+levels,175,205);
}
public String newColor()
{
  //generate the random color
  int rand = (int)(Math.random()*4);
  if(rand==0)
  {
    code+="r";
    return "r";
  }
  if(rand==1)
    {
```

```
    code+="b";
    return "b";
  }
  if(rand==2)
  {
    code+="g";
    return "g";
  }
  if(rand==3)
  {
    code+="y";
    return "y";
  }
  return " ";
}

public void reset(Graphics g)
{
  //reset the red square
  g.setColor(new Color(200,50,50));
  g.fillRect(0,0,200,200);
  //reset the blue square
  g.setColor(new Color(50,50,200));
  g.fillRect(200,0,200,200);
  //reset the green square
  g.setColor(new Color(50,200,50));
  g.fillRect(0,200,200,200);
  //reset the yellow square
  g.setColor(new Color(200,200,0));
  g.fillRect(200,200,200,200);
  drawLevel(g);
}

public void actionPerformed(ActionEvent e)
{
  //if start was pressed, start the game
  start.setBounds(500,500,50,50);
  begin = true;
  repaint();
}

public void mouseExited(MouseEvent e){}
public void mouseEntered(MouseEvent e){}
public void mouseReleased(MouseEvent e){}
public void mousePressed(MouseEvent e){}
public void mouseClicked(MouseEvent e)
{
  normal = false;
  guesses++;
  //find out which square the player clicked
  if(e.getX()<200 && e.getY()<200)
  {
    lastLetter = "r";
    guess+="r";
  }
  if(e.getX()>200 && e.getY()<200)
  {
    lastLetter = "b";
    guess+="b";
  }
  if(e.getX()<200 && e.getY()>200)
```

```
    {
      lastLetter = "g";
      guess+="g";
    }
    if(e.getX()>200 && e.getY()>200)
    {
      lastLetter = "y";
      guess+="y";
    }
    repaint();
    String codeSeg = code.substring(0,guesses);
    if(!codeSeg.equals(guess))
    {
      //game over!
      over = true;
      repaint();
    }
    else
    {
      //the player can proceed
      if(guesses==code.length())
      {
        guess = "";
        guesses = 0;
        normal = true;
        //next level!
        repaint();
      }
    }
  }
  public static void main (String[ ]args)
  {
    new IanSays();
  }
}
```

Figures 57-5 and 57-6 shows the end of a challenging round of Ian Says.

Customizing the game

Use auditory cues instead of colors as memory targets.

Alter the design of the game board from 4 squares to piano keys.

Add additional squares of different colors to the grid after each successful round.

Give players a time limit for each round.

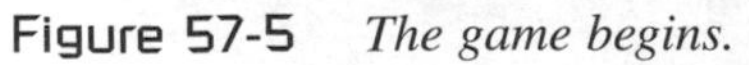

Figure 57-5 *The game begins.*

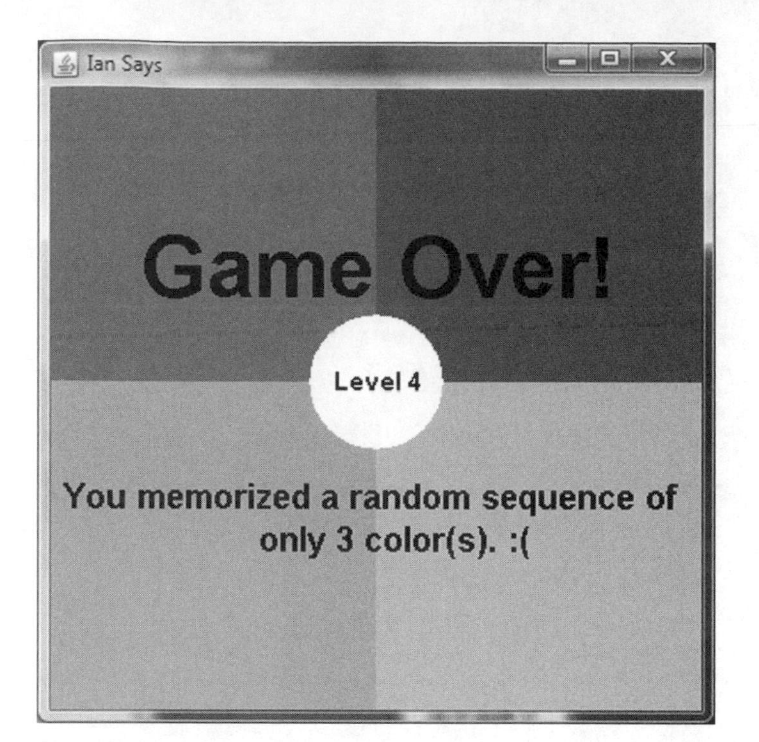

Figure 57-6 *Devastating loss at level 4. Darn it!*

Construct a finite number of levels and race against the clock to complete them.

In a test of eye-hand coordination, require right-handed players to use their left hand and left-handed players to use their right hand to respond.

Force players to delay their responses by "freezing" the board for several seconds.

Congratulations! You have mastered the art of programming video games.

You are clever, creative, and are a computer wizard!

You are a true EVIL GENIUS!!!

Index

Page references followed by f indicate an illustrative figure

A

B

C

G

I